A Universe of Atoms,
An Atom in the Universe

Springer

New York
Berlin
Heidelberg
Hong Kong
London
Milan
Paris
Tokyo

Mark P. Silverman

A Universe of Atoms, An Atom in the Universe

With 74 Illustrations

 Springer

Mark P. Silverman
Department of Physics
Trinity College
Hartford, CT 06106
www.trincoll.edu/~silverma

Cover illustration: Chris Silverman.

Library of Congress Cataloging-in-Publication Data
Silverman, Mark P.
    A universe of atoms, an atom in the universe / Mark P. Silverman.—2nd ed.
      p. cm.
    Rev. ed. of: And yet it moves. c1993.
    Includes bibliographical references and index.
    ISBN 0-387-95437-6 (acid-free paper)
    1. Physics. 2. Quantum theory. I. Silverman, Mark P. And yet it moves.
  II. Title.

QC21.3 .S55 2002
530—dc21

                                                         2002016006

ISBN 0-387-95437-6   Printed on acid-free paper

This book is a revised edition of *And Yet It Moves*,
published in 1993 by Cambridge University Press.

Printed in the United States of America.

9 8 7 6 5 4 3 2 1   SPIN 10865410

www.springer-ny.com

Springer-Verlag New York Berlin Heidelberg
A member of Bertelsmann Springer Science+Business Media GmbH

To
Sue, Chris, and Jen

atoms with consciousness . . .
matter with curiosity.
Stands at the sea . . .
wonders at wondering . . .
I . . . *a universe of atoms* . . .
an atom in the universe.

—Richard P. Feynman

Contents

Preface

Approximately 10 years have passed since the publication of *And Yet It Moves: Strange Systems and Subtle Questions in Physics*.[1] During this time, the book has done very well, being received favorably by both readers and reviewers. The exhaustion of the last printing has given me the opportunity to make revisions. The present volume, updated and expanded by three new chapters containing a total of 17 additional essays on a wide range of questions that I have explored in quantum mechanics, nuclear physics, thermodynamics, general relativity, and astrophysics, has been given a new title to reflect the broad thematic coverage and a new publisher (Springer-Verlag).

The Introduction (The Fire Within) that follows, based on the preface to the original edition, explains fully the purpose and content of this book. I wish to note briefly here, however, that time and the advance of physics have not dulled the scientific relevance of any of the essays. This book, like its predecessor, is not intended to be a popularization, a textbook, or a monograph of any field of physics. Rather, it is a personal account of the scientific underpinnings, motivations, lessons, and ramifications of some of the many fundamental physical problems that have engaged me throughout my career to the present. These are essays that anyone with an interest in contemporary physics can read, although it is certainly the case that the more serious the interest, the more meaningful will be the essays.

In the years following *And Yet It Moves*, I have written, besides the present book, three others, principally for physicists, teachers, and students, more specifically focused on those parts of my researches concerned with quantum interference phenomena,[2] classical optics and electromagnetism,[3] and quantum electrodynamics and atomic physics,[4] respectively. As was the case with my previous books, there is again in this one an underlying concern with physics education. By that I mean not merely the transmission of facts and formulas, but a communication of the delight of scientific exploration, the ultimate exercise of human curiosity and ingenuity, without which any science, especially physics, becomes lifeless and dull, however competently its technical details are taught.

In matters of science and education it is to my wife, Dr. Susan Brachwitz, and to my son, Chris, and daughter, Jennifer, that I owe the greatest debt of gratitude. Besides the full-time occupations of university teaching and research, I have had, together with my wife, the serious responsibility and privilege of instructing our children from infancy onward in our own home-based school. ["Home," however, has ranged over the globe from New Zealand to Finland with many points in between.] It was I, however, who received the most instruction, for what I have come to understand about the nature of learning and the stimulation of interest in science, I have learned with Susan's help from seeing our son and daughter evolve into young adults in an atmosphere supportive of their natural instincts to explore, discover, and create.

I would also like to thank Chris, an artist and computer scientist, for the lovely design of the cover of this book, and to express my gratitude as well to my editor and long-time friend, Dr. Thomas von Foerster, and my production editor, Terry Kornak, for guiding this book safely through all the shoals of production, and to Professor Michael Berry (Bristol) for his enthusiastic reception of *And Yet It Moves* and helpful comments for the new edition.

Mark P. Silverman
Tall Pines Institute
July 2001

Notes

1. M. P. Silverman, *And Yet It Moves: Strange Systems and Subtle Questions in Physics* (Cambridge University Press, Cambridge, 1993).
2. M. P. Silverman, *More Than One Mystery: Explorations in Quantum Physics* (Springer-Verlag, New York, 1995).
3. M. P. Silverman, *Waves and Grains: Reflections on Light and Learning* (Princeton University Press, Princeton, 1998).
4. M. P. Silverman, *Probing the Atom: Interactions of Coupled States, Fast Beams, and Loose Electrons* (Princeton University Press, Princeton, 2000).

Introduction: The Fire Within[1]

As a child, for as far back as I can recall, I always wanted to be a physicist—a nuclear astrophysicist or cosmologist, in fact. I am not exactly sure why, for I knew no one like that within my family or circle of acquaintances. I suspect that aspiration was owed largely to Eddington, Hoyle, Jeans, and especially Gamow, whose popular books I read avidly. Only rarely have the students whom I have taught over the past thirty five years heard of these people or of their books. Sometimes, out of sheer perversity, or perhaps genuine curiosity, I would remark to a class, "What? You mean you never read *One, Two, Three . . . Infinity* or *Mr. Tompkins in Wonderland?*" But the students would only look at one another with wry smiles, as if to confirm their suspicions that physicists are strange people and that they, unfortunately, got stuck with an especially peculiar one. Wonderland, indeed!

As a graduate student, I never studied astrophysics or cosmology, although, in recent years, my research and publications address key issues in these areas. Perhaps those same books that fired my imagination with the marvels of the physical world may have also led me to believe that the most fundamental mysteries of physics were largely exhausted.

I began my scientific odyssey in the field of medicine as part of a group researching malfunctions of the immune system. Finding experimentation on animals personally distasteful, however, and myself little inclined to constant preoccupation with disease, I changed to biochemistry with the heady, through misguided, notion of answering the question posed by Schrödinger's influential book, *What Is Life?* It was a profound disappointment, therefore, to end up on a project to analyze the nitrite content of corn.

I took up organic chemistry next. Predicting the outcome of complex chemical reactions by flipping electrons around pentagonal and hexagonal rings had a certain aesthetic appeal to me—at least on paper. In reality, however, "Molecules do what they damn please," as one professor told my wife when she was a graduate student at Harvard some years later. Having passed unscathed through more syntheses with toxic and explosive precursors than I now care to remember, I decided one day to push my luck no longer. Disaffected with a field that seemed to lack fundamental principles and in which I, quite literally, saw no future for myself, I turned to physical chemistry.

I enjoyed physical chemistry for a while, investigating molecules with electron and nuclear magnetic resonance, until I realized what it was about the subject that interested me most. It was physics. So I switched one last time and returned to the passion of my youth. I have remained a physicist ever since and have no regrets at all for the circuitous path that finally brought me back—or almost back—to the career I decided upon as a child. If anything, the diversity of experiences has made me a better scientist.

Even during the years of "wandering" before I rediscovered what it was I wanted to do with my life, I never actually abandoned the study of physics. I took physics courses at the university, although I have little recollection of anything noteworthy about them. It was not that those courses were necessarily ill-taught. At best, they conveyed well enough the mathematical or mechanical skills required for solving physics problems. But something essential was missing. No instructor ever addressed the question of why physics problems were worth solving or what made physics sufficiently interesting so that one would want to study it at all, let alone devote a lifetime to it. Not once prior to graduate school—and even then only rarely—can I recall a professor expressing personal interest in the abstractions on the classroom blackboard or the apparatus used for demonstrations. Sometimes I wonder how many potential physicists may have perished in lecture halls of universities and colleges for want of a larger vision of what physics was all about.

Fortunately for me, I did not need to rely on formal instruction for motivation. I loved to learn, although I did not particularly care to be taught—at least not in the traditional manner of lecturing and testing that deprived a person of the pleasure of discovery. I already knew from childhood many reasons why physics was interesting. I needed only to know that there were still wonderful things to learn and to do, and this I gradually discovered in the same way as before, by reading widely.

My vision of physics took shape under the tutelage of Galileo, Newton, Fresnel, Maxwell, Einstein, Bohr, Heisenberg, Schrödinger, Fermi, Dirac, and a score of others whose writings I struggled through. It was perhaps not the most efficient way to learn, for there was much I did not understand until much later; at times, I understood nothing at all. But what I did absorb was priceless: a sense that in those written words and mathematical relations were ideas of fundamental importance—deep ideas that with further effort I would one day be able to comprehend. The pages spoke as if the authors, themselves, were present. In this way, my passion for physics survived tiresome and seemingly pointless classroom analyses of falling projectiles, rolling cylinders, and swinging pendula that many a hapless student bore somnolently to the end of his final exam—and then promptly forgot.

*   *   *

What kind of physicist did I finally become? In an age when science is infinitely fragmented, its practitioners highly specialized, and experimentalists and theoreticians likely to find themselves on different floors, if not altogether in separate buildings, I hope it will appear neither coy nor audacious to give the reply I. I. Rabi gave when asked to classify himself: "I am just a physicist."[2] The German chemist Wilhelm Ostwald, who was much interested in the subject of scientific creativity, divided scientists into classicists, who systematically bring to perfection one or a few discoveries, and romanticists, who pursue a multitude of ideas, albeit incompletely. I rather like the colorful and sympathetic distinction drawn between these two dispositions by educator Gilbert Highet in his book *The Immortal Profession*[3]:

Will you decide (as Swift put it) to resemble a spider, spinning out endless webs from its own vitals, or a bee, visiting flower after flower and extracting a different sweetness from each of them? Will you be like those individualists one sees out west in Colorado and Wyoming, who dig their own little vertical mine shafts into the earth, and spend the rest of their days extracting ore from the same small vein? Or will you be a wandering prospector, trying first this mountain range and then that, never working out a single lode but always adventuring farther forward?

One has but to scan the employment notices in science periodicals to realize in an instant what type of scientist is sought the most today by academia, industry, or government. Nevertheless, for what it is worth, I confess unabashedly to being a romanticist who has spent years happily adventuring in whatever "mountain range" I found interesting. Unbound to any one field or to any one machine, I am attracted by problems, whether of an experimental or theoretical nature, that are conceptually intriguing, even if at the time I may be alone in thinking so.

The essays in this book are based on some of the research with which I have instructed and entertained myself over the past few decades. Touching on topics drawn from quantum mechanics, atomic and nuclear physics, electromagnetism and optics, gravity, thermodynamics, and the mechanics of fluids, these essays are about different physical systems whose behavior has stimulated my curiosity, provoked in me surprise, and challenged my imagination. There are strange processes for which no visualizable mechanism can be given; processes that seem to violate fundamental physical laws, but which, in reality, do not; processes that are superficially well understood, yet turn out to be subtly devious. The essays address specific questions or controversies from whose resolution emerge lessons of general significance.

For example, does an atomic electron move? How would one know? Would an "antiatom" fall upward? Is the vacuum really empty? Can an atom be larger than a blood cell? If it were, would it behave like a miniature planetary system? Can a particle be influenced by an electric or magnetic field that is not there—that is, through which it does *not* pass? How is it possible for randomly emitted particles to arrive preferentially in pairs at a detector—or, conversely, to avoid one another altogether? What constitutes a random process anyway? Could watching decaying atoms emit light in London have an effect on the corresponding radiative decay in New York? Does a "right-handed" light beam interact differently with matter than a "left-handed" light beam? How can light get brighter by rebounding from a surface (without violating the conservation of energy)? Is a basketball changed for having been turned 360°? Perhaps not, but what about an electron? Could one tell the difference between an electron that has jumped out of a quantum state and then back again and an electron that has never jumped at all? Is there really such a thing as a "Maxwell demon"? No?—then how is one to account for a simple hollow tube that blows hot air out one end and cold air out the other? And one of the grandest mysteries of all: Where or what is the 95% of the mass of the Universe that is "missing"?

Broadly regarded, there is a common theme that runs through the various chapters: the mystery and fascination of motion—whether it be the movement of an electron, the flow of air, or the propagation of a light wave. It is the strange behavior of what are often enough more-or-less familiar systems—at least to physicists—that brings to my mind the famous words of Galileo ("Eppure si muove") adopted as the title of the first edition of this book—words that signified to me not a mutter of defiance, but rather an expression of wonder and awe.

Although no mathematics beyond elementary calculus is used here, this book is not intended to be a popularization of any aspect of contemporary physics. Neither is it designed to be a textbook or monograph. I hope, of course, that the reader may find the collection of essays instructive, but my objective is not so much to teach physics as to communicate, through discussion of personally meaningful investigations, that the study of physics can be intensely interesting and satisfying even when one is not addressing such ultimate questions as the origin and fate of the universe. What follows, then, is essentially one scientist's personal odyssey in physics.

Admittedly, it may seem somewhat presumptuous to believe that one's own work would necessarily interest and instruct others, and for the encouragement to think thus, I have friends and colleagues throughout the world to thank. Indeed, one of the strongest impressions that a life in science has made upon me is the transcendence of common scientific interests over national, ethnic, racial, and religious differences that somehow seem to pose such barriers to social intercourse in other walks of life. I am often reminded of Sir Humphry Davy's and Michael Faraday's peregrination through France in the early 19th century, visiting French laboratories and factories and meeting with French scientists, at a time when France was convulsed with war. "It is almost impossible for an inhabitant of the twentieth century to believe," wrote Faraday's biographer L. Pearce Williams,[4] "that a party of English citizens could go about their ordinary affairs in the middle of an empire locked in a struggle to the death with England without the slightest inconvenience."

But I believe it. Under circumstances less dramatic perhaps, but nonetheless evocative of the experiences of Davy and Faraday, I have myself gone to my mailbox more than once to find—for example, from the (former) Soviet Union and Eastern European nations during the "cold war" or from Iran during the "hostage crisis"—a friendly letter opening up a scientific dialogue or extending an invitation to visit and lecture. Where else, but in science, I have often thought, would it be so natural and proper for total strangers half a world apart to exchange letters telling of their deepest interests. More than one scientific adventure began, in fact, with my opening or writing such a letter.

To someone like me, who has been for most of his professional life simultaneously a physicist and a teacher, the pursuit of physics is an activity intimately coupled with education. One conducts scientific research ideally to learn new things, and that inquiry is somehow incomplete until shared. Teaching science to others is, in effect, sharing the fruits of discovery made, not by oneself alone, but by some of the most creative people who have ever lived. It is not simply occupational parochialism that fosters my belief that science, in general and physics, in particular, are much more than merely the source of better technology and higher lifestyles, but rather a precious intellectual legacy to pass on to future generations. And yet, as anyone knows who keeps abreast of the current state of education in America, Britain, and elsewhere, science is one of the subjects least understood or favored by the general public. The enormous divergence between the public perception of science and the profoundly interesting and important heritage that scientists know it to be, should be a matter of great concern. I have chosen to end this book, therefore, with an essay not on physics, but on the teaching of physics—or, more generally, on why science is worth knowing and how it might best be learned.

Finally, it is a pleasure to acknowledge many delightful, far-ranging conversations and shared experiences with the following colleagues, some of whom are coauthors on papers issuing from the projects herein described: Professor Jacques Badoz of the Ecole Supérieure de Physique et Chimie Industrielles (Paris), Professor Ronald Mallett of the University of Connecticut (Storrs), Professor Geoffrey Stedman of the University of Canterbury (Christchurch), Mr. Wayne Strange of Trinity College (Hartford), Dr. Akira Tonomura and Dr. Hiroshi Motoda of the Hitachi Advanced Research Laboratory (originally in Tokyo, but subsequently relocated to Hatoyama), and my son, Chris R. B. Silverman, who, at the time I am writing this, is a student at Trinity College. It is not only those kinds of motion prescribed by physical laws that elicit wonder, but also, in a metaphorical sense, the extraordinary exchange of ideas and people that characterise the scientific enterprise itself.

Mark P. Silverman
Tall Pines Institute
July 2001

Notes

1. Based on the Preface to *And Yet It Moves: Strange Systems and Subtle Questions in Physics*, Cambridge University Press, New York, 1993.
2. J. S. Rigden, *Rabi*, Basic Books, New York, 1987, p. 8.
3. G. Highet, *The Immortal Profession*, Weybright and Talley, New York, 1976, pp. 62–63.
4. L. Pearce Williams, *Michael Faraday*, Simon and Schuster, New York, 1971, p. 36.

CHAPTER 1

The Wirbelrohr's Roar

With all due respect to Robert Boyle, there is a "spring" to the air that the venerable Irish physicist never dreamed of some three centuries ago when he introduced his fellow natural philosophers to the effects of pressure.[1] Air is not merely compressible; it can course and caper through appropriate devices in such ways as to please the ear and titillate, if not confound, the intellect. I learned that first hand from playing.

Most people I have encountered, for whom physics is anything but relaxation, could hardly imagine "physics" and "play" in the same sentence—except, perhaps, one denying their equivalence. Yet, the same laws that govern the erudite matters to which physicists give their attention also apply to recreation. Indeed, sometimes nature's subtlest wiles may be invested in the simplest child's toy. One of my favorite science photographs[2] shows Wolfgang Pauli and Niels Bohr hunched over the ground observing the behavior of a Tippetop, a curious little object that, shortly after being spun on its wide bottom, flips 180° and spins on its narrow handle. As far as I know, there may still be no consensus as to how it works.

When I think about the topic of physics "toys," I find it striking how often the phenomena that puzzle and amuse us involve the element of spinning. As a child, I was ever entranced by a small gyroscope precariously perched at the end of my finger or horizontally suspended by a loop of string around the rotation axis in apparent defiance of the laws of gravity. The gyroscope still fascinates me even though, as a physicist, I understand how it works. My own children, when they were young, were intrigued by a "one-way" spinner, also known as a celt, which I frequently borrowed from them for use in lectures on chiral asymmetry. It is a 4-in. piece of plastic (bearing the words "Turn on to Science") shaped like the hull of a clipper ship with just the slightest inequivalence between port and starboard sides. Spin it counterclockwise and it turns freely; spin it clockwise and it soon wobbles vehemently, stops, and rotates in the opposite sense!

A celt is startling to behold and by no means trivial to explain. In fact, a partially satisfactory explanation was provided by the cosmologist Hermann Bondi only some hundred years after this remarkable behavior was first reported.[3] Bondi's paper is not bedtime reading.

I have myself often fashioned a "two-way" spinner from a wooden pencil by carving a row of notches along its length and affixing a propeller (a popsicle stick works well) to the eraser with a pin. Stroke the notches with another pencil, and the propeller spins either clockwise or counterclockwise depending on a subtle manipulation by the stroker. How does ostensibly linear motion rotate the propeller? In some way, of course, the strokes must generate elliptical vibrations in the pencil, but the details are hardly obvious.

If the motion of solids, with relatively few degrees of freedom, can be puzzling, one can only begin to imagine the paradoxical possibilities that arise when fluids are admitted. Consider, for example, the simple radiometer found in many a museum gift shop. Illuminated by bright sunlight, the four vanes (black on one side, white on the other) spin wildly about a vertical shaft inside a highly evacuated bulb. James Clerk Maxwell, I have been told, was ready to discard his electromagnetic theory of light upon learning that the vane spun the "wrong" way—the wrong way, that is, if one assumes the vanes are driven by light pressure. It is not light, however, but residual gas that lies at the heart of the matter, although exactly how is still, more than a hundred years later, a question for discussion.[4]

Some years ago, while living in Japan, my family and I encountered near a train station in the town of Hakone the eerie strains of a most unearthly symphony. There, about twenty meters in front of us, a dozen or so Japanese children, whose number was quickly augmented by my own, were feverishly grabbing long flexible colored plastic tubes from the stand of a streetside vendor and twirling them furiously above their heads like lariats. The burst of tones that emerged from each musical pipe, designated "The Voice of the Dragon" by a sign in English, soared and dropped with rotational speed over what seemed like a good portion of the range of a flute. I have never forgotten this loud, wavering, rich-toned chorus of "dragon voices." Despite the outward simplicity of the toy, the details of its sound production are by no means trivial, and the efforts to understand it provided both my students and me worthwhile lessons in physics as well as much entertainment.[5]

However, when it comes to gases, nowhere are the intriguing effects of rotational motion as counterintuitive, I think, as in the case of the "Wirbelrohr."

*   *   *

I am not an avid reader of science fiction, and, in fact, except for the "classics" by such writers as Jules Verne and H. G. Wells, generally avoid this genre of literature altogether. My late father-in-law Fred, however, who for many years was a machinist at the AT&T Bell Laboratories, was a science fiction enthusiast with subscriptions to a number of such magazines spanning at least four decades. I recall in particular one visit to his home when, in the course of a chess game, we began discussing some of the strange devices he had constructed for engineers during his employment at Bell.

Perhaps the strangest device that he had ever made, however, he made for himself, he told me. I asked him what it looked like, and he replied that it was extremely simple: a hollow tube shaped like a **T** with no moving parts of any kind. Upon my inquiring as to what it did, Fred cocked his head, and I could see—or at least imagine I saw— a gleam in his eyes and the faint trace of a sardonic smile beneath his bushy white beard. It was a Wirbelrohr, he explained; you blew into the stem, and out one end of the cross-tube flowed hot air while cold air flowed out the other. I laughed; I was certain he was teasing me. Although I had never head of a Wirbelrohr, I recognized a Maxwell demon when it was described.[6] I asked my father-in-law whether he invented the device, to which Fred replied that he first read about it in one of his science fiction magazines. "Yes indeed!," I thought to myself, and the look on my face undoubtedly conveyed my incredulity as if my thoughts were audible. He insisted that it worked, and when it worked really well, the cold air could freeze water and the hot air could fry an egg!

I saw from Fred's expression that he was not teasing me. My father-in-law was from Switzerland; he was no physicist, but his skill in making things was exceptional. I had often thought to myself that he could make anything—although I meant, of course, anything real. Maxwell demons were, as far as I knew, imaginary. My curiosity was thoroughly aroused, all the more because I happened to be teaching a course in thermodynamics that same semester.

To my great disappointment, Fred had kept no record of the device he made, nor was he able to recall exactly when or from what magazine he obtained construction drawings. After all, he built the device some thirty years earlier. Nevertheless, having never discarded a single volume of his science fiction library, Fred promised that, as time permitted, he would search for the intriguing story. At the end of the visit, I returned home excited, but by no means convinced that the Second Law of Thermodynamics should be omitted from my lectures.

Two weeks later, a copy of the desired article arrived in the mail.[7] There, sandwiched between the last page of 38,000 Achnoid alien carbon people without brain chords and the first page of encephalo-

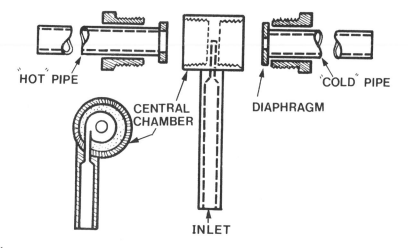

Figure 1.1. Schematic diagram of a Wirbelrohr or vortex tube. Room-temperature compressed air enters the inlet tube, spirals around the central chamber, and exits through the "hot" pipe with unconstrained cross section or through the "cold" pipe, whose aperture is covered by a diaphragm.

graphic analysts led by a powerful mental mutant during the dying days of the First Galactic Empire (two undoubtedly gripping tales that vividly reminded me once again why I rarely read science fiction) were the anatomical details of a Maxwell demon.

My father-in-law had certainly told the truth (as I said, he was Swiss). He, in fact, did more than that; he machined in his basement workshop a working model, which I received from him shortly afterward. The exterior was more or less just as he had described it: two identical long thin-walled tubes (the crossbar of the **T**) were connected by cylindrical collars screwed into each end of a short section of pipe that formed the central chamber; a gas inlet nozzle (the stem of the **T**), shorter than the other two tubes but otherwise of identical construction, joined the midsection tangentially (Figure 1.1). Externally, except for a throttling valve at the far end of one output tube to control air flow, the entire device manifested bilateral symmetry with respect to a plane through the nozzle perpendicular to the cross-tubes.

Only someone with the lung capacity of Hercules could actually blow into the stem. Instead, the nozzle was meant to be attached to a source of compressed air. Taking the Wirbelrohr to my laboratory, I looked skeptically for a moment at its symmetrical shape before opening the valve by my work table that started the flow of room-temperature compressed air. Then, with frost forming on the outside surface of one tube, I yelped with pain and astonishment when, touching the other tube, I burned my fingers!

*   *   *

*Thermo*dynamics is different from any other dynamics in physics; in fact, the very word "thermodynamics" is a misnomer. Whereas the term *dynamics* ordinarily embraces the idea of a system evolving in time under the action of specific forces (e.g., electrodynamics, hydrodynamics, aerodynamics, and chromodynamics[8]), the classical theory of thermodynamics is a study of systems in thermal equilibrium—systems, that is, whose macroscopic thermal properties are temporally unchanging. How some physical system has come to be in a state of equilibrium, or how much time is required for the system to go from one equilibrium state to another when external conditions are changed, is outside the principal concern of thermodynamics; for a problem in *this* area, contact your local specialist in kinetics.

To those unfamiliar with the subject, it may seem that, by excluding from its domain the intricate details of specific interactions, thermodynamics must necessarily be a weak and ineffective science compared with the other dynamical siblings in the family of physics. This, however, is not the case at all. In autobiographical notes that he merrily designated his "obituary," Albert Einstein, a man who spent a lifetime developing physical theories, wrote[9]

A theory is the more impressive the greater the simplicity of its premises is, the more different kinds of things it relates, and the more extended is its area of applicability. Therefore the deep impression which classical thermodynamics made upon me. It is the only physical theory of universal content concerning which I am convinced that, within the framework of the applicability of its basic concepts, it will never be overthrown. . . .

The strength of thermodynamics, as emphasized by Einstein, lies in its close and simple ties to experiment and observation. Let the whole edifice of chromodynamics—and therefore the theory of matter itself—collapse because quarks turn out to be nonexistent; the conclusions of thermodynamics will remain as sound as ever.

The vast latticework of thermodynamic interrelations rests principally upon two major laws, the First and the Second. The First Law is a generalized statement of energy conservation and is to be found in one form or another in all dynamical theories of physics. In short, energy can be transformed—from mechanical to electrical, or from electrical to heat, for example—but it cannot be created from "nothing" or destroyed. The Second Law, however, which can be expressed in a variety of seemingly inequivalent ways, is unique to thermodynamics. In its sum and substance, the Second Law affirms the essential irreversibility of natural or spontaneous processes.

The more tangible versions of the Second Law, attributable in one form to Lord Kelvin and Max Planck and in another to Rudolph Clausius, reveal the historical roots of thermodynamics in the practical problems of machine-making. According to Kelvin and Planck there can be no process the sole result of which is to absorb heat and convert it into work. "But what about the steam engine?" someone is bound to ask. True, it operates by heating water to steam which, upon expansion, does work, but that is not the *sole* result; heat is also discarded to the environment. Neither the steam engine nor any other engine converts 100% of the absorbed heat into work without in some way changing the state of the rest of the world (including, possibly, itself). According to the different perspective of Clausius, there can be no process the sole result of which is to transfer heat from a cooler to a hotter body. Now, before one is tempted to assert that a refrigerator does exactly that, let him recall again the restricting condition. A refrigerator does take heat from a cool body and pumps it to a hotter body, usually the ambient air, but only upon the input of work in the form of electrical energy. The foregoing two statements of the Second Law are completely equivalent; it is not difficult to show that violation of one implies violation of the other.

Whether it is a tribute to the indomitable spirit, or simply the perversity, of human nature, the interdictions posed by the Second Law have been a red flag before the eyes of many a bullish inventor. Energy is something most people, rightly or wrongly, believe they understand at least to some degree—and the thought of building a device that generates more energy than is employed to run it is probably not seriously entertained except by those entirely ignorant of all science. The content of the Second Law, however, which embodies highly abstract notions for what is otherwise so concrete a science, rests less easily on the mind. It stands as a challenge to the ingenious as well as the ingenuous.

The frustrating thing about the Second Law is that it forbids processes for which energy remains conserved and which one might naïvely hope can be made to work—somehow. But they cannot. A coin dropped from a height above a tabletop falls down and heats up a little upon impact. No one, I suspect, has ever witnessed a coin spontaneously rise up against the force of gravity at the expense of its own internal thermal energy thereby suffering a drop in temperature. In either case, mechanical and thermal energy can be made to balance, but the process occurs in one direction only. In a similar way, the outcome of setting a hot coin on top of a cold one is a final state of two lukewarm coins. One could wait, as they say, "until Hell freezes over" before the time-reversed process, whereby the lower coin becomes perceptibly colder by spontaneously transferring heat to the upper coin, ever occurs, even though the total energy of the two-coin system is again unchanged. The foregoing hyperbolic remark actually serves a purpose: It emphasizes an essential part of the unique quality of the Second Law vis-à-vis all other physical laws: its statistical validity.

Consider, for example, the First Law: the conservation of energy. Is it conceivable that although energy appears to be conserved in processes involving macroscopic amounts of matter, violations nevertheless occur from time to time on an atomic level? In the early 1920s, as physicists struggled to make sense of the structure of atoms and the nature of light, Niels Bohr, Hendrik Kramers, and John Slater published a paper (the notorious B-K-S paper) rejecting Einstein's light-quantum hypothesis and holding to the view that the principles of energy and momentum conservation cannot be strictly applied to individual interactions.[10] Highly controversial, the B-K-S paper elicited much discussion within the physics community. Einstein and Pauli scathingly criticized it; Schrödinger, by contrast, was fascinated by it. In the end, however, the B-K-S theory was decisively refuted by experimental studies of the Compton effect, the scattering of light by free electrons. If the conservation of energy and momentum applied only to bulk matter averaged over time, and not to individual quantum processes, then there would be a non-negligible probability than an illuminated electron could recoil in any direction whatever. Within the limits of experimental precision, constrained ever more tightly by new methods and increasing advances in technology, every reliable measurement consistently revealed that individual interacting pairs of electrons and photons strictly conserved both energy and momentum.[11]

That was probably the last time leading physicists seriously entertained the thought that the basic conservation principles of dynamics were only statistically valid. When in the 1930s the weak decay of elementary particles seemed to reveal violations of energy and momentum conservation, Pauli knew to look for an alternative explanation and predicted the existence of an elusive new particle, the neutrino.

Before the underlying statistical basis of thermal phenomena was clearly understood, some—Clausius, for example—regarded the Second Law to be rigorously valid in all domains of experience. The proscription that no process can, as a sole result, convert heat to work with perfect efficiency was interpreted strictly to mean *no* process ever. Perception of the Second Law as a manifestation of a law of large numbers was probably first recognized by James Clerk Maxwell, whose pioneering statistical studies of the distribution of particle velocities and associated colligative phenomena would have marked him as a master theoretical physicist even if he had never formulated the laws of electromagnetism.

Although the connection may not be obvious, the previous two formulations of the Second Law are equivalent to yet another formulation, more fundamental, in my view, as it is readily amenable to interpretation within the framework of the atomic theory of matter. This third version is expressible in terms of the abstract concept of entropy, which, in addition to energy, is one of the basic properties (or state variables) of an equilibrium thermodynamic state.[12]

From the macroscopic perspective of thermodynamics, the *change* in entropy (which is what one actually measures) associated with the transformation of a system from one equilibrium state to another is related to the heat absorbed or released. From an atomic perspective, however, heat is the energy exchange between physical systems as a result of random molecular motion. In fact, the temperature of a sample of matter in thermal equilibrium with its environment is linearly proportional to, and therefore a measure of, the average molecular kinetic energy. Correspondingly, the entropy of the sample is a measure of the disorder of molecular motion—that is, the number of distinctly different ways in which the particles can be distributed over allowed quantum states and yet still give rise to the specific macroscopic properties (e.g., pressure, temperature, volume) exhibited by the sample. The higher the entropy, the greater the disorder.

Looked at statistically, then, the entropy change for a transformation between equilibrium states is a measure of the relative probability of finding the molecules of the system in the microscopic (quantum) states compatible with the final macroscopic equilibrium state of the sample compared with finding them in the microscopic states compatible with the initial equilibrium state. With this in mind, one can express the third version of the Second Law as follows: In any spontaneously occurring process, the entropy always increases, unless the process is reversible, in which case the entropy change is zero.

Imagine a box divided into two sealed compartments of equal volume—one containing a gas, the other vacuum. Between the two compartments is a removable partition. When, by some external means, the partition is removed, the gas spreads into all of the available volume until the gas pressure is uniform throughout the box. Wait as long as you please, the gas will never return to the original compartment. What, never? Well, hardly ever! For all the molecules to move in such a way as to recreate a vacuum in the second compartment would require a highly improbable configuration of molecular velocities. Suppose, as symmetry would suggest, the probability of finding a gas molecule in one side or the other of the original partition is $\frac{1}{2}$; then the probability that all N molecules spontaneously and independently diffuse to the same side is $(\frac{1}{2})^N$. At room temperature (20°C) and 1 atm pressure, a sample of gas initially confined to $1\,cm^3$ contains about 2.5×10^{19} molecules.[13] Therefore, compared with finding the gas uniformly spread throughout the entire available volume, the probability that all N molecules retreat to the initial compartment is roughly

$$P(N) \sim \frac{1}{10^{10^{19}}}.$$

The denominator of this fraction is the number 1 followed by 10 billion billion zeros. Technically, the probability is not exactly zero, but it is certainly indistinguishable from zero for all practical purposes.

I do not know the equation of state for Hell, but if one considers the latter "frozen over" when, statistically speaking, not a proton remains in the observable universe,[14] I can illustrate how small the above number is or, conversely, how large the reciprocal number is. Contemporary theories of the elementary particles predict that baryon number is not rigorously conserved, and therefore protons should decay to positrons, among other things. Experiment currently places the proton lifetime in excess of 10^{33} years. Assuming that there is any validity at all to the expectation of a finite proton lifetime, let us be generous and set it as 10^{40} years (or about 3×10^{47} s). Having on occasion seen the number of protons in the observable universe set at about 10^{80}, I shall again be lavish and estimate the proton count at 10^{100} (after all, what are a few zillion protons more or less?). If at some moment there are N_0 protons, the number $N(t)$ remaining at a time t later is given by the exponential decay law [to be discussed further in Chapters 4 and 8; see Eq. (4.2b)]. On average, there will be one proton left after a time interval

$$t_1 = T \ln(N_0) \sim 10^{50} \text{ s}, \tag{1.1a}$$

where T is the mean proton lifetime. Wait another 10^{40} years, or a total time still on the order of 10^{50} s, and there is a fair chance that even that last proton will have decayed. Hell is now completely disintegrated, let alone frozen.

How long must one wait for the gas molecules to evacuate the second compartment? Let us assume—because it is simplest to do so and because other models will hardly make any difference in the final results—that the molecules can be treated as spherical balls of some specified radius and mass. To be concrete, let the mass be that of the proton (1.67×10^{-24} g) and the radius be on the order of the Bohr radius (1×10^{-8} cm). It follows from the kinetic theory of gases that under the equilibrium conditions of room temperature (20°C), 1 atm pressure, and a volume of 1 cm^3, the molecules in the gas move at a mean speed of about 10^5 cm/s and undergo roughly 10^{28} collisions per second.[15] A complete rearrangement of the approximately 10^{19} molecular velocities should therefore occur about once every 10^{-9} s. However, only one in $10^{10^{19}}$ rearrangements is likely to yield the desired configuration. Thus, the time interval for the molecules to return to the first compartment would be

$$10^{-9} \text{ s} \times 10^{10^{19}} \sim 10^{10^{19}} \text{ s}. \tag{1.1b}$$

Note that $10^{19} - 9$ in the exponent is still about 10^{19}. The number $1/P(N)$ is *so* large that the timescale for complete molecular rearrangement obtained from any reasonable model of molecular collisions remains insignificant in comparison.

The time expressed in relation (1.1b) exceeds the putative lifetime of all matter in the universe [estimated in relation (1.1a)] to such an extent that the gas molecules in the container will have long since crumbled to photons and neutrinos before totally evacuating the second compartment. Actually, the container, itself, would no longer exist.[16]

By contrast, if the box originally contained only one molecule, the likelihood of "all" of the gas being found in the original compartment is clearly 50%, and one could expect this configuration to recur over the time interval required for the molecule to traverse the length of the container, namely in about $1\,cm/(10^5\,cm/s)$ or about $10\,\mu s$.

Looked at from the perspective of probability, the Second Law represents, not an absolute interdiction, but rather a continuum of possibilities. When few particles are involved, the behavior of the system is invariant under time reversal—that is, processes can occur in either direction—in keeping with the fundamental equations of motion (such as Newton's second law or the equations of Schrödinger and Dirac) that do not distinguish an "arrow" of time. When, however, the numbers of particles involved are unimaginably huge, the spontaneous transformation of a system proceeds in that direction for which the resulting molecular configuration is overwhelmingly probable, the direction in which entropy increases.

Having understood the statistical nature—and wishing to illustrate the limitations—of the Second Law, Maxwell, noted for his incisive intellect and playful spirit, proposed a mechanism that has since become an integral part of thermodynamic lore[17]:

[The Second Law] ... is undoubtedly true as long as we can deal with bodies only in mass, and have no power of perceiving or handling the separate molecules of which they are made up. But if we conceive a being whose faculties are so sharpened that he can follow every molecule in its course, such a being ... would be able to do what is at present impossible to us.

And so was born the famous (or perhaps infamous) Maxwell demon. What could such a demon do?

Now let us suppose that ... a vessel is divided into two portions, A and B, by a division in which there is a small hole, and that a being, who can see the individual molecules, opens and closes this hole, so as to allow only the swifter molecules to pass from A to B, and only the slower ones to pass from B to A. He will thus, without expenditure of work raise the temperature of B and lower that of A, in contradiction to the second law of thermodynamics.

At the time of its enunciation in the early 1870s (at the end of an elementary textbook on heat), Maxwell's little "being" elicited little interest in several of the major thermodynamicists then alive. Clausius responded that the Second Law did not concern what heat could do with the aid of demons, but rather what it could do by itself. Ludwig Boltzmann, who contended with Clausius for priority in deriving the Second Law from mechanics, also side-stepped the problem by arguing that in the absence of all temperature differences characteristic of thermal equilibrium, no intelligent beings could form. However, to discard Maxwell's demon as merely frivolous is to miss an essential point seized upon by later physicists; namely whether or not an intelligent intervention (not necessarily a demon's) can exploit in some way the naturally occurring thermodynamic fluctuations within a system to circumvent the Second Law.

By about 1914, it was already quite clear that no inanimate mechanisms could do this. Although phenomena such as Brownian motion and critical opalescence showed clearly that substantial fluctuations in the thermodynamic properties of bulk matter in thermal equilibrium can be made to occur,[18] such fluctuations would also affect any mechanism devised to operate Maxwell's "trap door" in a way that admitted or rejected molecules selectively. Moreover, the smaller the mechanism, the stronger would thermal fluctuations act upon it, and, correspondingly, the more uncontrollable would be the outcome.

The final loophole, however, that of a device operated by intelligent beings, was eliminated by the nuclear physicist Leo Szilard, whose broad interests also embraced major contemporary issues in the life sciences. In what is now regarded as a classic paper[19] relating the concepts of physical entropy and information, Szilard argued that any intelligent being, even a demon, would have to make a measurement of some kind in order to exploit naturally occurring fluctuations; the very act of measuring would result in an entropy production sufficient to prevent violation of the Second Law. The idea was carried further some twenty years later when Leon Brillouin[20] demonstrated more concretely that a Maxwellian demon, working in an isolated system in thermal equilibrium, could not see the molecules. Bathed in a surrounding sea of isotropic blackbody radiation, the demon could never distinguish one molecule from another without recourse to his own source of illumination—and this additional light would generate an increase in entropy.

All of this, of course, has not ended the discussion of Maxwell's demon. Nevertheless, from the time of Maxwell's proposal around 1871 to the present, no one has ever found or constructed a functioning demon, and it is probably accurate to state, as did Nobel laureate thermodynamicist Percy Bridgman, "that the entire invention of the demon is most obviously a paper and pencil affair."[21]

So, what about the Wirbelrohr?

*   *   *

I withdrew my fingers quickly, shut off the air supply, and stared anew at my father-in-law's present. When frost at the cold end melted and the temperature of the hot end dropped, I dismantled the device, half expecting to see some diabolical little creature inside smiling at me. Actually, it was clear at the outset that the Wirbelrohr could never have functioned as a Maxwell demon (i.e., in violation of the Second Law). The mere fact that the Wirbelrohr had to be fed compressed air signified that initial work was done on the gas. Nevertheless, how the Wirbelrohr converted work into such a striking difference in temperature was a mystery to me.

With the few parts of the Wirbelrohr laid out on my table, I understood better the significance of the German name "Wirbelrohr," or vortex tube. The heart of the device is the central chamber with a spiral cavity and offset nozzle. Compressed gas entering this chamber streams around the walls of the cavity in a high-speed vortex. But what gives rise to spatially separated air currents at different temperatures? Regarding the pieces closely, I recognized immediately what had hitherto escaped my attention when I had only the story from the science fiction magazine (which scarcely made an impression on me as long as I thought it could be a hoax). Although there were indeed no moving parts of any kind, the internal geometry of the device belied the outward bilateral symmetry. The symmetry was broken by the placement in one cross-tube of a small-aperture diaphragm that effectively blocked the efflux of gas along the walls of the tube, thereby forcing this part of the airflow to exit through the other arm whose cross section was unconstrained.

The glimmer of a potential mechanism dawned on me. Had the incoming air conserved angular momentum, the rotational frequency of air molecules nearest the axis of the central chamber would be higher—as would also be the corresponding rotational kinetic energy—than peripheral layers of air. However, internal friction between gas layers comprising the vortex would tend to establish a constant angular velocity throughout the cross section of the chamber. In other words, each layer of gas within the vortex would exert a tangential force upon the next outer layer, thereby doing work upon it at the expense of its internal energy, at the same time receiving kinetic energy from the preceding inner layer. Energy would consequently flow from the center radially outward to the walls generating a system with a low-pressure, cooled axial region and a high-pressure, heated circumferential region. Because of the diaphragm, the cooler axial air had to exit one tube (the cold side), whereas a mixture of axial and peripheral air exited the other (the hot side).

The presence of the throttling valve on the hot side now made sense. If the low pressure of the air nearest the axis of the tube fell below atmospheric pressure, the cold air would not exit at all; instead, ambient air would be sucked *into* the cold end—which is what I found to be the case when the valve was fully open. By throttling the flow, pressure within the central chamber was increased sufficiently so that air could exit both tubes. Thus, this simple, yet ingenious, device transferred energy within its working fluid (air) by means of a mechanism incorporating no moving parts except for the fluid itself.

However, even if no demon was at work, did the Wirbelrohr violate— or come close to violating—the Second Law? Because it involved a complex, turbulent fluid flow, the operation of an actual vortex tube could not be described strictly by thermodynamics alone. Nevertheless, with some simplifying assumptions, I was able to calculate the entropy change incurred by passage through the Wirbelrohr of a fixed quantity of gas of known initial temperature and pressure. Under what is termed adiabatic conditions (i.e., with no heat exchange with the environment), the Second Law requires that the entropy change of the gas, alone, be greater than or equal to zero. The resulting mathematical expression, augmented by the equation of state of an ideal diatomic gas and the conservation of energy (First Law of thermodynamics), yields an inequality

$$x^f \left(\frac{1-fx}{1-f} \right)^{1-f} \geq \left(\frac{p_f}{p_i} \right)^{2/7} \quad \left(x \equiv \frac{T_c}{T_i} \right) \tag{1.2}$$

relating the temperature (T_c) of the cold air flow to the initial temperature (T_i) and pressure (p_i) of the compressed air, the fraction (f) of a gas directed through the cold side, and the final pressure (p_f) of the ambient gas (taken to be 1 atm). From the First Law, the temperature (T_h) of the hot air flow can be expressed in terms of T_c and T_i.

By setting the expression for the entropy change equal to zero, I could calculate the lowest temperature that the cold tube should be able to reach if the gas flow were an ideal reversible process. The result was astonishing. With an input pressure of 10 atm and the throttling valve set for a fraction $f = 0.3$, compressed air at room temperature (20°C) could, in principle, be cooled to about –258°C, a mere 15°C above the absolute zero of temperature! The corresponding temperature of the hot side would have been 80°C. Clearly, the actual performance of the vortex tube, whose operation was by no means a reversible process, was far from any limitation posed by the Second Law. That did not make it any the less fascinating.

Intrigued to know more about the tube, I returned to the obviously nonfiction science fiction article that Fred sent me and tracked down the couple of references provided therein. The first experimental demonstration of a vortex tube seems to have been reported in 1933 by a French engineer, Georges Ranque.[22] Because at the time the device was the subject of a patent application, Ranque provided no drawings or quantitative analysis. Nevertheless, I was pleased to find from the general principles he enunciated that I had arrived at a broad explanation largely coincident with his own.

Little more was apparently heard of this device until about thirteen years later when, after the Second World War, detailed experimental investigations of German physicist Rudolph Hilsch came to the attention of an American chemist, R. M. Milton of Johns Hopkins University, who had Hilsch's work published in English.[23] In Hilsch's hands, proper selection of the air fraction f (approximately $\frac{1}{3}$) and an input pressure of a few atmospheres gave rise to an amazing output of 200°C at the hot end and −50°C at the cold end. Hilsch, who was the one (not my father-in-law) to coin the term "Wirbelrohr," used the tube in place of an ammonia precooling apparatus in a machine to liquefy air.

What alerted the author of the science fiction article to the existence of Hilsch's work was an initially brief report in the news section of an American chemical engineering journal[24] in 1946. The information was apparently furnished by Milton, who had visited Hilsch's laboratory and brought back (or perhaps constructed later) a small model of the vortex tube. Milton, according to the journal report, was not satisfied with the interpretation of Hilsch and Ranque that frictional loss of kinetic energy produced the radial temperature distribution. Upon requesting journal readers to submit their own interpretations, the reporter soon found himself inundated by a flood of letters from all over the world, a few excerpts of which appeared in a second report, also in 1946. Then, signing off cheerily with the hope that the information might provide a solid basis for further investigation, the reporter ceased all mention, as far as I knew, of the vortex tube.

Left with a farrago of explanations and a slim collection of old references, I looked wistfully at my Wirbelrohr. Did anyone really know how it worked?

*   *   *

Faced with other more pressing matters, I put the tube aside, except for occasional classroom demonstrations. Some time afterward, when the Wirbelrohr was all but forgotten, I experienced one of those serendipitous twists of fate that make life interesting. Standing in a corridor of a convention center and biding my time between sessions that interested me at a physics conference, I scanned a pile of papers strewn over a nearby table. Suddenly, one of the papers, an abstract of a talk to be given (or quite possibly already given) at a different scientific society than the one then convening, caught my eye; in its title I saw the words "Ranque–Hilsch Effect." Upon returning home, I wrote to the first author of the abstract, who kindly sent me a copy of his papers.[25] What was proposed therein, supported by experiment, was a mechanism far different from anything that I had seen proposed before.

According to a story often told in connection with Wolfgang Pauli, an eccentric genius whose acerbic criticism could be devastating (Ehrenfest called him "Die Rache Gottes"—the wrath of God), Pauli presented a new theory of elementary particles before an audience including Niels Bohr. Bohr, by contrast known for his gentle qualities, could nevertheless rise to the occasion when critical remarks were required. "We are all agreed that your theory is crazy," he allegedly replied. "The question which divides us is whether it is crazy enough to have a chance of being correct. My own feeling is that it is not crazy enough."[26] Although the paper that I read was certainly not crazy, it seemed to me sufficiently strange and original to have a chance of being correct.

With a loud roar, air rushes turbulently through the Wirbelrohr, just as it does through a jet engine or a vacuum cleaner. Buried within that roar, however, is a pure tone, a "vortex whistle" as it has been called, that emerges from the selective amplification of background noise. Although high-pitch whistles are often associated with the swirling flow of gas in turbomachinery with rotating shafts and blades, the vortex whistle can be produced as well by the tangential introduction and swirling of gas in a stationary tube. It is this pure tone or whistle, whose frequency increases with the velocity of swirling—and hence with the pressure of the compressed air—that is purportedly responsible for the spectacular separation of temperature in a vortex tube.

The Ranque–Hilsch effect is a steady-state phenomenon (i.e., an effect that survives averaging over time). How can a high-pitch whistle—a sound that, depending on air velocity and cavity geometry, can be on the order of a few kilohertz—influence the steady (or, in electric terms, the dc) component of flow? The answer, so the authors contended, was by "acoustic streaming." As a result of a small nonlinear convection term in the fluid equation of motion, an acoustic wave can act back upon the steady flow and modify its properties substantially. In the absence of unsteady disturbances, the air flows in a "free" vortex around the axis of the tube; the speed of the air is close to zero at the center (like the eye of a hurricane), increases to a maximum at around mid-radius, and drops to a small value near the walls of the tube. Acoustic streaming, however, deforms the free vortex into a "forced" vortex within which the air speed increases linearly from the center to the periphery. Acoustic streaming and the production of a forced vortex, rather than mere static centrifugation, engender the Ranque–Hilsch effect.

The experimental test of this hypothesis could not have been any more direct. Remove the whistle—and *only* the whistle—and see whether the radial temperature distribution remains. To do this, the authors first monitored the entire roar with a microphone and sent the resulting electrical signal to a signal analyzer that decomposed it into composite frequencies, of which the discrete component of lowest frequency and largest amplitude was identified as the vortex whistle. Next, they enclosed the central cavity of the Wirbelrohr inside a tunable acoustic suppressor: a cylindrical section of Teflon with radially drilled holes serving as acoustic cavities distributed uniformly around the circumference. Inside each hole was a small tuning rod that could be inserted fully (i.e., until it touched the outer shell of the Wirbelrohr) to close off the cavity or could be withdrawn incrementally to make the cavity resonant at the specified frequency to be suppressed.

To simplify their experimental test, the authors sealed off one output of the vortex tube and monitored with thermocouples the temperature difference between the center and periphery of the cavity (which was effectively equivalent to monitoring the temperature difference between the two output tubes). In the absence of the suppressor, an increase in the pressure of the compressed air produced, as I had noticed when experimenting with my own vortex tube, a louder roar and greater temperature difference. When, however, the acoustic cavity was adjusted to suppress only the frequency of the vortex whistle (leaving unaffected the rest of the turbulent roar), the temperature *difference* plunged precipitously at the instant the corresponding input air pressure was reached (Figure 1.2). In one such trial, the centerline temperature jumped a total of 33°C from −50°C to −17°C. With further increase in air pressure, the frequency of the whistle rose and, as it exceeded the narrow band of the acoustic suppressor, the temperature difference began to increase again. Additional evidence came from a striking transformation in the nature of the flow, itself, discernible with a touch of the hand. Before the frequency of the vortex whistle was suppressed—and while, therefore, a significant radial temperature separation was produced in the tube—the exhaust air swirled rapidly near and outside the tube periphery in the manner expected for a forced vortex. Upon suppression of the whistle, however, the forced vortex was also abruptly suppressed; now quiescent at the periphery, the air rushed out close to the centerline.

So the "demon" in the Wirbelrohr did not merely roar—it whistled, blowing hot and cold air simultaneously out different sides of its mouth. Thermodynamic analysis has shown that the Ranque–Hilsch effect is not particularly efficient at producing cold air. I have esti-

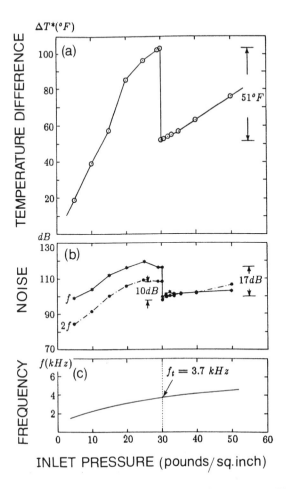

Figure 1.2. Relation of the vortex whistle to the Ranque–Hilsch effect. Increasing the pressure of the compressed air raises the frequency (f) of the vortex whistle [graph (c)]. When the frequency of the whistle corresponds to the tuned frequency (f_t) of the acoustic suppressor, the contribution of the whistle (and its first harmonic at $2f$) to the acoustic noise is greatly diminished [graph (b)], the temperature at the centerline of the central chamber jumps upward, and the temperature difference between what is effectively the "hot" and "cold" pipes falls precipitously [graph (a)]. The temperature difference grows again as f is made greater than f_t. (From M. Kurosaka *et al.*, AIAA/ASME 3rd Joint Thermophysics Conference, 1982, Paper AIAA-82-0952.)

mated the coefficient of performance, defined as the amount of heat removed from a mole of gas divided by the amount of work done on the gas, to be ideally just under unity (e.g., about 0.9), whereas the corresponding performance of a reversible machine (a Carnot heat pump) operating under the same conditions is about 5. The actual performance of Hilsch's tubes ran at or below 0.2. But, who could be so crass as to talk about the efficiency of a remarkable phenomenon?

For all I know, the case of the mysterious Wirbelrohr is largely closed, although, science being what it is, future versions of that device may yet hold some surprises in store. I have sometimes wondered, for example, what would result from supplying a vortex tube, not with room-temperature air but with a quantum fluid, like liquid helium, free of viscosity and friction.

The exorcism of the demon in the Wirbelrohr will not, I suspect, dampen one bit the ardor of those whose passion is to challenge the Second Law. Despite the time and effort that has been frittered away in the past, others will undoubtedly try again. On the whole, such schemes are bound to fail, but every so often, as in the case of Maxwell's own whimsical creation, this failure has its positive side: when, from the clash between human ingenuity and the laws of nature, there emerge sounder knowledge and deeper understanding.

1.1. Wirbelrohr Follow-up

In the years following publication of *And Yet It Moves*, I had again put the vortex tube aside and concerned myself with other matters. Always curious, however, I could not help wondering, in the midst of preparing this edition, whether acoustic streaming as an explanation of the hot and cold air flows has stood the test of time, or whether other mechanisms may have been proposed and tested. Having now at my disposal a remarkable information resource—the Internet—which was not available to me when I first began to write *And Yet It Moves*, I typed into my search engine a few keywords and sat back in anticipation of what I would find.

I did not have long to wait—0.05 s to be exact. The very first item on the relatively short list of returns bore a somewhat cryptic URL[27] with truncated, tantalizing excerpts of text promising an interesting result. I clicked the hypertext title and up popped an official-looking memorandum directed to a scientific news group concerned with fusion. Fusion? What could possibly connect the vortex tube to nuclear fusion? I looked further and was even more surprised. The memorandum originated from no less a bulwark of national security than the Naval Undersea Warfare Center (NUWC) of the United States Navy. "Eureka," I thought, "the U.S. Navy plans to power its nuclear submarines with vortex tubes!" I ran my eyes rapidly down the page to see the details of this exciting (and thoroughly unworkable) undertaking and, from the haunting familiarity of the text, realized immediately that my speculation was conceived too hastily. There, on the screen before me, was the text of *my own book*, five laser-printed pages full of material from this very chapter, complete with diagrams!

Apart from the thought, natural to an author, that the dissemination of his book over the Internet without permission must constitute some sort of copyright infringement, I stared in amazement at the screen, wondering why the U.S. Navy was citing my discussion of the Wirbelrohr. In answer to my silent query, I read further:[28] "The following may be relevant to the Potapov device. It contains excerpts from 'And yet it moves . . . strange systems & subtle questions in physics,' by Mark P. Silverman, ..."

I never heard of "the Potapov device," but it did not take more than a few mouse clicks to learn all that I needed to know about it. According to one source,[29] it was a water-heating device developed in Moldavia by Dr. Yu. S. Potapov "reported to produce a heat output up to 3 times greater than the energy required to drive it." But why stop at a mere 300% efficiency? Another more exuberant source[30] proclaimed that "Potapov's devices input several kilowatts of electricity into a centrifugal water pump . . . and gets [sic] out reportedly 400% to 1000% excess power in hot water!" Moreover, the device is available commercially and "hundreds upon hundreds of satisfied customers have ratified the technology in the marketplace!" I presume that meant that many Moldavians bought the device, but there is no mention of what they thought after trying it. Based on experiments reported in the first source, however, I can readily guess. "The Potapov device," the experimenters reported ruefully, "did not show any evidence of over-unity performance in our tests. We can find no explanation for the failure of this Potapov device to perform as reported (300% over-unity)."

I can suggest an explanation: The device does not work, has never worked, and will never work, and the report of "over-unity performance"—pseudoscience jargon for getting out more energy than one puts in—is either a deliberate fabrication or inept self-delusion. No device (like an engine or a pump) operating cyclically (i.e., returning to its original state after an operating cycle) produces more energy than it receives; to do so would violate the First Law of Thermodynamics. Moreover, the Second Law is even more restrictive; it prescribes what fraction of input energy can at best for given circumstances be converted into useful work—and this fraction is always less than 1. The only way a device could release more energy than it receives would be by tapping into the chemical or nuclear potential energy of its working material (a noncyclic process), which, in the case of the Potapov device, is water.

Water (H_2O) contains hydrogen atoms and the fusion of hydrogen atoms into helium (the process that powers the Sun) releases an enormous quantity of energy. For example, the fusion of deuterium (^{2}H) and tritium (^{3}H), two isotopic forms of hydrogen, to form helium (^{4}He) generates a neutron and 17.6 MeV (million electron volts) of energy.[31] [By contrast, the chemical combination of two hydrogen atoms to form a hydrogen molecule (H_2) releases only about 27 eV (i.e., a million times less energy).] It is indeed wonderful to contemplate the production of vast quantities of energy by hydrogen fusion, except that the Potapov device, operating (as I understand it) off room-temperature water, could never do this. For hydrogen atoms to get close enough to fuse, they must overcome the repulsive electrical force or Coulomb barrier between them, and this requires a mean kinetic energy per particle of about 10 keV (thousand electron volts), or a temperature of about 100 million degrees Centigrade.[32]

There are many out there in "cyberspace," I have found, who do not like the laws of thermodynamics, or other physical laws for that matter. They regard them not as limitations imposed by an indifferent Nature, but as barriers constructed by an arrogant scientific priesthood for the purpose of thwarting their wishes. Their diversity embraces all kinds of irrational belief—the denial of biological evolution, the denial of an ancient Earth, or the espousal of countless invalid schemes for generating energy out of nothing. Among the latter is a large subculture devoted to the exploitation of "cold fusion," the generation of nuclear fusion at temperatures close to room temperature, usually (although not exclusively) by various kinds of electrochemical reactions. All attempts by creditable laboratories to reproduce such claims have, to my knowledge, failed.

There is a certain irony to the ending of this chapter. I began, in effect, with my father-in-law's search through his science fiction collection for an article he once read concerning the vortex tube, and, as a consequence of my own internet search for the vortex tube, I have found the NUWC message with extensive excerpts from this chapter embedded in websites touting "new energy," "new science," and so forth. The only difference between science fiction and "new science" is that the authors of the former *know* they are writing fantasy. The authors of the latter, I suspect, do not want to know. As one such site proclaims,[33] it "is a big nasty nest of 'true believers' . . . and skeptics may as well leave in disgust."

Driving out skeptics, however, will not change the reality of the physical world. Neither the Potapov device nor any other room-temperature water pump is going to generate more energy than it receives, or perform work with an efficiency greater than that permitted by the Second Law. If you do not believe me, just ask one of those 38,000 Achnoid mutants without brain stems . . . or was it one of those encephalographic carbon aliens without adenoids . . . or . . . whatever.

Notes

1. R. Boyle, *New Experiments, Physico-mechanical, Touching the Spring of the Air and Its Effects*, (Oxford, 1660). A discussion of these experiments may be found in R. Harré, *Great Scientific Experiments*, Oxford University Press, New York, 1983, pp. 74–83.

2. This picture is reproduced in *Niels Bohr: A Centenary Volume*, edited by A. P. French and P. J. Kennedy, Harvard University Press, Cambridge, MA, 1985, p. 177.

3. H. Bondi, The Rigid Body Dynamics of Unidirectional Spin, *Proceedings of the Royal Society of London* **A405** (1986) 265. A qualitative description of the origin of the celt's behavior is given by J. Webb, Torque of the Devil, *New Scientist* (26 July 1997) 35.

4. Maxwell did not remain confused over the radiometer effect for long, but addressed its mechanisms in a seminal paper, On Stresses in Rarified Gases Arising from Inequalities of Temperature, *Philosophical Transactions of the Royal Society of London* **A170** (1879) 231; reprinted in *The Scientific Papers of James Clerk Maxwell*, Volume 2, edited by W. D. Niven, Dover, New York, 1952, pp. 681–712.

5. I write about my investigation of the "dragon tube" in *Waves and Grains: Reflections on Light and Learning*, Princeton University Press, Princeton, NJ, 1998, Chapter 14.

6. A comprehensive guide to the literature on Maxwell demons is given by H. S. Leff and A. F. Rex, Resource Letter MD-1: Maxwell's Demon, *American Journal of Physics* **58** (1990) 201.

7. A. C. Parlett, Maxwell's Demon and Monsieur Ranque, *Astounding Science Fiction* (January 1950) 105–110.

8. Quantum chromodynamics refers to the strong interactions of elementary particles (quarks) mediated by a type of charge whimsically designated "color."

9. A. Einstein, Autobiographical Notes, in *Albert Einstein, Philosopher–Scientist*, Volume 1, edited by P. A. Schilp, Open Court, La Salle, IL, 1969, p. 32.

10. N. Bohr, H. A. Kramers, and J. S. Slater, The Quantum Theory of Radiation, *Philosophical Magazine* **47** (1924) 785.

11. Although quantum mechanics allows for energy and momentum *non*-conserving virtual processes (such as the ephemeral creation and annihilation of particles in the vacuum), these violations occur over time intervals too short to be revealed directly by experiment; their existence is inferred from theory.

12. The term "energy" derives from a Greek root meaning "work." The term "entropy," introduced by Clausius, derives from a Greek root signifying "transformation."

13. This follows from the ideal gas law: (Pressure)(Volume) = Nk (Absolute Temperature), where Boltzmann's constant k is 1.38×10^{-16} erg/K, and the equilibrium conditions, expressed in suitable units, are 1 atm pressure = 10^6 dyne/cm^2, room temperature = 293 K, and volume = 1 cm^3.

14. Without protons, there can be no neutrons bound in atomic nuclei; free neutrons decay (to protons, electrons, and antineutrinos) with a mean

lifetime of about 15 minutes. Positrons (from proton decay) and electrons would presumably combine and mutually annhilate. There should be nothing left, then, except electromagnetic radiation and neutrinos. Whether hot or cold, such a universe ought to quality as a hellish place.

15. The mean speed v (technically, the root-mean-square speed) of molecules of mass M can be estimated by equating the mean molecular kinetic energy $\frac{1}{2}Mv^2$ and mean thermal energy $\frac{3}{2}kT$; thus, $v \sim \sqrt{kT/M}$. A single molecule of cross-sectional area A sweeps out a volume Av per second of travel within which occur about $N(Av/V)$ collisions with other molecules in the container of volume V. Thus, the total rate at which molecular collisions occurs is approximately N^2Av/V per second.

16. To say that the lifetime of a proton is so many years does not mean that one must necessarily wait that long for the particle to decay. It *might* decay within the next minute, although the enormous lifetime indicates that that is highly improbable. The inference of the time interval for evacuation of the second compartment must be interpreted in the same statistical way—that is, the molecules *could* all leave within the next few moments, but most likely they will not.

17. J. C. Maxwell, *Theory of Heat*, 8th ed., Longman, Green, and Co., London, 1885, pp. 328–329.

18. Brownian motion refers to the random movement of small particles in a fluid (e.g., pollen grains in water) as a result of the spatially nonuniform impacts by the molecules of fluid. Critical opalescence is the onset of a milky appearance in an initially transparent fluid at a temperature and pressure close to those for which a phase change occurs. Large fluctuations in the density, and therefore in the refractive index, of the fluid lead to substantial light scattering at all wavelengths, hence the whitish appearance.

19. L. Szilard, Über die Entropieverminderung in einem thermodynamischen System bei Eingriffen intelligenter Wesen, *Zeitschrift für Physik* **53** (1929) 840. The paper is reprinted together with an English translation in *The Collected Works of Leo Szilard: Scientific Papers*, edited by B. T. Feld and G. W. Szilard, MIT Press, Cambridge, MA, 1972, p. 103.

20. L. Brillouin, Maxwell's Demon Cannot Operate: Information and Entropy. I, *Journal of Applied Physics* **22** (1951) 334.

21. P. W. Bridgman, *The Nature of Thermodynamics*, Harvard University Press, Cambridge, MA, 1941, p. 161.

22. G. Ranque, Expériences sur la Détente Giratoire avec Productions Simultanées d'un Echappement d'air Chaud et d'un Echappement d'air Froid, *Journal de Physique et Radium* **4**(7) (1933) 1125.

23. R. Hilsch, The Use of the Expansion of Gases in a Centrifugal Field as Cooling Process, *Reviews of Scientific Instruments* **18** (1947) 108; translation of an article in *Zeitschrift der Naturwissenschaft* **1** (1946) 208.

24. R. L. Kenyon, Maxwellian Demon at Work, *Industrial Engineering and Chemistry* **38**(5) (1946) 5; The Demon Again, *Industrial Engineering and Chemistry* **38**(12) (1946) 5–14.

25. M. Kurosaka, Acoustic Streaming in Swirling Flow and the Ranque–Hilsch (Vortex Tube) Effect, *Journal of Fluid Mechanics* **124** (1982) 139; M. Kurosaka, J. Q. Chu, and J. R. Goodman, Ranque–Hilsch Effect

Revisited: Temperature Separation Traced to Orderly Spinning Waves or "Vortex Whistle," Conference of the American Institute of Aeronautics and Astronautics, 1982.

26. W. H. Cropper, *The Quantum Physicists*, Oxford University Press, New York, 1970, p. 57.

27. The Uniform Resource Locator (URL) is the address of each article of text, graphic, sound, or video on the World Wide Web.

28. bernecky@starbase.nl.nuwc.navy.mil (W. Robert Bernecky), 1 July 1995.

29. http://www.eden.com/-little/yusmar/potapov.txt

30. http://www.planetarymysteries.com/energy/ie.html

31. One electron volt, or 1 eV, is the energy acquired by an electron falling through a potential difference of 1 V. An energy on the order of 10 eV is required to ionize a hydrogen atom; this is roughly the energy scale at which chemical and biological processes occur. The energy equivalent of the electron mass (from Einstein's relation $E = mc^2$) is on the order of a million electron volts (MeV); this is roughly the threshold beyond which nuclear processes occur.

32. In general, the Kelvin scale must be employed in thermodynamic relationships. For example, the mean kinetic energy per particle is given by $\frac{3}{2}kT$, in which T is the Kelvin temperature and $k = 1.38 \times 10^{-23}$ J/K = 8.63×10^{-5} eV/K is Boltzmann's constant. Temperatures on the Kelvin (K) and centigrade (°C) scales are related by K = °C + 273, but with a number of the order of 100 million, the addition of 273 hardly matters.

33. http://www.eskimo.com/-billb/weird/wvort.html

Musical Bottles, Flying Balloons, and Hot Stoves: The Uncommon Physics of Common Things

2.1. The Good Sound of Coke™: Physical Modeling by Analogy

In the lighthearted, madcap African satire, *The Gods Must Be Crazy*, a Coke bottle, nonchalantly tossed from the cockpit of an airplane, landed in the midst of an isolated Bushman family never before exposed to the familiar commodities of "civilization." Of the many uses the family found for this mysterious "heaven-sent" gift, among the most pleasing was that of a musical instrument. (As the story unfolded, however, there were other *less* pleasing attributes—and the resourceful Bushman went to great lengths to return the gift and recover his peace of mind.) By teaching courses based on what I have called "self-directed learning"[1] —the radical proposition that students learn science better when striving to answer questions that arise out of their own curiosity—I have often been led to explore imaginative avenues of physics that would not likely have occurred to me had it not been for the curiosity of some student. In this way, I, together with a student colleague (E. R. Worthy), likewise came to realize that a Coke bottle—or, more precisely, about ten bottles containing different volumes of water—does indeed make a splendid instrument. Yet, surprisingly, for so superficially simple a structure, the tones of the bottle are by no means easily accounted for. For my student and me, as for the Bushman, the Coke bottle has not been drained of all its mystery.

To exploit the musical properties of the Coke bottle (or any other bottle) as an acoustic resonator, one must determine the relationship between the fundamental frequency f and the length of the air column ℓ. Despite the overall cylindrical symmetry of the bottle, the problem is a challenging one—and within the elementary physics literature that my student and I surveyed, we encountered no discussion of the issues involved beyond the standard geometric depiction of axial standing waves in tubes of constant cross section.[2] As shown in Figure 2.1 for the case of a tube sealed at one end (like the Coke bottle),

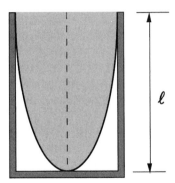

Figure 2.1. Fundamental mode of a single open-ended organ pipe.

each longitudinal mode has a displacement node at the closed end and (to good approximation) a displacement antinode at the open end. The lowest-frequency mode, therefore, has a wavelength $\lambda = \frac{1}{4}\ell$, from which it readily follows that the fundamental frequency is

$$f = \frac{v_s}{4\ell} \qquad (2.1)$$

where v_s is the speed of sound, which is about 344 m/s at 1 atm and 20°C.

From the shape of a Coke bottle, illustrated in Figure 2.2a, one might think that the "organ pipe" of Figure 2.1 would serve as a useful model for predicting the fundamental frequency. This, however, is far from the case. Nevertheless, a relatively simple approach that avoids solving the differential equations of wave theory can be made by analogy between an acoustic resonator and the ordinarily more familiar elements of ac circuit analysis. A comprehensive wave analysis of acoustic systems leads to equations of the same form as those of ac circuit theory when the lengths of individual components are small compared with a wavelength of sound. Justification of this assertion is by no means trivial, but is demonstrated in advanced books on theoretical acoustics.[3] From such a comparison, we find the following: (1) the gauge pressure (the difference between actual and equilibrium air pressures) in the acoustic system corresponds to the voltage at a point in the circuit; (2) the air flow (volume/time) through an orifice corresponds to the electric current; (3) a short narrow tube (a constriction) of length ℓ_c and cross section S_c is equivalent to an inductance (termed the analogous inductance)

$$L_a = \frac{\rho \ell_c}{S_c} \qquad (2.2a)$$

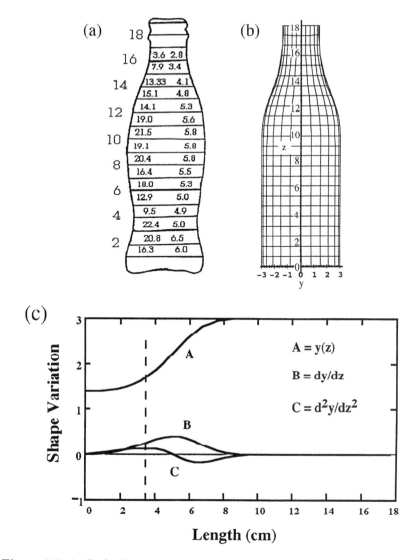

Figure 2.2. (a) Scale drawing of a Coke bottle; the three sets of numerals designate (left to right) the inside air column length (in cm), the water volume of cylindrical segment (in cm³), and the bottle diameter (in cm). (b) Generation of a Coke bottle from a hyperbolic tangent curve. (c) Variation with length along the bottle of the generatrix and its first and second derivatives.

in which ρ is the mass density of air ($\sim 1.2\,\text{kg/m}^3$ at $1\,\text{atm}$ and $20°\text{C}$); (4) a broad tube (a tank) of length ℓ_t and cross section S_t is equivalent to the analogous capacitance

$$C_a = \frac{S_t\,\ell_t}{\rho v_s^2};\tag{2.2b}$$

(5) radiation of sound (of angular frequency ω) from a constriction opening into free space constitutes an analogous terminal resistance

$$R_a = \frac{\rho \omega^2}{2\pi v_s}. \tag{2.2c}$$

One further refinement is necessary to make the model correspond more closely to reality. Because the antinode of a standing wave in a tube actually lies a little above the open end, we should replace ℓ_c in Eq. (2.2a) by an effective length

$$\ell_e = \ell_c + 0.8\sqrt{S_c} \tag{2.3}$$

that depends on the size of the opening. The correction, which is not necessary for the (much larger) tank, shows that even a flat aperture ($\ell_c = 0$) contributes an analogous inductance.

With the preceding relations, a wide variety of acoustic systems (bottles, horns, reed instruments, strings, loudspeakers, etc.) can be modeled in terms of their electrical counterparts. Now, let us examine the Coke bottle.

To an approximation sufficient for the purposes of this discussion, the Coke bottle (8 fluid ounces) of Figure 2.2a comprises a short neck inserted into a longer tank, a structure known as a Helmholtz resonator. Blowing across the mouth of the bottle excites the air inside to vibrate, but only those vibrations at the resonant frequencies of the bottle are amplified. Unless the bottle is "overblown," it is principally the fundamental tone that one hears, and it is this tone alone that we want to predict. If we neglect energy dissipation at the open end and at the walls, the bottle can be modeled by the ac circuit of Figure 2.3 containing a capacitor (of capacitance C) and inductor (of inductance L) in series. A series LC circuit exhibits a complex impedance[4] $Z = i(X_C - X_L)$ in which

$$X_C = \frac{1}{\omega C} \tag{2.4a}$$

is the capacitive reactance and

$$X_L = \omega L \tag{2.4b}$$

is the inductive reactance for an applied harmonic signal of angular frequency ω. If the capacitive and inductance reactances are equal, then the impedance of the circuit vanishes. From Eqs. (2.4a) and (2.4b), it follows that this resonance condition occurs at the frequency

$$f = \frac{\omega}{2\pi} = \frac{1}{2\pi\sqrt{LC}}. \tag{2.5}$$

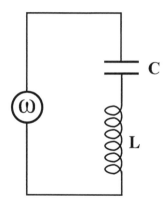

Figure 2.3. Diagram of the series LC circuit to which the Coke bottle, modeled as a Helmholtz resonator, corresponds.

Substitution of relations (2.2a) and (2.2b) and effective length (2.3) into Eq. (2.5) leads to the following expression for the fundamental frequency of a cylindrical bottle:

$$f = \frac{v_s}{2\pi}\sqrt{\frac{S_c}{S_t \ell_e \ell_t}} = \frac{v_s a}{2\pi b}\frac{1}{\sqrt{(\ell_c + 0.8a\sqrt{\pi})\ell_t}} \qquad (2.6)$$

with a and b the radii of the mouth and the base of the bottle, respectively. Because a, b, and ℓ_c are fixed parameters for a particular bottle, Eq. (2.6) expresses the fundamental frequency as a function of the variable tank length $\ell_t = \ell - \ell_c$, where ℓ is the full length of the air column. Thus, in marked contrast to Eq. (2.1) for the constant-diameter organ pipe in which $f \propto \ell^{-1}$, the fundamental of the bottle should vary as $(\ell - \ell_c)^{-1/2}$.

Now, the Coke bottle, whose radius varies smoothly from mouth ($a \sim 1.4\,\mathrm{cm}$) to base ($b \sim 3\,\mathrm{cm}$), is not, strictly speaking, a Helmholtz resonator, which, technically, comprises two joined tubes each of constant radius. How, then, is one to decide where the constriction ends and the tank begins? A good rule, supported by examination of the equations characterizing wave propagation in the bottle, is as follows: Take ℓ_c to be the distance from the mouth to the point where the second derivative of the bottle shape is maximum. Briefly, the wave equation for sound produced by the bottle differs from the comparable equation for an organ pipe only by a term containing this second derivative. Although small in magnitude and effectively applicable only over a small segment of the bottle length, this term is responsible for the marked difference in acoustic behavior between the bottle and the organ pipe.

As an illustration, look at Figure 2.2b, which depicts a mathematical simulation of the Coke bottle obtained by rotating the curve (with radius y and length z in cm)

$$y(z) = 1.6 \tanh\left(\frac{(z-18)^3}{216}\right) + 1.4 \qquad (2.7)$$

about the symmetry axis. Apart from the acoustically unimportant shallow indentation near the base in Figure 2.2a, the generatrix (2.7) provides an accurate representation of the size and shape of the Coke bottle. Equation (2.7) was obtained by trial and error, guided by the "principle of simplicity" to select the simplest curve that makes a smooth transition between the mouth and base. The desired sigmoid shape almost cried out for a hyperbolic tangent; the third power of the argument best reproduced the curvature of the bottle in the critical region where the constriction joins the tank. Figure 2.2c shows the variation with length of the generatrix (curve A) and its first (curve B) and second (curve C) derivatives. The location of the positive maximum value of curve C establishes that $\ell_c \sim 3.5$ cm.

Upon substituting into Eq. (2.6) the preceding Coke-bottle parameters and the speed of sound in air at room temperature, there emerges the final relation

$$f = \frac{1089}{\sqrt{\ell - 3.5}} \text{ Hz} \qquad (2.8)$$

for fundamental frequency f (Hz) as a function of air column ℓ (in cm).

$$* \qquad * \qquad *$$

So, what is the "sound" of Coke? To test the predictive accuracy of our model, Eqs. (2.6)–(2.8), my student and I measured the frequency of the tones obtained by blowing across the mouth of a Coke bottle filled to different levels of water. In keeping with the spirit of a home-based project to be performed with apparatus more or less readily available outside the physics laboratory, the resonant frequencies of the bottle were measured by means of a guitar tuner calibrated against a well-tuned piano. Water levels were sought for which the tuner registered standard notes, which were then converted to the corresponding frequencies. Heights were measured to within 0.1 cm, and the experimental uncertainty in frequency is estimated from the intervals of the guitar tuner to be less than $2^{1/48} - 1$ times the frequency of middle C, or approximately 4 Hz.

Figure 2.4, which gives results for both the Coke bottle and a right-circular cylinder closed at one end, summarizes the outcome of these experiments. One readily sees that the observed frequencies of the Coke bottle bear out very well the ac circuit resonator model and that a model of the bottle as an organ pipe is in thorough disagreement with experiment even though the curvature of the Coke bottle is relatively small (as shown in Figure 2.2c).

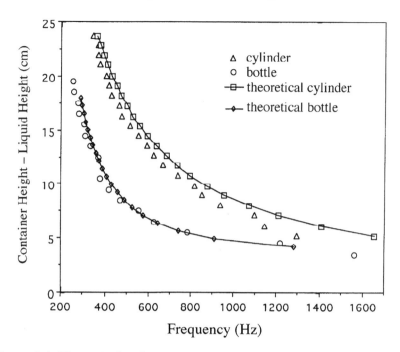

Figure 2.4. Theoretical and experimental variation of air-column length with frequency for both a Coke bottle and a right-circular cylinder closed at one end. At the scale shown, the uncertainties in length and frequency are smaller than the plotting symbols.

For the reader interested in the musicality of the bottle, the following table records the water levels (in cm) required to produce notes needed for various popular tunes[5]:

C_4	1.5 cm	C_5	10.2 cm
D_4	3.4	D_5	11.1
E_4	5.4	E_5	12.1
F_4	6.4	F_5	12.4
G_4	8.3	G_5	13.1

Although a general discussion of environmental effects on the tones of the Coke bottle would take us too far afield, temperature (T) is sufficiently important to consider briefly. All other parameters remaining unchanged, the increase in sound velocity v_s with T would raise the pitch of the bottle, as shown explicitly in Eqs. (2.1) and (2.6). However, raising T causes both the glass container and water contents to expand, thereby changing the length of the air column. Because the volume coefficient of expansion of water (2.1×10^{-4} per °C) is greater than that of glass ($\sim 1.1 \times 10^{-5}$ per °C for Pyrex), an increase in temperature should lead to a shorter air column and, therefore, to a higher pitch. In our experiments, we measured the variation in frequency as a function of temperature and found, overall, an increase of 22 Hz (approximately the interval of one note) over the range of 85°C—a change corresponding to a net rise in water level of about 1 cm.

Because a Coke bottle is not of uniform diameter, the change in water level (and therefore the length of the air column) engendered by a given volume expansion will have greater consequence where the bottle is narrow than where it is wide. Thus, temperature variations will affect more severely those pitches in the higher octaves than in the lower ones.

*   *   *

Although much of the physics that excites a student's imagination may often pertain to exotic realms far from daily reality (like black holes, time travel, and the fate of the universe), there is also a certain satisfaction that comes from being able to understand the behavior of familiar objects. Learning physics, I believe, is greatly facilitated when teachers can convey an appreciation for the power of general physical principles to account for what students frequently experience, yet rarely understand.

It may turn out as well—and such is the case with the Coke bottle— that, for all its familiarity, a superficially simple object hardly worth noting may pose a daunting challenge. In such instances, the use of analogy provides a powerful strategy. If science, as Nobelist Peter Medawar has written, is the "Art of the Soluble,"[6] then the *art* of that art is modeling, the capacity to exploit threads of commonality between outwardly dissimilar systems to arrive at a partial understanding of a complex and puzzling phenomenon. Moreover, in this art of modeling, what best serves as a model system can be surprising, at least to the uninitiated. Without prior experience, very few students—even physics graduate students—would look at a Coke bottle and see a resonant LC circuit rather than the structurally closer counterpart of an organ pipe.

However, analogy is not identity—and what a model omits from first consideration may yet prove decisive to deeper inquiry. The study of the humble Coke bottle is by no means a closed book. For example, if the bottle were of soft plastic, then a gentle deformation by squeezing would damp out the fundamental tone. Why? Is this the ineluctable consequence of broken cylindrical symmetry? No, for hard-glass bottles of highly elliptical cross section render strong fundamental tones. (Try blowing across an empty maple syrup bottle.) The simple resonator model does not explain this.

On the other hand, one might ask why the LC circuit model works as well as it does. In a puzzling reversal of expectations, I was initially astonished to discover that supposedly more sophisticated mathematical models of acoustic resonators that treated the bottle as a continuous transmission line with no arbitrary division between neck and tank predicted fundamental tones *less* accurate than those of the cruder resonator model with lumped-circuit elements. How can this be? It must suffice to say here only that to understand variable-diameter resonators like a Coke bottle in all their complexity constitutes a study of three-dimensional waves, including both radial and longitudinal modes of air vibration as well as the elastic properties of the vessel walls. The investigation is a fascinating one, but not recommended for the mathematically fainthearted.

From the perspective of experiment, however, the ready availability of personal computers with microphones and sound-analyzing software makes it possible to explore the acoustic properties of bottles and other simple resonators in great detail. It is an excellent way to learn about waves and vibrations in familiar systems with mysteries yet to be explored.

2.2. Comedy of Errors: What Every Aeronaut Needs to Know

As a physics teacher, I have often pointed out—to motivate a captive audience that would not likely have been sitting before me had not medical school or other professional school requirements loomed over them—that there is survival value to learning physics. To go unarmed into a technologically complex world without the slightest understanding of the universal laws and fundamental principles that make such a world possible is to be as naked and helpless as our paleolithic ancestors must have been before lightning and thunder. That, at least, was how the rhetoric went—and I cannot say with conviction that the majority of students found it convincing. However, here at last, is an indisputable example—drawn from no less a bastion of journalistic integrity than the newsletter of the American Physical Society—that awareness of physics could convey a degree of protection against self-destructive acts of ignorance.[7]

The case at hand is that of the unfortunate Californian who longed to float leisurely some 10 m above his back yard, eating sandwiches and drinking beer, until such time as he chose to descend. To realize his dream, he purchased 45 weather balloons, which he inflated with helium and attached to his lawnchair, secured by a tether to the bumper of his jeep. Then, having provisioned his lawnchair with the necessary snacks and a pellet gun with which to pop the balloons to effect his descent, the enterprising aeronaut released the tether—whereupon (according to the news report) he streaked like a rocket into the sky, reaching equilibrium, not at 30 feet as intended, but at 11,000 feet!

There, he drifted cold and frightened for 14 hours until he was noticed by the pilot of a passing jetliner. (Now, the plight of the hapless man is, in reality, no laughing matter, but can you imagine what must have gone through the mind of the air traffic controller to whom the pilot reported having passed someone with a pellet gun in a lawnchair at 11,000 feet?) Eventually rescued by the crew of a helicopter, the physics-deficient flier was arrested for having flown his lawnchair into the air-approach corridor of Los Angeles International Airport.

The *APS NEWS* report of this adventure reached me at a most propitious moment, my physics class having just completed its study of fluids and begun to examine the properties of ideal gases. There was a lesson—indeed several—to be learned from this adventure and, not being one to waste an opportunity, I promptly made it the focus of the following day's lecture. With the data provided in the news article— plus a modicum of creative modeling—a physics-savvy person can predict with adequate accuracy the height at which his or her lawnchair would settle (and would therefore know enough at least to throw in a down jacket and thermos of hot tea along with the sandwiches and beer). There *is* survival value to the study of physics!

Let us examine this vital issue.[8]

*   *   *

The Barometer Story: Model One

I designate by m the mass of the balloons and load and by V the volume of displaced air of density ρ. By Archimedes' principle, it follows that the balloons come to rest at an altitude h such that the total weight of the suspended objects is balanced by the buoyant force B, where

$$B = \rho V g = mg. \tag{2.9a}$$

Thus, the density of the air at h must equal the mean density (total mass/total volume) of the objects:

$$\rho = \frac{m}{V}. \tag{2.9b}$$

Although the news report did not give the mass and volume explicitly, enough information is furnished to allow a not-unreasonable estimate. First, the total mass. Taking account of all pertinent items, I would assign masses as follows:

Aeronaut	85 kg
Lawnchair	20 kg
45 Balloons	10 kg
Six-pack of beer + pellet gun + sandwiches	5 kg

for a total $m = 120\,$kg. The aeronaut may seem a bit portly, but then I inferred from the news report that he drank a lot of beer. I have also assumed that the lawnchair is of the sturdy wooden variety and not a flimsy aluminum one.[9]

Regarding the displaced volume, the report specifies only that, when fully inflated, the radius of a balloon exceeded two feet. Based on a weather balloon I cherished as a child and the fact that the lawnchair ascended precipitously, I estimate the maximum radius to be closer to three feet. This leads to a total volume $V = 144\,$m^3. In arriving at this value, I have assumed that, once filled to capacity at ground level, the balloons do not inflate further upon rising (for, to proceed otherwise, I would need information about the elastic properties of the balloon material, and the problem would become virtually intractable at the elementary level).

From Eq. (2.9b) and the preceding assumptions, the question then becomes: At what height above ground is the air density $\rho = 120\,$kg$/144\,$m$^3 = 0.83\,$kg/m^3? Recall that at ground level, where the pressure is $P_0 = 1\,$atm $\sim 10^5\,$N/m^2, the corresponding density of the air (at room temperature $T \sim 293\,$K) is, to good approximation, $\rho_0 = 1.2\,$kg/m^3. Thus, $\rho/\rho_0 \sim 0.69$.

The simplest (albeit approximate) method of attack is to apply what I call the "Barometer-Story formula," named for a delightful essay that I habitually read to my class whenever we study fluids.[10] Written by a physics teacher (who I am quite willing to believe may have actually had the experience related in the essay—but this I do not know), the story describes the response of a bright student asked on an examination to "Show how it is possible to determine the height of a tall building with the aid of a barometer."

Wearied by college instructors trying to tell him what to think, the student came up with numerous methods—all sound but impractical and altogether intentionally irrelevant to the particular point the teacher wanted to test—with the consequence, of course, that he received a zero for that question. For example, tie a barometer to the end of a cord, swing it as a pendulum, and determine the value of g at ground level and at the top of the building. "From the difference between the two values of g," said the student, "the height of the building can, in principle, be calculated." You get the picture. The essay is short, hilarious, and satisfying (at least to me and my class), for in the end the student triumphs. I highly recommend it to physics teachers; one of my own students confided afterward that he will now go to his grave knowing the barometer formula, whereas had he encountered it merely as an end-of-chapter exercise, he would have already forgotten it.

From the familiar form of the ideal gas equation of state

$$PV = nRT \tag{2.10a}$$

(with temperature T expressed in Kelvin), the number of moles per volume (n/V) can be readily eliminated in favor of the gas density to yield

$$\rho = \frac{MP}{RT},\qquad(2.10b)$$

in which M is the formula or molar mass (traditionally termed the molecular "weight," although this is a misnomer.) For air, with an approximate composition (accurate enough for our purposes) of 75% N_2 and 25% O_2, the gram molecular weight is $M \sim 29\,g$. R, the universal gas constant, is 8.2 J/mol K.

If we assume for the present that the temperature of the atmosphere is constant (i.e., independent of height), it follows from Eq. (2.10b) that the density is linearly proportional to pressure and, therefore,

$$\frac{\rho(h)}{\rho_0} = \frac{P(h)}{P_0}.\qquad(2.11)$$

The difference in air pressure between ground level and height h is simply the weight of a column of air of length h and unit cross-sectional area, or

$$P(h) = P_0 - \rho_0 g h\qquad(2.12)$$

if, as an additional approximation, I now take the air to be incompressible. [Equation (2.12) is the pressure–height relation that the physics teacher sought from the recalcitrant student in the Barometer Story.]

Strictly speaking, Eqs. (2.11) and (2.12) are inconsistent with one another, for the density of the gas cannot both change and be constant in the same problem. However, because the variation in density is already accounted for in Eq. (2.11), it is not too crude an approximation over a sufficiently small change in altitude to assume constant density for the evaluation of $P(h)$. How small is sufficiently small? With insertion of Eq. (2.12) into Eq. (2.11), the resulting expression itself suggests an answer:

$$\rho(h) \sim \rho_0 \left(1 - \frac{\rho_0 g h}{P_0}\right) = \rho_0 \left(1 - \frac{h}{h_0}\right).\qquad(2.13)$$

The approximation should be valid for altitudes low compared with the characteristic height

$$h_0 \equiv \frac{P_0}{\rho_0 g} \sim 8600\,m.\qquad(2.14)$$

I mention, in anticipation of what is to follow, that Eq. (2.13) is, in fact, a series expansion to first order in h/h_0 of the exact expression for the density variation of an isothermal atmosphere.

Substitution of Eq. (2.13) into Eq. (2.10b) leads to

$$h = h_0\left(1 - \frac{\rho}{\rho_0}\right) \sim 2700\,\text{m} \sim 8800\,\text{ft} \qquad (2.15)$$

as the equilibrium height of the lawnchair. This is somewhat lower than the reported height, but then we did not have to work too hard to get the answer—and, in any event, the outcome is orders of magnitude beyond what the aeronaut *thought* his elevation would be based on no quantitative reasoning at all.

However, we can work a little harder and do a little better.

The Isothermal Atmosphere: Model Two

Under the previous assumption that the temperature of the air remains constant (let us say at room temperature $T = 293\,\text{K}$), it is not difficult to derive the exact variation of density ρ with altitude z. Figure 2.5 shows the pertinent dynamical details. A cylindrical plug of gas of cross section A and height Δz remains in static equilibrium if the upward force of the air, $P(z)A$, on the bottom of the plug balances the sum of the downward force of air, $P(z + \Delta z)A$, on the top of the plug and the downward force of gravity, $\rho g \Delta z$, at the center of mass of the plug, leading to the well-known barometric equation

$$-\rho g = \frac{P(z + \Delta z) - P(z)}{\Delta z} \xrightarrow{\Delta z \to 0} \frac{dP}{dz}. \qquad (2.16)$$

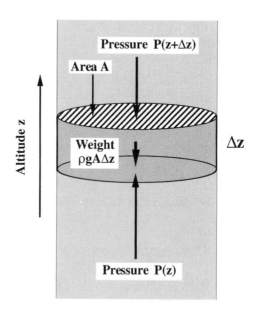

Figure 2.5. Diagram of forces on a cylindrical section of air within an isothermal atmosphere in static equilibrium.

Replacing pressure P in Eq. (2.16) by expression (2.10b) for density ρ leads to

$$\frac{d\rho}{dz} = -\left(\frac{Mg}{RT}\right)\rho = -\frac{\rho}{h_0} \tag{2.17a}$$

or, equivalently,

$$\frac{d\rho}{\rho} = d\ln\rho = -\frac{dz}{h_0}, \tag{2.17b}$$

which is readily integrated between $z = 0$ and $z = h$ to yield the exponential solution

$$\rho(h) = \rho_0 e^{-h/h_0}. \tag{2.18}$$

Note that the characteristic height $h_0 = RT/Mg$ in Eq. (2.17a) is precisely the same quantity as the h_0 in Eq. (2.14); this readily follows from use of Eq. (2.10b).

The exponential function, as discussed in the previous chapter, arises in two different, but equivalent, ways: (1) as the solution to a differential equation whenever the variation in a quantity is proportional to the remaining quantity [e.g., $d\rho \propto \rho$ in Eq. (2.17a)], and (2) as the limit

$$e^x = \lim_{n \to \infty}\left(1 + \frac{x}{n}\right)^n \tag{2.19}$$

of a sequence of terms as the index n approaches infinity. Substituting into Eq. (2.18) the approximation $e^x \sim 1 + x$, obtained by terminating the sequence (2.19) at $n = 1$, generates the result, Eq. (2.15), of Model One.

From the exact solution (2.18), the equilibrium altitude reached by the aeronaut is found to be

$$h = h_0 \ln\left(\frac{V\rho_0}{m}\right) \sim 3100\,\text{m} \sim 10{,}300\,\text{ft}, \tag{2.20}$$

which lies quite close to the 11,000-ft altitude reported in the news article.

However, with yet more effort, we can obtain an even more reliable answer. And it is *worth* the effort, for we are about to encounter something unexpected and counterintuitive.

The Adiabatic Atmosphere: Model Three

Although the prediction of Eq. (2.20) is good, the assumption that the temperature of the Earth's atmosphere remains the same at all heights is not valid. I can recall a number of transoceanic flights in which the cruising altitude of the aircraft and the outside temperature were simultaneously displayed over the cabin entrance. At roughly five miles high, the air temperature had fallen to approximately −20°C. If the temperature varied linearly with altitude and the ground was close to +20°C (room temperature), the preceding observation would imply a rate dT/dh of about −8°C/mile or −5°C/km. This is actually very close to the linear variation of −6.5°C/km recorded by atmospheric scientists over the approximate 12–16-km extent of the troposphere, the lowest layer of Earth's envelope of air.[11]

Since the height of the troposphere greatly exceeds the reported equilibrium altitude of the aeronaut, let us adopt the constant rate $dT/dh = -6.5$°C/km and explore the consequences of a model with linear variation in temperature. (The reason for designating this an "adiabatic atmosphere" will be made clear shortly.) Because it is often useful to work with dimensionless ratios when solving a problem, I will introduce a second characteristic height z_0 defined by the temperature–altitude relation

$$T(z) = T_0\left(1 - \frac{z}{z_0}\right) \qquad (2.21)$$

with T_0, the temperature (293 K) at ground level. From the requirement that $dT/dh = -T_0/z_0 = -6.5$°C/km, it follows that $z_0 \sim 45{,}000$ m.

Substitution of Eq. (2.21) into the barometric equation (2.16) leads to a differential equation

$$\frac{d\rho}{\rho} = d\ln\rho = -\left(\frac{1}{h_0} - \frac{1}{z_0}\right)\frac{dz}{1 - (z/z_0)}, \qquad (2.22)$$

which, at first glance, may seem complicated, but in reality is quite straightforward to integrate, for it involves the exact differential of a natural logarithm on both sides. Note, too, that if we let z_0 increase without bound, the atmosphere again becomes isothermal [see Eq. (2.21)] and the right-hand side of Eq. (2.22) reduces to the defining relation, Eq. (2.17b), of Model Two. For finite z_0, however, integration of Eq. (2.22) from $z = 0$ to $z = h$ yields a power-law expression

$$\rho = \rho_0\left(1 - \frac{h}{z_0}\right)^{(z_0/h_0)-1} \qquad (2.23)$$

Although the mathematical forms of solutions (2.23) and (2.18) are outwardly quite dissimilar, their kinship becomes apparent when the representation of an exponential as a limiting process [Eq. (2.19)] is again recalled. If the parenthetical expression on the right-hand side of Eq. (2.23) were recast as $[1 - (h_0/z_0)(h/h_0)]^{(z_0/h_0)-1}$, then it would have the approximate value of e^{-h/h_0} if z_0/h_0 were sufficiently large so that

−1 in the exponent could be neglected. For the parameters pertinent to our problem, the actual value of this ratio is

$$\frac{z_0}{h_0} = \frac{Mg}{R|dT/dz|} \sim 5.3 \tag{2.24}$$

and is independent of the choice of ground-level temperature T_0.

As our final estimate of the aeronaut's altitude h, the inversion of Eq. (2.23) leads to

$$h = z_0\left(1 - \frac{m}{\rho_0 V}\right)^{\frac{1}{(z_0/h_0)-1}} \sim 3700\,\text{m} \sim 12,100\,\text{ft}, \tag{2.25}$$

which also accords well with the reported facts (and is probably closer to the true altitude if our assumptions regarding m and V are accurate).

For purposes of comparison, Figure 2.6 illustrates the variation in air density with altitude for both the isothermal and adiabatic atmospheres.

However, something does not seem quite right here. Look at the numerical outcome in the preceding equation. It is *larger* than the estimate derived from Eq. (2.20) for an isothermal atmosphere. Yet, the air temperature is now falling with altitude. Should we not expect the density of colder air to be greater than that of warmer air—and therefore the aeronaut to level off at a *lower* altitude than if the atmosphere remained at room temperature all the way up? This curious feature is brought out strikingly in Figure 2.6. At any fixed value of the relative air density ρ/ρ_0, the linear-temperature curve lies to the right of the constant-temperature curve (i.e., at greater altitude) over the entire extent of the troposphere ($\sim$0–15 km).

There is no calculational error. A cursory examination of the barometric equation of motion shows the resulting behavior to be indeed possible. Because $P \propto \rho T$, the derivative dP/dz in the barometric equation (2.16) leads to two terms: one, deriving from $d\rho/dz$, reduces the air density with increasing altitude, but the other term, arising from dT/dz, carries the opposite sign and thereby causes the density to fall off at a slower rate than that of the isothermal atmosphere. It is these two opposing actions that lead to the coefficient $1/h_0 - 1/z_0$ in Eq. (2.22). How can that be? What went awry?

Nothing went awry. Rather, we have rediscovered a seminal property of air—indeed of any fluid—heated from below: It rises (and sometimes in startling ways). A graphic example of this behavior, first explained by Lord Rayleigh[12] and today still a subject of intensive investigation, is the Rayleigh–Bénard effect, the self-organization of convection cells within a short column of fluid confined between two planar barriers, the lower maintained at the greater temperature.

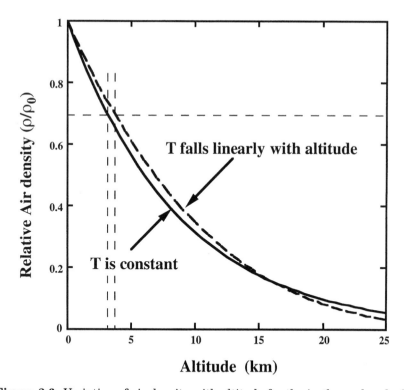

Figure 2.6. Variation of air density with altitude for the isothermal and adiabatic atmospheres. The horizontal dashed line marks the relative air density $(m/V)/\rho_0 \sim 0.69$ at the height at which the aeronaut settles. The two vertical dashed lines denote the corresponding altitudes. Note that the aeronaut levels off at a greater altitude in the adiabatic atmosphere than in the isothermal atmosphere.

Earth's atmosphere provides another example, less startling perhaps than the phenomenon studied by Bénard and Rayleigh, but no less interesting—and certainly far more significant in its overall impact on all of us. It is this convective flow in the atmosphere that bathes us in sea breezes by day and land breezes by night and rattles us unnervingly with atmospheric turbulence during our air flights.

Were the atmosphere left unperturbed for a sufficiently long time, it would eventually assume the quiescent state of thermal equilibrium, the density of each gas component falling exponentially with height. But such is not the case. Incessantly agitated under a negative temperature gradient, air is continually transferred from one part of the atmosphere to another. However—and this is the crucial feature—because the conduction of heat in gases is very slow, the atmosphere is never permitted to assume the equilibrium distribution we have discussed in Model Two. Instead, before an element of gas newly arrived at some location can adjust its temperature to that of its surroundings, it is again moved away. The distribution of the atmosphere, therefore, is determined by the condition that an element of gas, on being moved from one place to another, takes up the requisite pressure and volume in its new position without there being any loss or gain of heat by conduction.[13]

The foregoing process by which a quantity of gas undergoes a change in pressure, volume, and temperature without exchanging heat with the environment is termed "adiabatic" from the Greek word for "impassable." An ideal gas undergoing an adiabatic process satisfies the constitutive relations

$$PV^\gamma = \text{constant}, \tag{2.26a}$$

$$T\rho^{\gamma-1} = \text{constant}, \tag{2.26b}$$

in which $\gamma = c_P/c_V$ is the ratio of the molar specific heat at constant pressure to the molar specific heat at constant volume. For a diatomic gas (or mixture of diatomic gases like air), γ is expected on the basis of the equipartition theorem of classical physics to be $\frac{7}{5} = 1.4$.[14]

Had we known to begin with the adiabatic relations (together with the ideal gas equation of state and the barometric equation), we could have deduced the linear dependence of temperature on altitude rather than adopt it as an empirical fact. By casting the resulting expressions into forms comparable to Eqs. (2.21) and (2.23), we could then relate the heat-capacity ratio γ to our ratio of characteristic heights z_0/h_0 and thereby predict the rate of temperature fall through the chain of connections:

$$\frac{z_0}{h_0} = \frac{Mg}{R|dT/dz|} = \frac{\gamma}{\gamma-1} = \frac{c_p}{R}, \tag{2.27a}$$

$$\frac{dT}{dz} = -\frac{g}{c_P/M} = -\frac{\text{Gravitational field strength}}{\text{Heat capacity per unit mass}}. \tag{2.27b}$$

Insertion of the classical value $\gamma = 1.4$ into Eq. (2.27a) gives $dT/dz = -10°C/km$, which is not too far from the actual rate of $-6.5°C/km$.[15] The discrepancy may be attributable to the fact that, in reality, Earth's atmosphere is an extremely complex system, affected in no small way by the irregularities of the planet's surface and the reflectivity of the clouds.

It is precisely such complexity, however, that makes the physical world so interesting and therefore the physicist's capacity to interpret it in terms of a few basic laws and simple models so remarkable. The predicament of our aeronaut aside, perhaps it is not so much the survival value of physics that is worth emphasizing after all, but the intrinsic pleasure and satisfaction that comes with understanding.

2.3. Cool in the Kitchen: Radiation, Conduction, and Newton's "Hot Block" Experiment

The rate at which an object cools down gives valuable information about the mechanisms of heat loss and the thermal properties of the material. In general, heat loss occurs by one or more of the following four processes: (1) conduction, (2) convection, (3) evaporation, and (4) radiation.

In *conduction*, heat is transferred through a medium by the collisional encounters of thermally excited molecules vibrating about their equilibrium positions or, in the case of metals, by mobile, unbound electrons; only energy, not bulk matter itself, moves through the material. *Convection*, by contrast, refers to the transfer of heat through the action of a moving fluid; in free or natural convection, the motion is principally the result of gravity acting on density differences resulting from fluid expansion. *Evaporation* entails the loss of heat as a consequence of loss of mass, the faster-than-average molecules escaping from the free surface of a hot object, thereby removing kinetic energy from the system. Last, *radiation* involves the conversion of the kinetic and potential energy of oscillating charged particles (principally atomic electrons) into electromagnetic waves, ordinarily in the infrared portion of the spectrum. From the perspective of classical physics, charged particles moving periodically about their equilibrium positions (or indeed undergoing any kind of acceleration) radiate electromagnetic energy.

Although the physical principles behind the four mechanisms lead to different mathematical expressions, it is widely held that if the temperature of a hot object is not too high, then the decrease in temperature in time follows a simple exponential law, an empirical result historically bearing Newton's name. But how good an approximation to reality *is* Newton's law—and what in any event determines whether the temperature of the hot object is too high? Furthermore, although Newton's name is readily associated with his laws of motion, law of gravity, and various optical phenomena (e.g., Newton's rings, Newton's lens equation), it does not usually appear in discussions of thermal phenomena. Indeed, apart from this one instance, a search through a score or more of history of science books and thermal physics books at various levels of instruction produced but one other circumstance for noting Newton's name and that was his failure to recognize the adiabatic nature of sound propagation in air.[16] This historical footnote accentuates, however, the circumstance that Newton pursued his interests at a time long before the concept of heat was understood. He died in 1727, but the beginning of a coherent system of thermal physics might arbitrarily be set at nearly a hundred years later when Sadi Carnot published (1824) his fundamental studies on "the motive power of fire" (*La puissance motrice du feu*).

What, then, prompted Newton to study the rate at which hot objects cool, how did he go about it, and where did he record his work?

Let us look at the historical questions first. In stark contrast to Newton's other eponymous achievements, for which anyone desirous of knowing their origins could turn to such ageless sources as *Principia* or *Opticks*,[17] the paper recording the law of cooling is decidedly obscure. After much searching, I discovered a reprinting of this elusive work in an old and dusty physics sourcebook.[18] According to the author, William Francis Magie, the paper, "A Scale of the Degrees of Heat," was published anonymously in the *Philosophical Transactions* in 1701, although Newton was known to have written it.

Despite its obscurity, this is, like much of Newton's work, a fascinating paper. In contrast to what I expected, Newton's principal concern was *not* to nail down the precise formulation of another physical law, but rather to establish a practical scale for measuring temperature. By 1701, Newton, then about 60 years old, had long since completed the fundamental studies of his youth—motion, gravity, the calculus, spectral decomposition of light, diffraction of light, and much else—to take up the position of a British functionary. In 1695, he had been appointed Warden of the Mint and moved from Cambridge to London. It seems reasonable to speculate that Newton's concern with temperature and the melting points of metals was motivated by his responsibility for overseeing the purity of the national coinage.

All the same, the experiment was vintage Newton: clever use of the simplest materials at hand to carry out a measurement of broad significance.[19] Having selected linseed oil, which has a relatively high boiling point (289°C) for an organic material, as his thermometric substance, Newton presumed that the expansion of the oil was linearly proportional to the change in temperature. With this thermometer and a chunk of iron heated by the "coals in a little kitchen fire," Newton proceeded to establish what quite possibly was the first temperature scale by which useful measurements were made. He set 0 on his scale to be "the heat of air in winter at which water begins to freeze" and defined 12 to be "the greatest heat which a thermometer takes up when in contact with the human body." On this fixed two-point scale, the "heat of iron ... which is shining as much as it can" registered the value 192.

Having established the above points, as well as other intermediate values (e.g., 17: "The greatest heat of a bath which one can endure for some time when the hand is dipped in it and is kept still"[20]), Newton sought an independent procedure for confirming their validity. To do this,

... I heated a large enough block of iron until it was glowing, and taking it from the fire with a forceps ... I placed it at once in a cold place ... and placing on it little pieces of various metals and other liquefiable bodies, I noted the times of cooling until all these bodies lost their fluidity and hardened, and until the heat of the iron became equal to the heat of the human body. Then by assuming that the excess of the heat of the iron and of the hardening bodies above the heat of the atmosphere, found by the thermometer, were in geometrical progression when the times were in arithmetical progression, all the heats were determined.... The heats so found had the same ratio to one another as those found by the thermometer.

Thus Newton's law of cooling first saw light of day.

In fact, that small section above is *all* that Newton had to say about "Newton's law." Note that not once in the entire paper does Newton mention the word "temperature." At this time, the concepts of heat and temperature were poorly understood and confounded; Newton refers to both as "heat" (*calor* in Latin). Note, too, that nowhere does Newton mention the word "exponential" or give the equation of exponential form

$$T - T_0 = (T_m - T_0)e^{-kt} \qquad (2.28a)$$

(with rate constant k, ambient temperature T_0, and maximum temperature T_m) that explicitly shows the temporal variation synonymous with Newton's law. However, in verifying the points on his scale, Newton asserted that "the heat which the hot iron communicates in a given time to cold boldies ... is proportional to the whole heat of the iron," or, as we would express mathematically in current symbolism,

$$\frac{dT}{dt} = -k(T - T_0). \qquad (2.28b)$$

Equation (2.28a) is the solution to Eq. (2.28b), and from Eq. (2.28a), the reader will readily confirm that

$$\frac{T_1 - T_0}{T_2 - T_0} = \frac{T_2 - T_0}{T_3 - T_0} = \frac{T_3 - T_0}{T_4 - T_0} = \cdots = e^{k\Delta t} \qquad (2.28c)$$

where the temperatures $T_1, T_2, T_3, \ldots$ are all measured at equal intervals of time ($t_1 = \Delta t$, $t_2 = 2\Delta t$, $t_3 = 3\Delta t, \ldots$). This is the "geometrical progression" of temperatures (above the ambient temperature) when the times are in "arithmetical progression," which Newton assumed.

The law is simple and useful. But is it true? This question came to mind at a time when I was teaching my son Chris physics and calculus during his senior year of high school, and so we investigated the matter together.

*   *   *

"It is certain," wrote Benjamin Thompson (Count Rumford) at the opening of his own seminal paper on the flow of heat,[21] "that there is nothing more dangerous in philosophical investigations than to take any thing for granted, however unquestionable it may appear, till it has been proved by direct and decisive experiment." Thus inspired, Chris and I retired to our kitchen to test, as best we could, the law governing the cooling of a hot block of iron.

As a substitute for the block of iron and Newton's open kitchen fire (which surely would have invalidated our home insurance contract), we used, instead, an electric range and turned the right rear burner on HI so that it "was shining as much as it can." The ambient temperature was measured to be 25.5°C with a mercury-in-glass thermometer, which we also used to calibrate a digital thermocouple thermometer[22] placed in contact with the burner. The glowing burner registered 456°C, which would appear to be somewhat cooler than Newton's kitchen fire.[23] All the same, it was hot enough to test Newton's law.

Turning the range off, we simultaneously activated a stopwatch and recorded the temperature of the burner at intervals of 1 min for a total of 35 min, at which time it approached ambient temperature closely enough to terminate the experiment. The temperatures, measured to a precision of 1°C for $T \geq 200°C$ and 0.1°C for $T < 200°C$, are plotted with circles in Figure 2.7. It is convenient and instructive to plot the data as dimensionless quantities. The vertical axis gives the ratio of the instantaneous temperature to the ambient temperature (all temperatures in Kelvin). The horizontal axis registers the time in units of a characteristic "radiation time" t_r, which in this experiment was found to be 25 min. The dashed line in the figure is the exponential curve (i.e., Newton's law) obtained as a least-squares fit to the data. The fitting procedure, performed with statistical software on a Macintosh computer, minimized the the sum of the squares of the deviations of a straight line from the natural logarithm of $T - T_0$, which, according to Eq. (2.28a), should be a linear function with slope $-k$ and intercept $\log_e(T - T_m)$.

It is clear that Newton's law does not represent the mechanism of heat loss very well. If not Newton's, then what law governs the physics at work here?

Under the conditions of this experiment—initially glowing solid iron in (for the most part) stationary air—the principal mode of heat loss is radiation until the reduced temperature (T/T_0) has fallen to about 1.2. The net rate (dQ/dt) at which a hot body immersed in an ambient medium of temperature T_0 loses energy by radiation is given by Stefan's law[24]

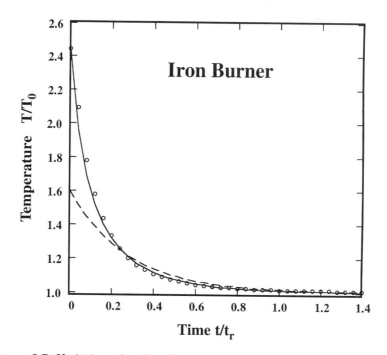

Figure 2.7. Variation of reduced temperature with reduced time for the cooling of an electric burner. Circles mark experimental points; the solid line represents Stefan's law (heat loss by radiation); the dashed line is an exponential fit (Newton's law). Radiation time $t_r = 25\,\text{min}$.

$$\left(\frac{dQ}{dt}\right)_{\text{rad}} = -\varepsilon\sigma A(T^4 - T_0^4), \qquad (2.29)$$

in which ε is the emissivity of the material, $\sigma = 5.67 \times 10^{-8}\,\text{W/m}^2\,\text{K}^4$ is a universal constant, A is the effective radiating area, and T is the absolute temperature. The first term on the right-hand side is the radiant power lost to the environment; the second term is the radiant power received from the environment. A general thermodynamic argument can be given (although not here) that the material parameter ε must be the same for both radiant emission and absorption. Note that the rate of energy loss is proportional to the fourth power of T, whereas in Newton's law [Eq. (2.28b)], it is proportional to the first power of T. [Under the present circumstances, Eq. (2.28b) corresponds to net cooling by conduction, as will be seen shortly.]

When an object radiates an amount of energy dQ, the drop in temperature dT depends linearly on the mass m and specific heat capacity c of the material:

$$dQ = mcdT. \qquad (2.30)$$

(There should be no confusion of c with the vacuum speed of light, which does not appear in this chapter.) Upon substitution of Eq. (2.30) into Eq. (2.29) and division of both sides of the equation by T_0, Eq. (2.29) takes the dimensionless form

$$\frac{dJ}{d\tau} = 1 - J^4 \qquad (2.31)$$

with reduced temperature $J \equiv T/T_0$ and reduced time $\tau \equiv t/t_r$, the characteristic radiation time referred to earlier being defined by

$$t_r = \frac{mc}{\varepsilon\sigma AT_0^3}. \qquad (2.32)$$

Equation (2.31) no longer explicitly contains material properties or physical constants and can be solved readily by separating variables, decomposing the right-hand side into a sum of rational terms

$$d\tau = \frac{dJ}{1-J^4} = \frac{1}{4}\left(\frac{dJ}{1-J} + \frac{dJ}{1+J} + \frac{2dJ}{1+J^2}\right)$$

and applying the elementary integration formulas for the natural logarithm and inverse tangent. This leads to the implicit relation for J:

$$\left(\frac{J-1}{J+1}\right) = \left(\frac{J_m-1}{J_m+1}\right)e^{-2(\arctan J_m - \arctan J)}e^{-4\tau} \qquad (2.33)$$

with $J_m = T_m/T_0$.

With a little additional effort, it is not difficult to reduce Eq. (2.33) to an approximate explicit relation for $J(\tau)$. Combine the two phase terms into a single phase by using the trigonometric identity[25]

$$\arctan x + \arctan y = \arctan\left(\frac{x+y}{1-xy}\right),$$

make the small-argument $(x < 1)$ approximations $\arctan x \approx x$ and $e^x \approx 1 + x$, and carry through the algebraic manipulations to isolate $J(\tau)$, obtaining

$$J = 1 + \frac{2(J_m-1)(J_m^2 - 2J_m + 3)}{(J_m^2+1)[(J_m+1)e^{4\tau} - J_m + 1] - 4(J_m-1)}. \qquad (2.34)$$

Applied to Eq. (2.33), the small-argument approximation implies that $(J_m - J)/(J_m^2 + 1) \ll 1$, which is best fulfilled when J is close to its maximum value. In other words, we would expect the relation (2.34) to describe radiative heat loss well and to become progressively poorer as the temperature approaches ambient temperature (in which case, radiation becomes secondary to conduction). By contrast, it is to be noted that when $\mathcal{J}$ is close to the ambient temperature ($\mathcal{J} \sim 1$), the radiative cooling law takes the form of Newton's law, for Eq. (2.31) becomes approximately

$$\frac{d\mathcal{J}}{d\tau} = (1-\mathcal{J})(1+\mathcal{J})(1+\mathcal{J}^2) \approx -4(\mathcal{J}-1). \tag{2.35}$$

Looking again at Figure 2.7, one sees that this expectation is indeed borne out. The solid curve, which closely matches the experimental points, is calculated from Eq. (2.34), with the radiation time $t_r \sim 25\,\text{min}$ the only adjustable parameter. Figure 2.8, in which $\log_e(\mathcal{J} - 1)$ is plotted against τ, shows the experimental results from another perspective. Clearly, the locus of experimental points (circles) is not linear. (Actually, the log of the log of $\mathcal{J}$ makes a nearly straight line.) The solid line, Eq. (2.34), follows the experimental points up to about $0.8t_r$ units of time, after which conduction sets in and pure radiation theory is no longer adequate. Note, however, that the log function greatly exaggerates what are actually small discrepancies between theory and experiment since (for any base a) $\log_a(x) \to -\infty$ as $x \to 0$. The dashed line is the least-squares linear fit leading to the exponential curve in Figure 2.7.

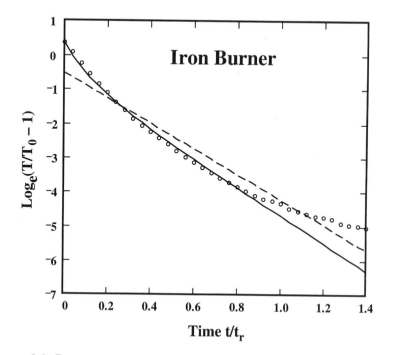

Figure 2.8. Representation of Figure 2.7 on a logarithmic scale.

From Eq. (2.32) and the empirical radiation time, we can estimate the emissivity of the burner. Employing the physical quantities $m = 0.263\,kg$, $A = 0.056\,m^2$, $T_0 = 299\,K$, $c = 448\,J/kg\,K$, and $t_r = 25\,min$ in Eq. (2.32), we obtain $\varepsilon \sim 0.91$, which is quite reasonable for an object with a blackened, oxidized surface. By comparison, the emissivity of soot is 0.95 and that of flat black paint is 0.94.[26]

How is it possible that Newton, who started out with an even higher temperature than we did, obtained "Newton's law" (i.e., an exponential decrease in temperature)? Actually, who can say with certainty that he did? His short paper contains no experimental record at all of the variation in temperature of the hot iron with time. He states, but does not demonstrate, that "the heat which the iron loses in a given time is proportional to the whole heat of the iron." Moreover, no information is given as to how Newton measured intervals of time—no mean task in an age when an inexpensive digital wristwatch (our own chronometer) did not exist.

Last and conceivably most significant, Newton did something with his hot block that we did not do with our burner: He removed it from the fire and "placed it . . . where the wind was constantly blowing." Newton did this specifically so that "equal parts of the air are warmed in equal times and carry away a heat proportional to the heat of the iron." Forced convection, which played no role in our own experiments, would have provided an additional cooling mechanism.

In any event, having satisfied ourselves that our own hot block experiment could be accounted for satisfactorily by Stefan's law rather than by Newton's law, we inquired next into the consequences of both conductive and radiative energy-loss occurring together. It is of particular interest to ascertain whether the effects of radiation are perceptible over a temperature range sufficiently low that heat loss is dominated by conduction and to determine whether, in fact, Newton's law provides a good model under these circumstances.

The rate at which a hot object initially at maximum temperature T_m loses heat by conduction across a region of thickness d bounded by a surface of area A is described adequately by the relation

$$\left(\frac{dQ}{dt}\right)_{con} = -\frac{k_T A}{d}(T - T_0) \tag{2.36}$$

where k_T is the coefficient of thermal conductivity of the material. Use of Eq. (2.30) in Eq. (2.36) to relate again dQ and dT results in the dimensionless equation

$$\frac{d\mathcal{J}}{d\tau} = 1 - \mathcal{J}, \tag{2.37}$$

where now (and for the rest of this chapter) we define the reduced time $\tau \equiv t/t_c$ in terms of the characteristic "conduction time"

$$t_c = \frac{mcd}{k_T A}.\qquad\qquad(2.38)$$

The form of Eq. (2.37) is precisely that of Newton's law, and the solution is

$$\mathcal{J} = 1 + (\mathcal{J}_m - 1)e^{-\tau},\qquad\qquad(2.39)$$

or, equivalently (in terms of original variables), Eq. (2.28a) with rate constant k identified with t_c^{-1}.

To test relations (2.37)–(2.39) on a system for which conduction ought ideally to be the only significant cooling mechanism, we cut a small rectangular block of white styrofoam and covered it with a thin wrap of aluminum foil for which the emissivity is very low ($\varepsilon \sim 0.02$). Highly reflective surfaces by definition do not absorb radiation, and poor absorbers make poor emitters, a fact that often seems paradoxical to those encountering it for the first time. We inserted the digital thermometer probe down the long axis of the block and set the block (fastened vertically to a chemical stand) into a pot of water. When the water was boiling vigorously and the display of the thermometer registered 100°C, we removed the block from the water, set the stand on the kitchen counter (in the absence of wind!), and recorded the temperature with resolution of 0.1°C in intervals of 1 min, as before. The experimental points are plotted with circles (upper dataset) in Figure 2.9. The solid line through the circles is the exponential curve calculated from Eq. (2.39) and leads to a conduction time $t_c = 5.3$ min.

That the value obtained for t_c is reaonable may be seen by substituting into Eq. (2.38) the appropriate parameters for our styrofoam block: $m = 0.02$ kg, $c = 1226$ J/kg K, $d = 0.006$ m, $A = 0.0132$ m^2, $k_T = 0.029$ W/m K. The theoretical result is 6.4 min.

In the second part of the experiment, designed to enhance the effects of radiation without changing any other property of the system, we simply painted the foil surface black, using flat lampblack paint which, to a large extent, is an oil emulsion of soot. That the blackening of the surface markedly affected the cooling rate is shown by the locus of diamond plotting symbols (lower dataset) in Figure 2.9.

It is important to note (although the graph does not show it) that an exponential fit to the "black" data is as poor as before. The dashed line in Figure 2.9 is an exponential curve parametrically adjusted (not fit) to match visibly well the overall pattern of data points. That even this attempt is poor can be seen in the logarithmic plots of Figure 2.10. Newton's law does not work particularly well here. How, then, can we account for these results? If not Newton's nor Stefan's, then what or whose law applies?

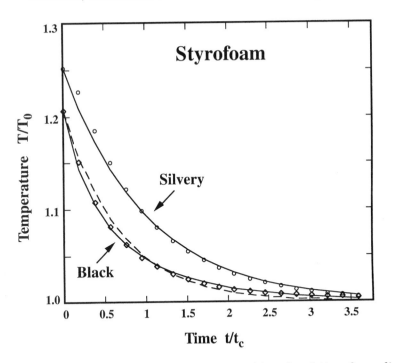

Figure 2.9. Variation of reduced temperature with reduced time for cooling of a block of styrofoam. Circles mark experimental points of white styrofoam covered with reflective foil; diamonds mark experimental points of the same object painted black. The upper solid line is derived from Eq. (2.39) with conduction time $t_c = 5.3\,\text{min}$; the lower solid line is derived from Eq. (2.42) with radiation parameter is $\gamma = t_c/t_r = 0.21$; the dashed line is derived from Eq. (2.39) with conduction time $t_c = 3.6\,\text{min}$.

By combining the radiation law (2.29) and the conduction law (2.36) together with the temperature–heat relation (2.30), one obtains the dimensionless cooling law

$$\frac{d\mathcal{J}}{d\tau} = (1+\gamma) - \mathcal{J} - \gamma \mathcal{J}^4, \tag{2.40}$$

in which the parameter γ is the ratio of the conduction and radiation times:

$$\gamma = \frac{t_c}{t_r} = \frac{\varepsilon \sigma d T_0^3}{k_T} \tag{2.41}$$

Although Eq. (2.40) may look more or less tractable, it cannot be integrated analytically to yield an exact closed-form expression. Nevertheless, it can be integrated numerically, and Figure 2.11 shows a

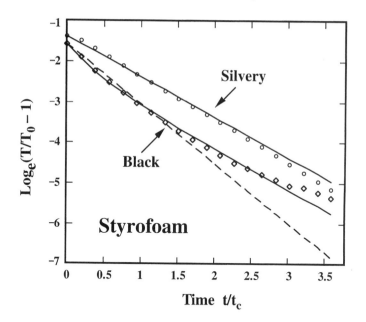

Figure 2.10. Representation of Figure 2.9 on a logarithmic scale.

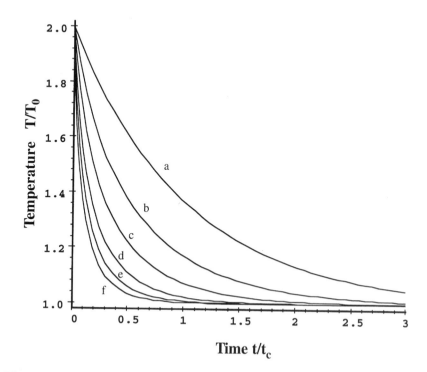

Figure 2.11. Numerically calculated cooling curves for a system with both conductive and radiative energy loss. The one adjustable parameter is γ which equals (a) 0 [no radiation], (b) 0.1, (c) 0.25, (d) 0.5, (e) 0.75, (f) 1.0.

sequence of cooling curves, solutions of Eq. (2.40) obtained with the symbolic computation software Maple, showing the transition from pure conduction with $\gamma = 0$ (i.e., Newton's law) to strong radiation $\gamma = 1$. A discussion of the numerical procedure, which is one of the Runge–Kutta methods, would take us too far afield but can be sought in appropriate reference books.[27]

It is also possible to derive an approximate solution to Eq. (2.40), which works quite well in the low-temperature regime (i.e., when $\mathcal{J}_m$ is not much in excess of 1). In that case, we treat the radiative part of Eq. (2.40) (i.e., the terms containing γ) by the approximation in Eq. (2.35) to obtain an exponential solution of the form of Eq. (2.28a), but now with rate constant $k = 1 + 4\gamma$. This approximation is then substituted back into Eq. (2.40) to obtain, after some patient effort, the interesting expression

$$\mathcal{J} = 1 + (\mathcal{J}_m - 1)e^{-\tau} \exp\left(-\frac{a}{k}\left(1 - e^{-k\tau}\right) \right) \qquad (2.42)$$

with $a = 4\gamma[1 + \frac{3}{2}(\mathcal{J}_m - 1)]$. Note that Eq. (2.42) involves the exponential of an exponential, a law ostensibly quite different from Newton's or Stefan's law.

The solid line through the "black" data in Figures 2.9 and 2.10 is the theoretical curve calculated from Eq. (2.42) using the same value of t_c as obtained for the "silvery" data, since the rate of heat conduction is determined by the conductivity of the styrofoam and should not be significantly affected by a thin layer of surface paint. Theory and experiment are in excellent accord when γ, the only adjustable parameter, takes the value 0.21. Then, from Eq. (2.41), with $T_0 = 301.1\,\mathrm{K}$, we find the emissivity of the block to be about $\varepsilon = 0.7$.

*   *   *

The investigations described in this section, as I have already noted, were initially undertaken as part of a high school physics course emphasizing the inclusion of meaningful research opportunities (in lieu of "cookbook" laboratory exercises) in accord with my philosophy of self-directed learning. By collaborating as partners in an endeavor of mutual interest, both student and teacher acquired some useful lessons in the workings of science and the intricacies of history.

Puzzled by frequent reference in math and physics textbooks to a law of Newton's of whose origin we knew nothing and by the apparent unquestioning credence with which the law was reported to hold widely, we tracked down Newton's paper. To our surprise we found that, far from demonstrating a physical law, the investigation of cooling, whose corroborative details Newton did not even bother to report, was to Newton solely an auxiliary procedure in the more important task of creating a practical temperature scale. A procedure, moreover, that, for all one can tell from the written account, was based on a mathematical relation that Newton merely assumed to be true.

As to the validity of Newton's law, however, our own kitchen experiments indicate that, where energy loss by radiation contributes significantly—even when the temperatures involved are relatively low—an exponential variation does not make a particularly good model. Exceptions to Newton's law are not hard to find. The cooling of a hot burner on an electric range is very well accounted for by Stefan's law. The cooling of a piece of black styrofoam—an object with high emissivity and low thermal conductivity—is accounted for by "the Silvermen's law" [if I may so call Eqs. (2.40) and (2.42)].

And therein lies perhaps the most important lesson of all: Abide Rumford's advice, and you cannot go too far astray for too long.

Notes

1. M. P. Silverman, Self-Directed Learning: A Heretical Experiment in Teaching Physics, *American Journal of Physics* **63** (1995) 495; Self-Directed Learning: Philosophy and Implementation, *Science & Education* **5** (1996) 357; Problem-Based Learning and Self-Directed Learning, *What Works II: Postsecondary Education in the 21st Century*, Penn State University, State College, PA, 1996.
2. See, for example, D. Halliday and R. Resnick, *Fundamentals of Physics*, Wiley, New York, 1988, pp. 425–428.
3. L. L. Berenek, *Acoustics*, McGraw-Hill, New York, 1954, pp. 128–143; P. M. Morse, *Vibration and Sound*, American Institute of Physics, New York, 1976, pp. 233ff.
4. In circuits that obey Ohm's law, the potential difference (voltage) across a circuit element and the current that flows through the element are linearly related, $V = IZ$, in which the coefficient of proportionality Z is called the impedance. If the element dissipates energy as heat, then Z is the familiar real-valued resistance R. However, the element may store energy in an electric field (or equivalently as charge on conducting plates) or in a magnetic field (or equivalently as current through a solenoid), in which case Z is a purely imaginary-valued capacitive or inductive reactance. A real circuit element may exhibit both resistance and reactance to varying degrees depending on the frequency of the electromagnetic signal it is carrying.
5. Two familiar songs played by my student before the class are

   (A) "Mary Had a Little Lamb":
   EDCDEEE DDD EGG EDCDEEE EDDEDC;
   (B) "Jingle Bells":
   EEE EEE EGCD E FFF FEEEE EDDE DG
   EEE EEE EGCD E FFF FEE GGFD C

   The designation C_4 is "middle C" on the tempered scale, nominally corresponding to a frequency of 261.6 Hz. Each succeeding half-tone (C#, D, D#,

E, etc.) in the octave between C_4 and C_5 is theoretically higher in frequency than the preceding half-tone by the factor $2^{1/12} = 1.0595$. Music, however, is not a precise science like physics, and a musician will play tones the way they sound best. The water levels that produce the notes listed in the text correspond only approximately to the frequencies of the tempered scale.

6. P. B. Medawar, *The Art of the Soluble*, Methuen, London, 1967.

7. 1997 Darwin Award Winner, *APS NEWS* **7** (January 1998) 7. The Darwin Awards, usually bestowed posthumously, celebrate the theory of evolution by commemorating the remains of those who have removed themselves from the human gene pool in spectacularly stupid ways. (The official web site is http://www.darwinawards.com.)

8. In a correction ["But It Was Such a Good Story . . ."] published several months later (March 1998), the *APS NEWS* noted that the reported incident actually occurred in 1982, that the flight lasted two hours (not fourteen), and that the lawnchair descended (without rescue by helicopter) onto power lines, blacking out a neighborhood for 20 minutes. A collection of news reports about the incident, which came to my attention long after this chapter had been completed, is available at the website http://www.markbarry.com/amazing/lawnchairman.html. There, one can see photos of the launch and hear a recording of the radio communication between the aeronaut and his ground crew. There is nothing to indicate that the balloonist was frightened; indeed, he was very much enjoying the flight.

9. Actually, it turned out to be the latter!

10. A. Calandra, The Barometer Story: Angels on a Pin, reprinted in *The Shape of This Century: Readings from the Disciplines*, edited by D. W. Rigden and S. S. Waugh, Harcourt Brace Jovanovich, New York, 1990, pp. 343–344.

11. M. Neiburger, J. G. Edinger, and W. D. Bonner, *Understanding Our Atmospheric Environment*, W. H. Freeman, San Francisco, 1973, p. 27; A. Maton et al., *Exploring Earth Science*, 2nd ed., Prentice-Hall, Needham, MA, 1997, p. 193.

12. Lord Rayleigh, On Convection Currents in a Horizontal Layer of Fluid, When the Higher Temperature Is on the Under Side, *Philosophical Magazine* **32**, Series 6 (1916) 529–546; reprinted in B. Saltzman, *Theory of Thermal Convection*, Dover, New York, 1962, pp. 3–20.

13. Sir James Jeans, *The Dynamical Theory of Gases*, Dover, New York, 1954, p. 335.

14. According to the equipartition theorem, a molecule in equilibrium with a thermal reservoir at temperature T has a mean molar energy $\frac{1}{2}RT$ for each dynamical degree of freedom. Thus, for a diatomic molecule with 5 degrees of freedom (3 degrees of translation along the x, y, and z axes; 2 degrees of rotation around the x and y axes if the molecule is aligned along the z axis), the mean internal energy is $\frac{5}{2}RT$ and, therefore, $c_V = \frac{5}{2}R$ and $c_P = c_V + R = \frac{7}{2}R$, from which follows $\gamma = \frac{7}{5}$. It is assumed that the temperature is sufficiently low that vibrational degrees of freedom are unexcited; otherwise, the numerical value of γ would be lower than 1.4. The equipartition theorem is a classical theorem that breaks down when the mean

thermal energy per molecule is comparable to the quanta of energy for transitions to excited states.

15. A more direct way to arrive at Eq. (2.27b) is to use the thermodynamic relation $dQ = c_P\, dT - V\, dP$ for a differential quantity of heat absorbed by an ideal gas. In the case of an adiabatic process ($dQ = 0$), the preceding equation, together with the barometric equation, yields dT/dz immediately.

16. Newton's calculation of the speed of sound in air lacked the specific heat ratio γ, which reflects the fact that heat cannot be exchanged between the sound wave and ambient medium within the duration of one period (reciprocal of the sound frequency). This error was subsequently corrected by Laplace.

17. Sir Isaac Newton, *Principia*, Motte's translation into English, revised by Cajori, University of California Press, Los Angeles, 1966; *Opticks*, based on the fourth edition, London, 1730, Dover, New York, 1952.

18. W. F. Magie, *A Sourcebook in Physics*, McGraw-Hill, New York, 1935, pp. 125–128. The paper is briefly cited—but with no mention of the law of cooling—in Richard Westfall's biography of Newton, *Never at Rest*, Cambridge University Press, London, 1980, p. 527. The citation is Scala graduum caloris, *Philosophical Transactions*, **22** (1700–1701), 824–829.

19. For additional discussion of Newton's experimental genius, see M. P. Silverman, *Waves and Grains: Reflections on Light and Learning*, Princeton University Press, Princeton, NJ, 1998), Chapter 5: Newton's Two-Knife Experiment.

20. If 12 degrees Newton (°N) corresponds to body temperature (37°C), then the hottest sustainable bath of 17°N corresponds to 52°C. Newton's value is actually quite good. While living in Japan, I experienced total immersion in the skin-searing temperatures of Japanese baths fed by hot springs. The hottest of such baths in Japan is said to be in the town of Kusatsu and is recorded at 57.8°C.

21. This paper ("Convection of Heat"), among other Rumford writings, is also to be found in Magie's Sourcebook (pp. 146–161).

22. Extech Model 421305 digital thermocouple thermometer; ambient operating range 0–50°C; measurement range −50°C to 1300°C; resolutions of 0.1°C and 1°C, depending on range.

23. Given the common origin (0°C for the freezing point of water) and linearity of the Newton scale, it follows that 37°C/12°N = 456°C/x°N or $x = 148$ on the Newton scale.

24. See, for example, R. A. Serway and J. S. Faugn, *College Physics*, 5 ed., Saunders, New York, 1999, p. 356.

25. This follows readily from the more familiar expression $\tan(a + b) = (\tan a + \tan b)/(1 - \tan a \tan b)$.

26. E. Hecht, *Physics: Algebra/Trig*, 2nd ed., Brooks/Cole, New York, 1997, p. 1020.

27. See, for example, A. Heck, *Introduction to Maple*, 2nd ed., Springer-Verlag, New York, 1997, p. 537.

CHAPTER 3

The Unimaginably Strange Behavior of Free Electrons

3.1. Variations on "The Only Mystery"

Strangeness is a relative thing. With varying degrees of sophistication, I have been thinking about physics for more than forty years now, and this has no doubt both strongly and subtly influenced how the world presents itself to my eyes. There are laws and principles as familiar to me as the names of my children; most people are unaware of them and would not believe them even if informed.

As I leave my office, I ball up one last piece of scrap paper and toss it into the recycling box. The paper follows a smooth parabolic arc as it lands. Were I to toss a rubber ball or a steel ball bearing in exactly the same way, I know that (barring air resistance) it would follow the same path in the same time. All objects, irrespective of mass, chemical composition, or any other physical property, fall at the same acceleration at the same location on the surface of the Earth. Galileo allegedly demonstrated this some four centuries ago. Yet, surveys of science "literacy" show that much of the American and British public readily subscribe to the Aristotelian notion that heavy objects intrinsically fall faster than light ones.

It is dark out when I reach my car to start for home. The Moon lies suspended above one of the campus sports fields like an enormous orange. I know, however, that it is falling toward the Earth with an acceleration roughly 1/3600 that of the wadded paper I tossed some moments earlier. I have no fear of being crushed for, although it is falling, the Moon will never reach the Earth—not in my lifetime at least, if at all. An inward radial attraction, in fact, is what makes the Moon go around the Earth in a circular orbit. Again, most people would find that thought strange. Like René Descartes, they imagine some force pushing the Moon tangentially around its orbit.

Upon reaching home, I apply the brakes and my car stops. If I did not apply the brakes, the car would eventually stop anyway (although not in a convenient location) because of friction. Excluding friction (and eventual obstacles), however, I know that the car would continue to move forward at a constant speed forever. "Move forward by itself forever?," I can hear one of my nonphysicist friends protest; "Impossible! You have to push or pull an object to keep it moving." Yet, even now, the Voyager probes, long since detached from the rockets that launched them, continue to penetrate unimpeded the void of interstellar space.

The various consequences of the laws of gravity and motion addressed above *are* in some ways strange, but not unimaginably so. They are features of the macroscopic world to which physicists have reconciled themselves and which they can understand in terms visualizable to the mind's eye. Newton, for example, illustrated some three centuries ago in the *Principia* how a sequence of increasingly wide parabolic arcs of a free-falling projectile leads naturally to the circular trajectory of an orbiting satellite. Much later, during the second decade of the 20th century, the mass independence of the law of free fall found its explanation in Einstein's general theory of relativity, which created the imagery of a conjoined space and time (space–time) warped by the presence of matter. The resulting contours of this incorporeal four-dimensional terrain constrain all matter to move along the shortest (actually, the extremal) paths or geodesics.

There is a qualitative difference between the tangible realm of classical physics, to which Newton's and Einstein's laws of motion and gravity belong, and the submicroscopic domain of the elementary particles and their composite structures. The principles governing the latter give rise to strange consequences that, at least to my satisfaction, have never been—and most likely can never be—adequately interpreted in terms of objects or processes drawn from the world of macroscale experiences. The behavior of such systems is unimaginably strange.

"A great physical theory . . . when it is confirmed, takes on its own impersonal existence in the course of time, becomes completely detached from its originator, and is finally received as self-evident."[1] So wrote the editor of a collection of Erwin Schrödinger's personal correspondence on wave mechanics. Having spent much of my professional life thinking about the intricacies of quantum physics, I am dubious that the theory will ever become self-evident (if, indeed, one can even characterize classical physics that way[2]). Certainly, quantum mechanics is no longer the novelty that it was when its foundations were being laid in the 1920s, and a seemingly endless supply of basic textbooks makes the subject common knowledge throughout the physics community. Nevertheless, familiarity with the fundamentals of quantum mechanics has not, by any means, exhausted the surprises to which these principles still give rise.

The attribute of the quantum world that is responsible in large measure for its strangeness is that the denizens of this world, the elementary particles for example, propagate from one point to another as if they were waves, yet are always detected as discrete lumps of matter. There is no counterpart to this behavior in the world we experience directly with our senses.

Imagine pouring a container of sand onto a flat plate with a small centrally located hole. A few centimeters below the bottom of the plate is a tiny movable detector that counts the number of sand particles arriving per unit of time. It would show, in accord with our expectations, that the greatest number of sand grains is registered directly under the hole, and that this number diminishes as the detector is moved transversely (i.e., parallel to the plate) away from the hole. Puncture another hole in the plate near the first one, and the sand pours through both in such a way that the total number of grains reaching the detector at any location is the sum of the number of grains reaching that point from each hole independently. In other words, opening up more holes can only *increase*, and never decrease, the total amount of sand reaching the detector per unit time at any location.

However, what if the experiment were performed with *electrons* rather than with sand? Quantum mechanics predicts that if the aperture size and separation are comparable in magnitude to the wavelength of the electron (or, depending on the experimental configuration, some other characteristic length parameter), the scattered electrons, like light waves, should give rise beyond the perforations to an undulatory interference pattern (Figure 3.1). For a plate with two identical rectangular apertures, the electron intensity, or number of electrons striking a unit area of the detector surface per unit of time, might be described mathematically by the expression

$$I(\theta) = 2I_0 \left(\frac{\sin^2 \beta}{\beta^2} \right)(1 + \cos 2\alpha). \qquad (3.1a)$$

Here, the deviation of the electrons from the forward direction is measured by the angle θ; I_0 is the contribution to the electron intensity from either aperture alone. The last factor in Eq. (3.1a) represents the *interference* between the components of the electron wave issuing from each aperture. The phase angle α on which the interference pattern depends is given by

$$\alpha = \left(\frac{\pi a}{\lambda} \right) \sin \theta, \qquad (3.1b)$$

where a designates the distance between the centers of the apertures and λ is the electron wavelength. The first factor in parentheses in Eq. (3.1a) describes the *diffraction* of the electron wave through a single aperture, let us say of width b. For the sake of simplicity, the aperture length is assumed to be much longer than the electron wavelength, in which case it will not significantly affect the passage of the electrons.

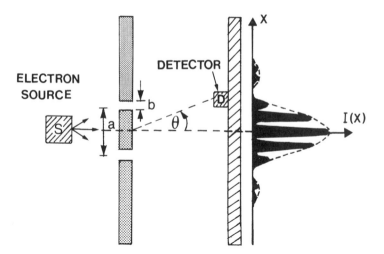

Figure 3.1. Two-slit electron interference experiment. An electron propagates like a wave from source S through the two slits to the detector D, but always registers as a discrete particle. The particle distribution $I(x)$ about the forward direction shows the oscillatory behavior of wave interference. The broken line which envelops the interference fringes is the single-slit electron diffraction pattern.

The phase angle β of the diffraction pattern (which envelops the interference pattern) is expressible as

$$\beta = \left(\frac{\pi b}{\lambda}\right)\sin\theta. \tag{3.1c}$$

Expressions (3.1a)–(3.1c), which characterize the diffraction and interference of ideally monoenergetic electrons (or monochromatic light) by a particular configuration of slits, are provided only as an example. Nevertheless, they serve to highlight important differences between doing the experiment with sand grains or with objects that behave like waves. At the central maximum of the particle distribution (i.e., in the forward direction $\theta = 0$), the electron intensity can be four times—not two times—the intensity from a single aperture. More generally, for a configuration of N slits, this intensity enhancement increases as N^2. At the locations of the interference minima, where 2α is an odd-integer multiple of π, no electrons are detected. Thus, opening up a second aperture has resulted in fewer electrons arriving at certain locations than with only one aperture open. When the aperture width and separation are sufficiently larger than the wavelengths, however, one sees from the expressions for α and β that outside of the forward direction, the amplitude of the diffraction pattern becomes very small and the fringes of the interference pattern become extremely narrow, eventually beyond the spatial resolution of the detector. The electrons, distributed nearly exclusively in the forward direction with a mean intensity of $2I_0$, would then seem to pass through the holes of the plate in much the same way as do grains of sand.

It is to be stressed that no electrons can be created or destroyed in this experiment; hence, the total number of electrons received by the detector at all locations must be equal to the total number that has passed through all apertures. Somehow, scaling the properties of the particles and the width of the apertures from the macroscopic size of a sand grain down to the ultrasmall size of an electron has radically altered the way in which the particles are distributed.

Richard Feynman, who had the good fortune to create his own version of quantum mechanics some twenty years after Schrödinger and Heisenberg developed theirs, characterized the wavelike interference of particles as

... a phenomenon which is impossible, *absolutely impossible*, to explain in any classical way, and which has in it the heart of quantum mechanics. In reality, it contains the *only* mystery.[3]

The fact that electrons, which always register at a detector like hard little balls of sharply defined mass and charge, give rise in large numbers to an interference pattern may be surprising, but this is not the core of the mystery to which Feynman referred. The real enigma unresolvable by any mechanism of classical physics becomes apparent only when the electron flux (another word for intensity or number of particles "flowing" through a unit area per unit of time) is reduced to such an extent that no more than one electron at a time passes through the apparatus. It is then that one must really come to grips with the implications of an electron wave.

The electron wave is *not* to be thought of as a water wave, sound wave, or any other wave in a medium which represents an actual physical displacement of matter; nor is it like a classical light wave composed of oscillating electric and magnetic fields which, though immaterial, is an expression of the classical electric and magnetic forces such a wave would exert on a unit electric charge. From the perspective of quantum mechanics, classical waves are composed of enormous numbers of elementary quantum excitations. For example, 1 watt of pure red light of wavelength 650 nanometers (nm) represents an emission of about 3×10^{18} quanta of light (or photons) per second. The wave characterizing the electrons is a *probability* wave; it allows one to calculate the probability of finding an electron within a given spatial region at a specified time.

Quantum mechanics does *not*, however, permit one to determine in which direction a *particular* electron (or any other elementary particle) that has been diffracted by some obstacle or aperture will eventually go. The arrivals of single electrons at the detector are random. Yet, according to theory, the random arrival of individual electrons in a sufficiently large number should build up in time the same interference pattern that would be engendered quickly by a large electron flux.

In the course of my research, I frequently investigated the quantum behavior of electrons theoretically. Although there is beauty and a measure of personal satisfaction in equations that reveal to the mind's eye striking new phenomena, the full implications of particle interference are so startling that they must be seen firsthand to be adequately appreciated.

In the mid-1980s, at a time when Japanese research laboratories were opening up to Western scientists, I had the pleasure of being invited to the Hitachi Advanced Research Laboratory (ARL) at Kokubunji, a part of the Tokyo prefecture. Created only a short time earlier in the midst of the already flourishing Central Research Laboratory (CRL) devoted principally to applied research, the ARL was to be a sort of hybrid Japanese-style Bell Labs and Princeton Advanced Institute concerned with fundamental studies. I was there in part to help the electron holography group under the direction of Dr. Akira Tonomura find novel ways to employ its craft.

The manufacture of electron microscopes is a specialty of the Hitachi Company, and central to the operations of the electron holography laboratory was a majestic state-of-the-art 150-kV field-emission electron microscope. The source of electrons is a sharp tungsten-cathode filament 10 nm wide (about the width of 100 atoms) from which electrons are drawn off by an electrostatic potential of a few thousand volts. One characteristic feature of the electron source deriving from the small tip size is the high degree of coherence of the electrons. "Coherence" is a much used word in physics, and even within the narrowed scope of electron microscopy, it has several connotations. It is effectively a measure of the extent to which electron interference can occur. For the present, let it suffice to say that the electrons produced by field emission can give rise under appropriate circumstances to several thousand interference fringes—an order-of magnitude improvement over other electron sources.

Shortly after my arrival at the ARL, it occurred to me that, by employing a sufficiently attenuated beam, the electron holography group could make a video showing, in real (or accelerated) time, the evolution of the electron self-interference pattern, one electron at a time. Such a film, I suggested, would be of much use to physics teachers. Unfortunately, the project would not be possible, I was told, because the Hitachi chief management was still skittish over funding a purely basic research laboratory and looked particularly askance at experiments for "classroom films." Somewhere along the line, however, the management had a change of mind, for the experiment I proposed was eventually done, and a five-minute black-and-white video cassette was prepared which captured one of nature's most amazing phenomena.[4] (The promotional advantage of "classroom films" as an aid to sales and recruitment did not go unnoticed for long. When I returned as visiting Chief Researcher to the ARL a few years later, Hitachi was in the midst of replacing the cassette with a full-length color and sound film on electron interference.)

According to the de Broglie relation expressing one facet of the dual wave–particle behavior of matter, the wavelength λ associated with electrons moving with linear momentum of magnitude p is given by

$$\lambda = \frac{h}{p} \tag{3.2a}$$

where h is Planck's constant ($\sim 6.6 \times 10^{-27}$ erg s). The second facet of the wave–particle duality is the Einstein relation

$$E = h\nu, \tag{3.2b}$$

relating particle energy E and wave frequency ν. In the Hitachi experiment, the wavelength of electrons emitted from the field-emission tip and accelerated through a potential difference of 50 kV was 5.4×10^{-3} nm (about one-tenth the Bohr radius of an unexcited hydrogen atom, and five orders of magnitude smaller than the wavelength of visible green light). With a kinetic energy of about 50 keV, the electrons moved relative to a stationary laboratory observer at a speed approximately one-half the speed of light ($c = 3 \times 10^{10}$ cm/s). Although fast by terrestrial standards, this speed v is still sufficiently below c that the Newtonian expression for momentum ($p = mv$) and kinetic energy ($K = \frac{1}{2}mv^2$) lead to a value of λ reasonably close to that obtained from the exact relativistic expression.

As an electron wave propagated through the barrel of the microscope (Figure 3.2), it was focused by electromagnetic lenses[5] and split by an electron biprism, a fine wire filament at a potential of about 10 V placed between two parallel plates at ground potential. The biprism served in place of the two apertures. At the lower end of the microscope, single electrons impinged on a fluorescent film which emitted about 500 photons for each electron into a fiber plate that channeled the photons through to an underlying photocathode. Electrons ejected from the photocathode by the incident light were accelerated to 3 keV and entered a multichannel plate, a sort of honeycomb detector and electron multiplier, by means of which the coordinates of the electron point image were determined with a position-sensing device. The arrival of an electron at a given channel was stored in an image proces-

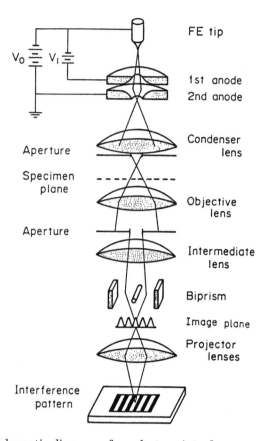

Figure 3.2. Schematic diagram of an electron interference experiment with a field-emission (FE) electron microscope. An electron beam drawn by a high potential from the FE tip and accelerated through various focusing devices is split by the biprism and recombined in the image plane. The classically inexplicable outcome is that interference fringes are formed even when only one electron at a time passes through the microscope. (Courtesy of A. Tonomura, Hitachi Advanced Research Laboratory.)

sor, and the accumulating electron image could be viewed in real time on a TV monitor.

Initially, there is no interference pattern; the detector simply registers, as expected, the random arrival of one electron at a time, each of which showed up on the monitor as a white dot. With the passage of time, these random arrivals build up the classical two-slit interference pattern of fringes (Figure 3.3). The element of periodicity is barely discernible after the arrival of a few hundred electrons. It is definitely present, although not distinct, after a few thousand (squinting helps). After several tens of thousands of electrons, the fringes stand out boldly; except for the wavelength scale, they are indistinguishable

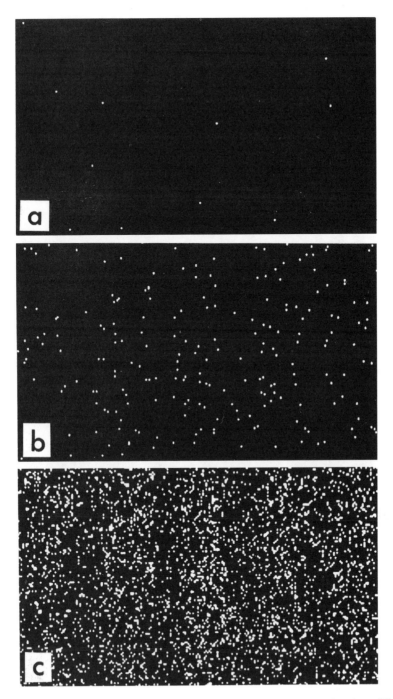

Figure 3.3. Development of an electron interference pattern in time. Electrons arrive at the detector at the rate of approximately 1000 per second with an average spatial separation of 150 km; the distance between source and detector is only about 1.5 m. The approximate number of recorded electrons in frames (a) to (e) are respectively 10, 100, 3000, 20,000, and 70,000. (Courtesy of A. Tonomura, Hitachi Advanced Research Laboratory.)

71

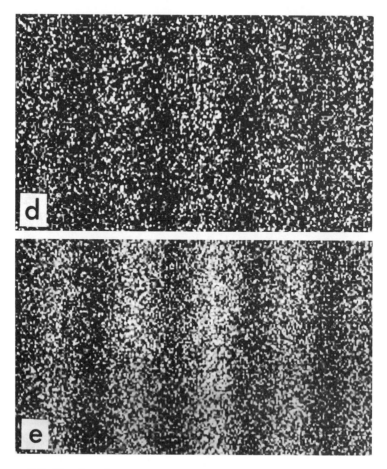

Figure 3.3. *Continued.*

from the fringes produced by laser light under comparable experimental conditions. No one with a sense of curiosity could view this film and, knowing the circumstances, not be profoundly puzzled by the behavior of matter on a subatomic scale.

How do the electrons "know" where to go? Can there be some sort of cooperative effect between electrons emitted at different times that somehow leads to their preferential arrival at some locations and avoidance of others? This is highly unlikely, for the conditions were such that from a classical perspective, one must regard the emissions as random, widely separated events. Let us consider a few relevant details.

Monochromatic waves, like frictionless surfaces and massless springs, are idealizations in physics; they cannot be produced by real sources that have been in operation for a finite length of time. The wavelike nature of the field-emission electrons is better represented by a wave packet. The length of the wave packet, designated the temporal coherence length ℓ_c, is a rough measure of the spatial extent (along the direction of propagation) within which one is likely to find the electron. The electron is not emitted from the cathode tip at a precisely known instant in time; rather, there is an uncertainty (the coherence time t_c) in the duration of the emission. The coherence length is then

$$\ell_c = vt_c, \tag{3.3a}$$

in which v is the mean electron speed. As a consequence of this temporal uncertainty, the energy of the electrons is also unsharp to an extent ΔE, where, according to one form of the uncertainty principle,

$$t_c \sim \frac{\hbar}{\Delta E}. \tag{3.3b}$$

The constant $\hbar$ (pronounced "h-bar"), ubiquitous throughout quantum physics, is defined as $\hbar = h/2\pi$. From the de Broglie relation (3.2a) and (nonrelativistic) expressions for the electron energy and linear momentum, one can derive the following formula for the coherence length:

$$\ell_c = \left(\frac{2E}{\Delta E}\right)\lambda. \tag{3.3c}$$

In the Hitachi experiment, the uncertainty in electron energy was about 0.5 eV. Estimated from relation (3.3c), the coherence length of the electron was then on the order of a micron, which, although small, is considerably greater than the electron wavelength and comparable in size to the diameter of the biprism filament. (One micron, or $1\,\mu$m, is 10^{-6} m, about the size of some bacteria.) The wave packet comprised $\ell_c/\lambda = 2 \times 10^5$ electron wavelengths and, therefore, represented a highly, although not perfectly, monochromatic electron beam.

How could one be sure that effectively only one electron at a time contributed to the interference pattern? By adjustment of the focal length of one of the lenses, the electron current reaching the detector was set to approximately 1000 electrons/s. Thus, one electron followed another at time intervals of about a millisecond. Moving at half the speed of light, the electrons were then separated from one another by about 150,000 m, or a distance on the order of 100,000 times the length of the electron microscope! Under these circumstances, an individual electron propagated from the source to the detector long before a succeeding electron was "born."

So the question remains: How can a coherent macroscopic pattern be systematically created by randomly arriving noninteracting particles? It may be of interest to note that biologists face an analogous problem in accounting for patterns of animal coloration. How, for example, do the cells along the thin strip at the growing edge of a mollusk shell create intricate shell designs that are millions of times larger than the cells themselves?[6] The answer must lie in cell inter-action—perhaps through diffusion of pigment-activating chemicals. No such interaction can adequately explain electron interference.

The Hitachi experiment was not the first of its kind (although it was the first that I was involved in personally), but rather one of the last and most conclusive in a line of analogous experiments dating back to just a few years after Einstein proposed the existence of photons. In 1909, in an experiment remarkable for its technological simplicity, the British physicist G. I. Taylor[7] photographed the shadow of a needle illu-minated by a light source so weak that, on average, only a few photons at a time impinged on the needle. After an exposure time of about 2000 hours the interference fringes of the diffraction pattern stood out as sharply as if a strong light source and much shorter exposure time had been employed. By contrast, the exposure time of the Hitachi experi-ment was about one hour.

The inadequacy of any explanation of interference phenomena based on the mutual interaction of electrons (or, as the case may be, photons) was noted by P. A. M. Dirac in his *Principles of Quantum Mechanics*, the bible of quantum mechanics for several generations of physicists.[8] According to Dirac (p. 9)

On the assumption that the intensity of a beam is connected with the proba-ble number of photons in it, we should have half the total number of photons going into each component. If the two components are now made to interfere, we should require a photon in one component to be able to interfere with one in the other. Sometimes these two photons would have to annihilate one another and other times they would have to produce four photons. This would contradict the conservation of energy.

One might add that for electrons, this would contradict the conserva-tion of electric charge as well.

If there can be no cooperative effect between electrons, and if the presence of an electron wave packet in some spatial domain corre-spondingly implies the probability of finding an electron there, it would seem that a given electron has to pass around both sides of the biprism wire simultaneously. Yet how can this be? The detector always regis-ters an electron as an entire massive particle; one would need to explain how an electron could fragment and recombine. The Hitachi team did not attempt to determine which path individual electrons took; had they done so, they would have found that an electron always passed to one side or the other, never to both simultaneously. This act of looking, however, would have destroyed the interference pattern. The electron distribution would then no longer have been oscillatory, but rather the same as that of the grains of sand.

One cannot (as Feynman says) "make the mystery go away by 'explaining' how it works." Nevertheless, there *is* a sort of explanation that stands as the central dogma of quantum mechanics; Dirac again expresses this clearly and succinctly in his *Principles* (p. 9).

The new theory [i.e., quantum mechanics] which connects the wave function with probabilities for one photon, gets over the difficulty by making each photon go partly into each of the two components. *Each photon then interferes only with itself. Interference between two different photons never occurs.* [Italics added by the author.]

Dirac addressed himself to the interference of photons, but the principle applies without qualification to electrons as well.

The italicized phrase above is very important, indeed essential, to the standard interpretation of quantum mechanics. The self-interference of an electron, by which is meant the interference of the split electron wave packet, can occur only if the two components of the wave packet can overlap. Thus, qualitatively speaking, the difference in the "optical path length" traversed by both components of the wave packet to a given point on the detector must not be much in excess of the coherence length ℓ_c if self-interference is to occur.

Quantum theory furnishes the means to calculate the properties of the interference pattern produced by a beam of electrons, but it provides no means to envision the actual path of an electron. The very idea of a path or trajectory in a case where single-electron interference occurs is largely rendered useless by the uncertainty principle.[9] The mechanism, if one can even employ the word, of how an electron interferes with itself is indeed a mystery. However, Feynman notwithstanding, this is not the only mystery. It is just the beginning.[10]

The self-interference of electrons is one manifestation of what is termed the wave–particle duality: the fact that "particles" like electrons evince wavelike properties, and "waves" like light evince particlelike properties. Examination of a diffraction or interference pattern does not reveal whether it has been made by electrons or by light (photons). This point is ordinarily deemed so obvious, once one accepts the wave–particle duality, that physics textbooks do not usually pursue it further. Nevertheless, electrons and photons are quite different. Electrons have mass, $m_e = 9.11 \times 10^{-28}$ g; photons, as far as is known, do not. (Examination of the torque on a toroid Cavendish balance due to the galactic magnetic field provides a conservative upper limit of the photon mass of about 3.6×10^{-49} g.[11]) Electrons are electrically charged; photons are neutral. All photons carry one unit (in terms of $\hbar$) of intrinsic angular momentum. The intrinsic angular momentum, or spin, of the electron is $\frac{1}{2}\hbar$. This seemingly small difference in intrinsic angular momentum is the basis for major qualitative differences in physical behavior. Photons are bosons (i.e., any number of them can be accommodated in a given quantum state); this, in essence, is the reason (from the standpoint of quantum mechanics) for the existence of classical light waves. Electrons are fermions, which signifies that at most only one electron can occupy a given quantum state; electrons cannot form classical waves.

Should not at least some of these properties—mass, charge, spin, statistics—affect the wave function and thereby distinguish electron from photon interference? They do—and at this point, the behavior of electrons becomes stranger still.

3.2. Electron Interference in a Space with Holes

The development of classical physics documents in many ways the triumph of the field concept, Faraday's insightful vision of the transmission of forces between matter by means of an invisible, yet pervasive, medium. Of the various fields discerned by physicists, those of gravity and electromagnetism are the most familiar and best understood. Gravity dominates the macroscale world of neutral matter, but it is many orders of magnitude intrinsically weaker than electromagnetism. Two electrons an arbitrary distance apart repel one another with an electrostatic force some 10^{42} times stronger than their mutual gravitational attraction. Ordinarily (although not always, as we shall see later), gravity does not have a significant impact on the quantum behavior of the elementary particles apart from those in highly collapsed, exotic systems like neutron stars and black holes. Let us concentrate here on electromagnetism and examine a quantum interference phenomenon arising from the existence of electric charge. Since light is electrically neutral, it is not expected to give rise to this effect.

All of the phenomena of classical electromagnetism follow from two sets of laws. On the one hand, there are Maxwell's equations, which describe the production of electric and magnetic fields from material sources of charge and electric current and from the spatio-temporal variation of the fields, themselves. Reciprocally, there is the Lorentz force law, which describes how the electromagnetic fields influence charged matter. Whether the Lorentz force is truly independent of Maxwell's equations is an interesting question, the answer to which depends essentially on what other assumptions one adopts about the properties of the fields. The point stressed here, however, is simply that (neglecting gravity) electrically charged particles interact with electric and magnetic fields; in the absence of such fields, classical physics provides *no* means by which the state of motion of charged particles can be perturbed. No E&M fields → no E&M force!

This remark is important because within the framework of Maxwell's theory, one customarily introduces, as a mathematical aid to the solution of problems, two auxiliary fields: the electromagnetic scalar and vector potentials, ϕ and $\mathbf{A}$ respectively. The electric and magnetic fields, designated $\mathbf{E}$ and $\mathbf{B}$, can be expressed in terms of the spatial and temporal derivatives of the potentials as follows[12]:

$$\mathbf{E} = -\text{grad } \phi - \frac{1}{c} \frac{\partial \mathbf{A}}{\partial t} \tag{3.4a}$$

$$\mathbf{B} = \text{curl } \mathbf{A} \tag{3.4b}$$

One could, therefore, represent the Lorentz force law, which takes the form

$$\mathbf{F} = e\left(\mathbf{E} + \frac{\mathbf{v}}{c} \times \mathbf{B} \right) \tag{3.4c}$$

for a single particle of charge e moving with velocity $\mathbf{v}$, in terms of these derivatives.

Nevertheless—and this again is essential to bear in mind—were it possible to envisage a region of space permeated *only* by the potentials yet *devoid* of all electric and magnetic fields, one would have to conclude from classical physics that a charged particle would experience no electromagnetic interaction in that region. For one thing, the electromagnetic potentials of a specified configuration of electromagnetic fields are not unique; they can be changed in certain prescribed ways by a mathematical procedure known as a gauge transformation[13] without in any way altering the electromagnetic fields, Maxwell's equations, and the Lorentz force law—and thus without changing the physical properties of the system. (A theory exhibiting this type of symmetry is said to be gauge invariant.) By contrast, real physical forces must be specified uniquely if classical physics is to lead to meaningful predictions.

The field configuration proposed above is not entirely a fanciful one. A very long current-carrying wire wrapped tightly to form a cylindrical coil (or solenoid) of finite radius produces an axial magnetic field in the interior region with a vanishingly small (as the length grows without bound) return magnetic field in the exterior region (Figure 3.4). Nevertheless, the exterior region is permeated by a vector potential field with equipotential surfaces that form concentric cylinders about the solenoid. The sense of circulation of the vector potential and the direction of the interior magnetic field depend on the sense of current flow through the windings. Although nature does not provide physicists with infinite solenoids (any more than with frictionless bearings), a real solenoid, to the extent that it is much longer than it is wide, can produce a magnetic field closely resembling the field of the ideal one. In any event, other geometrical configurations can be realized, and we will take up the practical details later.

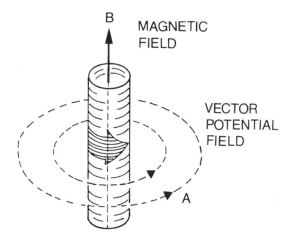

Figure 3.4. Static fields of an ideally infinitely long solenoid bearing a uniform current that circulates in the sense shown by the wide arrow. An axial magnetic field **B** fills the solenoid interior. A vector potential field **A** with cylindrical equipotential surfaces centered about the solenoid axis fills the space outside the solenoid (where the magnetic field is null).

What effect, if any, would such a solenoid have on charged particles if, to take a concrete example, it were placed midway between the two slits of the opaque partition employed in an idealized electron interference experiment (Figure 3.5)? The axis of the solenoid is oriented parallel to the plane of the partition (i.e., perpendicular to the page). It is to be understood that the solenoid is of sufficiently small diameter that it does not block the apertures and that one should neglect the diffraction that would occur at the cylindrical surface irrespective of the presence of the electric current and associated internal magnetic field. It is further assumed that the space accessible to the electrons is limited to the solenoid exterior where the magnetic field is null and only the vector potential exists.

Clearly, in view of what was said above about classical electromagnetism, no influence on the electron interference pattern would be expected. In the quantum world, however, the concept of force is not as fundamental as the concept of potential. Potentials can influence the phase of an electron wave function to produce phenomena for which there are no classical analogs. Spatially varying potentials usually give rise to some kind of force, even when that force plays no direct role in the interpretation of a physical effect. For example, the gravitational potential of the Earth influences the wave function of a particle moving *horizontally* (near the Earth's surface) rather than falling vertically in response to the gravitational force. This leads to a physically observable quantum interference effect (to be discussed

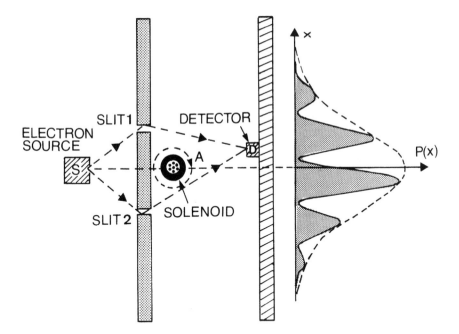

Figure 3.5. Schematic diagram of two-slit electron interference in the presence of a force-free vector potential field. The magnetic field inside the solenoid is directed perpendicularly into the page; the external vector potential field has a clockwise sense about the solenoid axis. Although the diffraction envelope remains undeviated from the forward direction, the interference fringes are displaced by a relative phase shift between the components of the electron wave issuing from slits 1 and 2. This phase shift is proportional to the magnetic field *within* the solenoid, the region from which the electrons are excluded.

later) that depends on the acceleration of gravity, but is *not* a direct consequence of the gravitational force, because the latter acts vertically downward only. In the present case, however, there is no electric or magnetic force at all in the region accessible to the electrons, and the ensuing effect is strange and thought-provoking even by the standards of quantum mechanics.

The first enunciation of what was eventually to become a major conceptual issue in the foundations of quantum mechanics was reported in 1949 by W. Ehrenberg and R. E. Siday[14] as something of an afterthought at the end of a long paper devoted to the correct determination of the refractive index in electron optics. The problem the authors addressed was of practical significance to the burgeoning field of electron microscopy, where one needed to be able to determine electron trajectories through focusing devices. (The concept of an electron trajectory is meaningful when the wavelike nature of the electron is not involved; electron propagation can then be described in what is effectively a geometrical optics limit.)

Realizing that the refractive index of an electron moving through the various focusing fields of an electron microscope "contains the vector potential and not the magnetic field strength," Ehrenberg and Siday concluded: "One might therefore expect wave-optical phenomena to arise which are due to the presence of a magnetic field but not due to the magnetic field itself, i.e. which arise whilst the rays are in field-free regions only." To emphasize their point, the authors even described a hypothetical two-slit electron interference experiment not unlike that proposed above—but with the source of the magnetic field (e.g., a solenoid) left unspecified; they correspondingly deduced that, for each increment of 3.9×10^{-7} gauss-cm^2 in magnetic flux between the two slits, the electron interference pattern would shift by one fringe.[15]

The physical quantity "magnetic flux," which shall be represented here by Φ, should bring to mind the image of magnetic field lines "flowing" through a surface. For a cylindrical region with constant axial magnetic field, such as the interior of the infinite solenoid, the flux is simply the product of the magnetic field strength B and the cross-sectional area (πR^2 for a solenoid of radius R). More generally, the magnetic flux of an arbitrary magnetic field through an arbitrary surface S is the surface integral

$$\Phi = \iint_S \mathbf{B} \cdot d\mathbf{S} \tag{3.5a}$$

where the dot or scalar product indicates that only the component of field perpendicular to the surface contributes.

As is well known from classical electromagnetism, the magnetic flux can also be deduced from the auxiliary vector potential field by means of a corresponding expression involving a contour or line integral

$$\Phi = \oint_C \mathbf{A} \cdot d\mathbf{l} \tag{3.5b}$$

completely around the magnetic field lines (like a string around a bundle of straw). The entire closed contour C may well lie in a region, such as the external region of the solenoid, where the magnetic field, but not the vector potential field, is null. In a prescient remark concluding their paper, Ehrenberg and Siday commented: "It is very curious that [there results] a phenomenon observable at least in principle with a flux; one expects a change in flux, but not steady flux, to have observable effects." The "change in flux" to which the authors referred recalls Faraday's law of induction and Maxwell's modification of Ampère's law, whereby a time-varying magnetic or electric flux engenders, respectively, electric or magnetic forces. These are processes well within the purview of classical physics. Thus, Ehrenberg and Siday's discovery was indeed "curious." Unfortunately, as so often occurs in science, the novelty of a discovery is unrecognized by one's contemporaries and lies fallow until rediscovered under more propitious circumstances.

The rediscovery took place independently ten years later. In a paper[16] regarded as a classic of quantum physics, Y. Aharonov and D. Bohm discussed in all its puzzling detail the strange phenomenon that bears their names. The Aharonov–Bohm (or AB) effect has been controversial in one way or another for over four decades—an extraordinary situation for a science like physics. Within the community of physicists interested in such matters, there are those (the majority) who believe that the effect is an essential consequence of quantum mechanics and that its observation has provided a fundamental confirmation of the theory. There are others who believe that the effect does not exist at all. And there are still others who, while admitting of the theoretical existence of the effect, are unconvinced that anyone has yet seen it. How can there possibly be such persistent divergence of opinion about the occurrence of a physical phenomenon?

At the core of the AB effect is the following characteristic of the electron wave function recognized by Dirac not long after the development of quantum mechanics. If $\psi_0(\mathbf{x}, t)$ is the wave function of an electron at some point $\mathbf{x}$ and time t in a space free of electromagnetic potentials (and consequently electromagnetic fields), then the wave function $\psi(\mathbf{x}, t)$ of the electron in the presence of a time-independent vector potential field at the same space–time location can be expressed in the form

$$\psi(\mathbf{x}, t) = \psi_0(\mathbf{x}, t) \exp\left(\frac{ie}{\hbar c} \int_{x_0}^{x} \mathbf{A} \cdot d\mathbf{l} \right). \tag{3.6a}$$

The line integral in the phase of the wave function is taken along a path P, largely arbitrary, that connects the point of origin of the electron motion $\mathbf{x}_0$ to the field point $\mathbf{x}$. Because both the path and the mathematical form of the vector potential are arbitrary, the phase of the wave function is not uniquely prescribed. Nevertheless, the wave function ψ of relation (3.6a) satisfies the quantum equation of motion (e.g., the Schrödinger equation for nonrelativistic electrons or the more general Dirac equation for relativistic electrons) when the vector potential is present, if ψ_0 is a solution when the vector potential is absent. The demonstration is quite straightforward, and, as far as I know, the above relation in itself scarcely raised any eyebrows *before* the AB paper pointed out unexpected physical consequences.

Note first of all that the indeterminate phase has no effect on measurements performed on an undivided electron beam, since the probability of finding an electron within some specified region, as well as the mean value of any physically observable property of the beam, depends on the expression $|\psi|^2 = \psi^* \times \psi$ from which the phase vanishes.

Consider, however, the two-slit interference experiment with the infinite solenoid. The electron wave function divides at the slits, with the component issuing from slit 1 propagating around one side of the solenoid and the component issuing from slit 2 propagating around the other side of the solenoid. After passage through the two slits and around the solenoid, therefore, the electron wave function comprises two terms

$$\psi(\mathbf{x}, t) = \psi_1(\mathbf{x}, t)\exp(iS_1) + \psi_2(\mathbf{x}, t)\exp(iS_2), \qquad (3.6b)$$

in which ψ_1 and ψ_2 are the wave functions that would issue from slits 1 and 2 in the absence of the vector potential field and S_1 and S_2 are the indeterminate phases of the form given in relation (3.6a). S_1 and S_2, however, are distinguished only the "path" taken by the electron. Actually, one can never know the path taken by the electron; all that really matters is that path P_1, to which phase S_1 is associated, lies on one side of the solenoid, and path P_2, to which phase S_2 is associated, lies on the other.

It is not difficult to show that the phase *difference* between the two components of the wave function is then

$$S_2 - S_1 = \frac{e}{\hbar c} \oint_C \mathbf{A} \cdot d\mathbf{l} = \frac{e\Phi}{\hbar c}, \qquad (3.6c)$$

where C is a *closed* contour about the solenoid. The integral therefore represents the magnetic flux Φ through the solenoid interior. The phase difference, in contrast to the phase of each component, is *not* indeterminate, but an experimentally accessible quantity. From relations (3.6b) and (3.6c), it follows that the electron intensity at a distance x (from the forward direction) along the axis perpendicular to both the solenoid and the incident electron beams takes the form (for two identical slits)

$$I(x) = I(0)\left[1 + \mathcal{V}\cos\left(a(x) + \frac{2\pi e\Phi}{hc}\right)\right]. \qquad (3.6d)$$

Here, $\mathcal{V}$ and $a(x)$ characterize the visibility (or contrast) and phase of the "ordinary" two-slit interference pattern in the absence of the current-carrying solenoid. (The single-slit diffraction factor is not included in the above expression since it is not relevant to the discussion at the moment.) The supplementary contribution to the phase that depends on magnetic flux is the AB effect. The flux-dependent term can be written as $2\pi(\Phi/\Phi_0)$, in which the constant $\Phi_0 = hc/e = 3.9 \times 10^{-7}$ gauss-cm^2 is one "fluxon," a fundamental unit of flux. If this expression really represents the outcome of the proposed experiment, then, as Ehrenberg and Siday first predicted, a change in flux by one fluxon should shift the pattern by one fringe.

Within two years of publication of the AB paper, several laboratories reported experimental confirmations of the effect. Nevertheless, the AB effect was puzzling in almost every way; neither the theoretical existence nor the experimental verification nor the authors' interpretation of the effect was readily accepted. In the words of Aharonov and Bohm[17]:

Although [our] point of view concerning potentials seems to be called for in the quantum theory of the electromagnetic field, it must be admitted that it is rather unfamiliar. Various of its aspects are often, therefore, not very clearly understood, and as a result, a great many objections have been raised against it. . . .

The point of view of the authors, embodied in the title of their seminal first paper, is that the presumed auxiliary electromagnetic potentials, even though they are indeterminate, are, in fact, more fundamental than the electromagnetic fields. At least initially, before the deep significance of gauge invariance to field theory was widely recognized, this view rested largely on the notion of causality. To be consistent with commonly understood ideas of cause and effect implicit, for example, in the principle of special relativity, interactions in physics must be local (i.e., a particle can interact only with the fields in its immediate vicinity). This perspective is expressed in the very formulation of physical laws as differential equations. In the AB effect, however, the only field at the site of an electron is the vector potential field (or, in variations of the effect, the scalar potential)—and this field is indeterminate.

To many, however, the interpretation that electromagnetic potentials are more basic than electromagnetic fields was (and perhaps still is) difficult to accept. After all, although the electrons may be subject to the laws of quantum mechanics, the fields are still the classical fields of Maxwell's electrodynamics. There is nothing in the AB effect that requires a quantum theory of electrodynamics, and classical electrodynamics can be formulated starting with either the fields or the potentials.

On the other hand, the alternative viewpoint, that the fields take precedence over (or are at least as fundamental as) the potentials, seemingly requires one to accept a most peculiar interpretation. Because the magnetic field is confined in a region of space inaccessible to the electrons, the particle–field interaction must occur *non*locally (i.e., by means of action at a distance). How can a magnetic field influence an electron that never passes through it?

One answer, maintained by a small minority, is that the whole issue is a tempest in a teapot: The AB effect does not exist except on paper.

The argument to support this view is linked to the nonuniqueness of the vector and scalar potentials. It is possible to find a gauge transformation for which the new (i.e., transformed) vector potential field *vanishes identically* in the region outside of the solenoid (or other current configuration). If this were indeed the case, the electrons could be made, by means of a purely mathematical manipulation, to pass through a region with neither a vector potential nor a magnetic field. Clearly, one would not expect any influence on an electron in that case; hence, the effect predicted for a nonvanishing vector potential must be fictitious.

This reasoning, however, is not sound. The type of gauge transformation at issue not only removes the external vector potential but effectively the *internal* magnetic field as well. It changes completely the physical system, and this is not permitted. Although there is wide latitude in the execution of gauge transformations, not every conceivable gauge transformation is an admissible one. A gauge transformation is a little like a change of coordinates; the selection of one coordinate system over another may afford more analytical convenience, but it must not change the physical system itself.

Ironically, the above point was already recognized in the 1949 paper of Ehrenberg and Siday, who posed the question: "One may ask if the anisotropy outside the [magnetic] field could not be avoided by an alternative value for A which also reproduces the field given. . . ." By "anisotropy," the authors meant the presence of the vector potential in the theoretical expression for the electron refractive index. A short demonstration showed that this was *not* possible and the authors concluded: "It is readily seen that no vector potential which satisfies Stokes' theorem will remove the anisotropy of the whole space outside the [magnetic] field. . . .' Stokes' theorem—the key to resolving the gauge transformation "paradox"—is the equality of relations (3.5a) and (3.5b). Expressed in words, the presence of a magnetic flux through a surface requires a nonvanishing vector potential field along some closed path bounding the surface. Any vector potential that does not satisfy Stokes' theorem for a specified magnetic field configuration is not acceptable. Apparently, the Ehrenberg and Siday paper was not widely read.

The AB effect, however, is a subtle one even for those who accept its existence. Indeed, what many physicists once thought (and perhaps still believe) the phenomenon to be is incorrect and violates basic physical principles! As depicted in the papers of both Ehrenberg and Siday and of Aharonov and Bohm, the phase shift engendered by the magnetic flux of the confined magnetic field redistributes the electron intensity out of the forward direction. So far, so good. Confusion arises, however, upon consideration of the actual manner of redistribution. In the words of Aharonov and Bohr, for example, the presence of the vector potential field of the solenoid has the following consequence: "A corresponding shift will take place in the directions, and therefore the *momentum of the diffracted beam*." (Italics added by the writer.) Is this in fact what the AB shift implies?

Recall at this point that although the two-slit interference pattern has, in principle, an indefinite lateral extent, the fringe contrast falls off rapidly outside the central region of the single-slit diffraction "envelope." [See relation (3.1a) and Figure 3.1.] For the fringes to be visible, the transverse coherence of the electron beam must extend at least over the width of the diffraction pattern (e.g., the region between the first two diffraction minima). This spatial or transverse coherence length ℓ_s (to be distinguished from the temporal or longitudinal coherence length ℓ_c, introduced earlier) is given to good approximation by the relation

$$\ell_s = \frac{\lambda}{2\delta}, \tag{3.7}$$

where δ is the initial angle of divergence (i.e., angular deviation from the forward direction) of the beam at its source. As an example, the field-emission beam employed in the electron self-interference experiment of Section 3.1 had a divergence angle $\delta = 2 \times 10^{-8}$ radian and a wavelength $\lambda = 5.4 \times 10^{-3}$ nm; thus, the transverse coherence length was $\ell_s = 0.014$ cm or $140\,\mu$m, two orders of magnitude larger than the temporal coherence length.

A common interpretation of the AB effect, expressed or implied in the expository literature, is that the shift in "momentum of the diffracted beam" refers to the shift of the diffraction pattern. Indeed, Feynman himself—one of the creators of quantum electrodynamics, the branch of physics that treats most comprehensively the interaction of particles and electromagnetic fields—had portrayed the AB effect as analogous to placing a strip of magnetic material (transparent to electrons) behind the partition with two slits; he showed that the resulting Lorentz force displaced the center of the diffraction pattern (as, in fact, it would).[18] This interpretation is *not* valid, however, for no such magnetic force is possible in a region ideally free of electric and magnetic fields. To represent the AB effect in this manner violates what is known as the Bohr correspondence principle.

Although quantum mechanics is a more comprehensive theory that classical mechanics, there must be some means of relating both the quantum and classical descriptions of a system under conditions where the latter theory is also applicable. This is the correspondence principle, first enunciated and widely used by Niels Bohr in the years before a consistent and complete theory of quantum mechanics was formulated. The principle can be implemented in a variety of ways of which one of the most common is to consider the limiting case of a quantum expression as Planck's constant h approaches zero. As h vanishes, the laws of the quantum world and the classical world become one; the electrons stream through the apertures like (charged) grains of sand.

The quantum-theoretic description represents this transition in the following way. First, the center of the diffraction pattern falls at the location to which the extant forces displace the corresponding classical particles in accordance with Newton's laws. Second, the quantum interference pattern oscillates infinitely fast, so that no real detector could reveal the fringes. Thus, in the limit of vanishing h, the AB effect must vanish and the diffraction pattern of the electron beam must be undisplaced. What, then, is the physical consequence of the AB effect in the real world where h is not zero?

Careful analysis of the Aharonov–Bohm effect would show that the magnetic flux shifts the two-slit interference pattern asymmetrically *within* the single-slit diffraction pattern (Figure 3.5). However, the center of the diffraction pattern, itself, is *not* displaced in keeping with the condition that no classical electromagnetic force acts on the electrons.[19]

The two-slit experimental configuration appears to be symmetric with respect to both slits. What determines whether the electron beam is displaced laterally toward slit 1 or toward slit 2? It is the direction of the magnetic field and, consequently, the sense of circulation of the vector potential that breaks the symmetry. The field within the solenoid can be oriented either "up" or "down"; a change in the field orientation would reverse the direction of fringe shift, even though the electrons do not directly experience the magnetic field.

There is no mechanistic explanation. The AB effect, since it is discernible only in the interference pattern (which vanishes as h approaches zero) and not in the diffraction pattern (which remains unchanged from that of a force-free electron beam), is a uniquely quantum mechanical phenomenon and, as such, beyond the visual imagery of classical physics.

Theory aside, what does experiment have to say about the matter; has anyone actually observed the AB phase shift? For a long while, interpretation of the few electron interference experiments reporting AB-type phase shifts in the presence of structures designed to simulate an ideal solenoid was somewhat ambiguous. One such structure, for example, was a very fine magnetized iron "whisker" less than $1\,\mu m$ in diameter. Unfortunately, to ensure that no magnetic field lines permeate the region accessible to the electrons—whereupon critics could argue that changes in the electron interference pattern derive from the familiar Lorentz magnetic force—was rather difficult.

In a series of experiments extending through the 1980s, my Hitachi colleagues took up the challenge.[20] Through a happy marriage of basic science and advanced technology, the condition of a confined magnetic flux was produced by fabrication—*not* of solenoids—but of tiny (about 10–20 μm in diameter) toroidal (doughnut-shaped) permalloy ferromagnets. Unlike a real solenoid, where the currents in the windings produce an external magnetic field, the toroidal magnetic field lines form circular loops within the interior of the toroid (around the hole). Electron microscopy itself could be used as a check for magnetic field leakage, thereby permitting imperfect toroids to be discarded. To make doubly sure that electrons would not penetrate the surface of the 20-nm-thick magnets, the Hitachi toroids were coated with a 300-nm-thick layer of niobium and a copper layer also on the order of a hundred nanometers in depth.

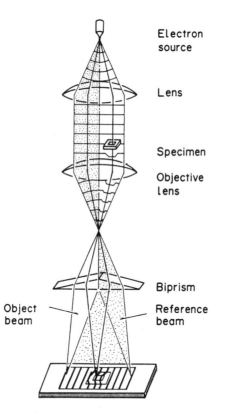

Figure 3.6. Schematic experimental configuration for observing the Aharonov–Bohm effect by means of electron holography. In the absence of the toroidal ferromagnet, the electron wave, split by the biprism and recombined in the image plane, generates straight, uniformly spaced reference fringes. With the toroid present, the portion of the electron wave diffracted by the toroid superposes with the reference wave at the image plane to generate a holographic image that preserves the phase (relative to the reference) of the electron wave function. (Courtesy of A. Tonomura, Hitachi Advanced Research Laboratory.)

To produce the AB effect, a shielded toroidal magnet was situated in the field-emission electron microscope above the electron biprism (Figure 3.6). In the imagery of classical physics, an electron in the beam could pass either around the outside of the toroid or through the central hole. Although the magnetic field geometry differs from that of the ideal solenoid, there is still a net magnetic flux through any surface bounded by these two types of classical trajectory. In the imagery of classical waves, the portion of the electron wave that propagates around the outside of the toroid serves as a reference in an experimental configuration analogous to that of optical holography; this reference wave is split at the biprism and gives rise to a pattern of interference fringes upon recombination at the image plane (a photographic film). The component of the electron wave that diffracts through the central hole of the toroid, however, should incur an AB phase shift relative to the reference and thereby produce fringe shifted with respect to the reference fringes by an amount depending on the magnetic flux winding through the toroid. Since the electrons cannot penetrate the shielded toroid, the projection of the toroid onto the film ought to appear as a solid black annulus (flat doughnut). Were the toroid not shielded, one would see within the body of the annulus the continuity of the outside reference fringes and the displaced fringes in the hole.

The experiment was duly conducted with the results as just described (Figure 3.7). But alas, the skeptics were unmoved. The toroids, so it was claimed, were not perfect, or at any rate not close enough to perfection, and the specter of the Lorentz force was again raised. Back to the drawing board (literally) went the Hitachi team.

To ensure beyond a reasonable doubt that the magnetic field of the toroid was adequately confined, the experimenters designed a low-temperature specimen stage to reduce the temperature of the toroid until the niobium layer becomes superconducting. A (Type I) superconductor displays the Meissner effect: Upon transition to the superconducting state, it will suddenly expel a pre-existing magnetic field from its interior. However, given the geometry of the tiny shielded toroids, expulsion of the magnetic field from the outer niobium layer is tantamount to confining it within the inner permalloy magnet.

There was one small potential problem, however: The use of superconductivity entailed the discouraging possibility that no fringe shift might take place at all! A feature of superconductors, not unrelated to the Meissner effect, is that the magnetic flux penetrating a superconducting loop is constrained to half-integer values of the fluxon hc/e. In other words, the AB phase shift in relation (3.6d) would take the form $2\pi(\Phi/\Phi_0) = 2\pi(n/2) = n\pi$, where n can be 0, 1, 2, 3, ... (One speaks of this as quantization of magnetic flux, but this should not be construed to imply a quantization of the magnetic field; the fields are entirely

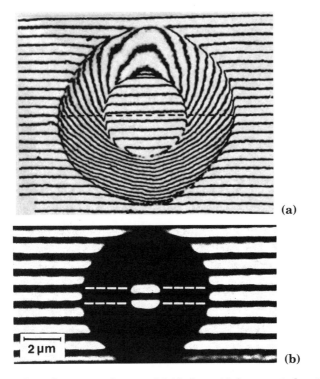

(a)

(b)

Figure 3.7. (a) Interferogram of an unshielded toroidal magnet showing the shift in fringes (i.e., lines of constant phase) for components of the electron wave passing outside the toroid or through the hole. Because the toroid is unshielded, electrons can also pass through the interior, allowing one to see the continuity of the fringes. (b) A superconducting layer over the toroidal magnet shields the outside region from magnetic field leakage and prevents, as well, electron penetration. The interferogram shows a 180° phase shift between the two components of the electron wave that diffract around the toroid or through the hole. (Courtesy of A. Tonomura, Hitachi Advanced Research Laboratory.)

classical.) If, for some reason, the Hitachi toroids all produced even integer multiples of the fluxon, then the AB phase shift would be an integer multiple of 360°, and therefore not observable. On the other hand, if toroids could be produced for which the quantized flux turned out to be odd-integer multiples of the fluxon, then phase shifts equivalent to 180° would occur, thereby giving rise to complete fringe reversal between the space outside the toroid and the central hole. This would be clearly observable.

Fortunately, nature was not so perverse as to deny the researchers the fruit of their hard efforts. Toroids of both types were produced, and the expected phase shift was observed under conditions approximating more closely than ever before the ideal confinement of the magnetic field. The AB effect exists!

Or does it? Can one point again to deviations from ideality that vitiate the conclusion? More largely construed, the query addresses a basic aspect of how science works.

The Meissner effect, as is well known, does *not* exclude a magnetic field completely from the interior of a superconducting material. An "evanescent" magnetic field (i.e., a field falling off exponentially with distance) can penetrate a superconductor to an extent (London penetration depth) that is ordinarily so small as to be negligible for a bulk substance. The penetration could, however, be significant for a thin film. The niobium layer covering the toroid is not a bulk material, but is it sufficiently thick? From the thickness of the layers and the precision of their technique for measuring fringe shifts, the Hitachi group estimated that the leakage flux outside a toroid must be far less than 1/200 of a fluxon. According to quantum theory, the ensuing Lorentz force should be negligible.

Well, what about electron penetration *into* the magnet? After all, 150-keV electrons are fairly energetic. Taking into account potentially relevant interactions between the electrons and the copper and niobium layers, the researchers estimated that about one out of every million electrons might penetrate the toroid sufficiently to experience a magnetic field. For a toroid $10 \,\mu$m in diameter and an electron flux of about $10^{-5} \,$A/cm^2, this amounts to some 50 electrons out of 50 million per second penetrating the toroid. Does this degree of imperfection invalidate the conclusion that the AB effect has been observed? Certainly not. There is no such thing, in my opinion, as a "definitive" experiment, an experiment so perfect or complete that it settles an issue for all time. There will always be deviations from ideal performance to which critics can point; there will always be new experiments that experimenters can make. The key concern, however, is whether or not the criticisms are valid and the improvements needed.

Within the framework of a mature science—indeed as a hallmark that a particular discipline is in fact a science—there must be objective ways for assessing the reliability of a given conclusion. It is insufficient as a mode of objection simply to list all manner of things that might conceivably be nonideal. To be taken seriously, a critic is obliged to demonstrate convincingly some relevant causal connection between the objection and the experimental outcome. In the present case, the evidence, both theoretical and experimental, for the existence of the AB effect is consistent and reproducible. To sustain prior objections at this point, one would need to explain how so striking a modification of the electron interference pattern could be produced by a minuscule trickle of penetrating electrons, or to demonstrate a flaw in the arguments leading to the presumed low level of penetrating electron flux, or to provide some alternative explanation (not involving the Lorentz force) of the observed phase shifts.[21]

For most physicists concerned with the issue, the AB effect is theoretically real and experimentally confirmed. Many probably already believed in it, if only on the basis of theoretical self-consistency, well before the Hitachi experiments. Nevertheless, the skepticism that motivated these experiments was, I think, beneficial. The ensuing research not only convincingly demonstrated the AB effect, but led, as well, to novel instrumentation and new discoveries concerning quantized magnetic flux. Skepticism in science, if not also in the arena of everyday life, is a healthy attribute to the extent that it reflects an open-minded willingness to be convinced by new facts. That is how science advances. What, finally, can one say about the interpretation of the AB effect regarding the fundamentality of electromagnetic fields or potentials? In quantum theory, the electromagnetic potentials are no longer merely secondary fields that facilitate computation; they are needed at the outset in order for the theory to be invariant under gauge transformations. Gauge invariance, once regarded as merely a curious feature of Maxwell's equations, has since been recognized as a most important symmetry to be maintained in all field theories. Indeed, this symmetry largely determines a priori the form of the interaction between particles and fields. (I discuss this point more comprehensively in Chapter 9.) Correspondingly, effects analogous to the AB effect are believed to occur, at least in principle, in areas nominally unrelated to electromagnetism such as gravity (general relativity theory) and the strong nuclear interactions (quantum chromodynamics). The interpretations of these effects are not always clear, nor are the experimental methods by which they might be observed. But that they are intrinsic to the theory and of fundamental significance is seemingly beyond doubt.

Recognition of the wider occurrence in physics of AB-like effects has led to reconsideration recently of the existence of an actual (i.e., electromagnetic) AB effect on light. Surprisingly, although such effects are not expected for completely neutral systems, the photon can nevertheless be influenced, at least in principle, by a magnetic flux. In classical electromagnetism, as a result of the linearity of Maxwell's equations, there is no mechanism by which light can interact with static electromagnetic fields or potentials (or with other light waves) in the absence of matter. Within the framework of relativistic quantum electrodynamics, however, a photon can, under appropriate circumstances, be transformed into an electron and positron pair whose brief lifetime is so short that, to within limits posed by the uncertainty principle, no physical law is violated. During their ephemeral existence, these oppositely charged particles (rather than the original photon directly) can interact with a vector potential field to give rise, after their subsequent mutual annihilation back to another photon, to light-scattering processes that depend on magnetic flux. The probability for the occurrence of such processes is extremely small, but that these processes exist at all highlights one of the seminal differences between classical and quantum electrodynamics.

The AB effect is a subtle one—at both the theoretical and experimental levels—and therein, in part, lies its great interest not only as a test of quantum mechanics and electrodynamics but also as a stark reminder that physics is a human activity characterized by intellectual ferment, struggle, and creativity. It is not simply a storehouse of equations, facts, and procedures whose straightforward application instantly produces "right" answers.

The confirmation of the AB effect does not, by any means, signify that the subject is an exhausted one. There are aspects to this phenomenon, as yet unexplored experimentally, that point to a physical reality stranger still than that revealed so far. One stands in awe at how devious and wonderful nature can be.

3.3. The Two-Electron Quantum Interference Disappearing Act

A long gray wall separates the Hitachi grounds from the rest of Kokubunji City. The familiar orange Chuo ("Middle Central") Line passes close by one side of the wall, taking commuters east to central Tokyo or west to outlying areas like the venerable city of Hachioji ("City of Eight Princes"). During my first visit to the Advanced Research Laboratory, I lived in Hachioji on a high hill overlooking the city and affording a memorable view of Mt. Fuji in the early morning hours when the air was clearest. Some years later, I climbed Fuji with a Hitachi colleague, starting at twilight in a solemn torch-lit procession of pilgrims, and caught a glimpse of the sunrise through a momentary parting of the thick curtain of mist that surrounded us. It was a moving experience. By then, I was living in Kokubunji within a short walk of the Laboratory.

Seen from outside the Hitachi wall, the gray tower of a company building looming up in the distance suggests just another industrial works. But inside, this mistaken impression evaporates before the extraordinary surroundings. Paths descend through wooded terrain to a large pond teeming with carp and lined with cherry and plum trees. Swans and ducks skim over the surface. A veritable botanical garden with a rich variety of trees, bushes, and flowers identified by small placards surrounds the pond and adjacent smaller pools. Footbridges connect the mainland to a few small wooded islands upon which here and there a stone lantern or Japanese shrine nestles unobtrusively. Beyond the pond is an extensive sports field with an outdoor amphitheater.

These gardens and woods served me well. In the early afternoon and evening of most days, I was wont to talk or jog around the pond, my head filled with thoughts of electrons, solenoids, and quantum mechanics. The Aharonov–Bohm experiments—as, indeed, all quantum interference experiments to date with free electrons—were performed effectively one electron at a time; that is, even with the brightest sources available, the observed interference effects were all manifestations of single-particle self-interference. Dirac's dictum aside, the possibility of quantum interference with two or more electrons intrigued me and I kept thinking about what types of effects could occur.

In the summer, around the time of the o-bon festival—a holiday somewhat analogous in spirit, but not in celebration, to the European All-Saints' Day—the ARL and CRL staff held a huge lawn party on the playing field in the late afternoon. Then, when the sky had darkened sufficiently so that the first stars appeared, the crown jewel of the day's activities would begin: the *hanabi* or "fire flowers," a spectacular display of fireworks lasting for almost an hour. I lay on my back in the soft grass watching burst after burst of brilliant particles and dreamily imagined them to be electrons shooting in all directions out of their source. What, I wondered, would two oppositely flying particles do if each encountered an AB solenoid at its own end of the sky?

The idea, although initially appealing, struck me after a few moments as uninteresting. Clearly, each particle would simply contribute to its own interference pattern of AB-shifted fringes. After all, once separated, the two particles go their merry way uninfluenced by one another. However, I had not reckoned on what surely must be one of nature's strangest attributes, a quantum mystery no less profound than that of self-interference (Feynman's "only" mystery). Some days later, when the inchoate images of celestial solenoids and *hanabi* electrons shooting through the heavens consolidated more soberly in my mind, I examined the problem systematically—and the results were surprising indeed.

To keep matters simple, imagine a compact source that produces wave packets with two electrons at a time. Like the fiery sparks of the *hanabi*, there is no preferred direction for electron emission. One electron can fly out in any direction whatever, as long as the other electron emerges simultaneously in the opposite direction. How is one to create such an electron source? Well, I am not sure; this is, after all, a *Gedankenexperiment*. Perhaps one can fabricate a double-tipped field-emission cathode that emits pairs of coherent electrons, one electron emerging from each end. Perhaps there are atomic processes involving the correlated excitation and ionization of two electrons. Or perhaps one can resort to the use of "exotic" atoms, atoms containing elementary particles other than the familiar ones (electron, proton, and neutron). If two electrons, for example, bind to a positive muon, there would result a muonic counterpart ($\mu^+ e^- e^-$) to the negative hydrogen ion H$^-$. A proton lasts forever (or at least many times the age of the universe); a muon, however, decays to other particles that flee rapidly from the scene of destruction. The decay of the muon in the $\mu^+ e^- e^-$ ion should leave two mutually repelling electrons. In any event, let us leave the technical details of a suitable electron source to the future.

Pick some point far to one side of the source (S) and place there an AB solenoid (with magnetic flux Φ_1) oriented perpendicular to the line between the source and the point; at a corresponding point on the opposite side of the source, locate a second AB solenoid (containing flux Φ_2) with its axis parallel to the first (Figure 3.8a). Let us suppose that the currents through the windings of the two solenoids circulate in the opposite sense so that the internal magnetic fields of the solenoids are antiparallel; it does not really matter—the principal results are not qualitatively changed.

Now, locate four electron "mirrors," as shown in the Figure 3.8a. If an electron—let us call it electron 1—heads in just the right direction to reflect from mirror M1 into a detector D1, then the companion electron—electron 2—will reflect from mirror M2' into detector D2. Correspondingly, if electron 1 reflects from mirror M1' into detector D1, then electron 2 reflects from mirror M2 into detector D2. The distance between source and mirror is the same for all four mirrors; likewise, the four (shortest) mirror–detector separations are all equal. In this way, there is no phase shift in the electron wave function arising from a difference in the electron optical path length. The two classically imagined paths of electrons 1 and 2 form complete loops about the respective solenoids S1 and S2, and it would seem then that all conditions for an AB interference effect at each detector are met.

It is to be assumed that the two solenoids (S1 and S2) and the two detectors (D1 and D2) are far apart from one another—indeed, so far apart that the observers may not even be able to see or communicate with one another. Perhaps the observer at D1 is not even aware that there is someone to switch on D2 and observe the other electron. From the perspective of observer 1, therefore, electrons are simply arriving regularly and being counted at D1. "Why should it matter if other electrons spewing out elsewhere?," he might ask. If an electron takes the path S–M1–D1, the wave function of the electron incurs the (nonunique) phase shift a_1, where, in accordance with relation (3.6a),

$$a_1 = \left(\frac{e}{\hbar c}\right)_{\text{Path S–M1–D1}} \int \mathbf{A}_1 \cdot d\mathbf{l}. \tag{3.8a}$$

(a)

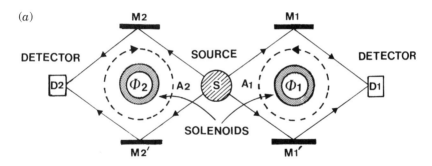

Figure 3.8. (a) Schematic diagram of an electron interference experiment with pairs of correlated electrons. For each electron emitted by the source S in a particular direction, there is a corresponding electron emitted in the opposite direction. One does not know, however, which paths (source → mirror → detector) the electrons take through the right and left interferometers. The count rate observed at the detector of one interferometer depends on the magnetic flux in the other interferometer through which the counted electrons could not possibly have passed.

Similarly, if the electron takes the path S–M1′–D1, there results a phase shift a_1':

$$a_1' = \left(\frac{e}{\hbar c}\right)_{\text{Path S–M1'–D1}} \int \mathbf{A}_1 \cdot d\mathbf{l} . \qquad (3.8b)$$

Corresponding phase shifts a_2 and a_2' involving the vector potential field $\mathbf{A}_2$ of solenoid S2 are incurred by the electron wave that propagates from the source S to D2 via mirrors M2 and M2′, respectively. Observer 1 does not concern himself with the effect of vector potential $\mathbf{A}_2$ in his vicinity because the electrons reaching detector D1 do not make a circuit around solenoid S2.

What does observer 1 predict will be the outcome? To the extent that he ignores the electrons emitted toward observer 2—and, for all he knows, may not even be counted—the first observer might reason as follows. The wave function—or at least the only part of it relevant to his own experiment—should be the sum of two probability amplitudes, one for each path the electron could take to detector D1. Thus, to within a constant factor, the net amplitude for arrival of an electron at D1 is

$$\psi(D1) \sim \exp(ia_1) + \exp(ia_1'). \qquad (3.9a)$$

Observer 1 would then deduce that the probability of an electron being received at D1 is proportional to $|\psi(D1)|^2$ or

$$P(D1) = \frac{1}{2}\left[1 + \cos\left(\frac{e\Phi_1}{\hbar c}\right)\right], \tag{3.9b}$$

which is, in effect, a special case of relation (3.6d). The normalization factor $\frac{1}{2}$ assures that the maximum probability is unity or 100%. We have also make use of Stokes' law, discussed earlier, which in the present case requires that

$$a_1 - a_1' = \frac{e\Phi_1}{\hbar c}, \tag{3.9c}$$

$$a_2 - a_2' = \frac{e\Phi_2}{\hbar c}. \tag{3.9d}$$

Observer 2, reasoning in a similar way that his experiment is independent of that of the distant observer 1, would deduce an analogous expression involving flux Φ_2. Together, the two observers would infer the following joint probability for an electron to be received at both D1 and D2:

$$P(D1, D2) = P(D1)P(D2) = \frac{1}{4}\left[1 + \cos\left(\frac{e\Phi_1}{\hbar c}\right)\right]\left[1 + \cos\left(\frac{e\Phi_2}{\hbar c}\right)\right]. \tag{3.9e}$$

The results may seem satisfying, for the probability inferred by each observer depends only on the magnetic flux of the solenoid in "his" part of the universe. The only problem is that the predicted outcomes [relations (3.9b) and (3.9e)] and the whole mode of thinking are *incorrect*.

The two electrons are not emitted independently since their "paths" (actually their linear momenta) are correlated; according to quantum theory, this correlation persists, no matter how far apart the electrons travel. If the observer at D1 determined the path of arrival of an incoming electron, he would know without having to make a measurement (if he were aware of the correlated emission) the path taken by the other electron. Of course, if he *did* determine the electron path, there would no longer be any quantum interference. Nevertheless, according to the hypothetical conditions of the *Gedankenexperiment*, an electron necessarily arrives at D2 if an electron arrives at D1, and one must determine the joint probability of electron detection at the outset.

The amplitude that one electron arrives at D1 via mirror M1 and therefore that the other arrives at D2 via mirror M2' is proportional to the product of the phase factors for each route:

$$A(M1, M2') \sim \exp(ia_1)\exp(ia_2'). \tag{3.10a}$$

Likewise, the amplitude that one electron arrives at D1 via mirror M1' and that, therefore, the other arrives at D2 via mirror M2 is

$$A(M1', M2) \sim \exp(ia_1')\exp(ia_2). \tag{3.10b}$$

The wave function $\psi(D1, D2)$ of the detected two-electron system is proportional to the sum of the above two amplitudes, and the joint probability of electron detection, given by $|\psi(D1, D2)|^2$ now takes the form

$$P(D1, D2) = \frac{1}{2}\left[1 + \cos\left(\frac{e(\Phi_1 - \Phi_2)}{\hbar c}\right)\right]. \qquad (3.10c)$$

From the perspective of classical physics, the above hypothetical experiment with correlated (as opposed to independent) electrons poses a curious dilemma in several ways. First, until one gets used to it, the idea of the AB effect, itself, where electrons are affected by passing *around*, and not through, a magnetic field is rather curious. But in the present configuration, the signal detected by one observer also depends on the distant magnetic field around which "his" detected electrons have *not* passed. Or, phrased differently, it depends on the electrons that go to the *other* observer even though the latter can be arbitrarily far away! Suppose, for example, that the magnetic flux is the same within the two solenoids; irrespective of the magnitude of the flux, the electrons, according to the correct relation (3.10c), would arrive at both detectors with 100% probability. By contrast, if the flux through S1 had been set to product a phase shift $e\Phi/\hbar c$ of 180°, then observer 1, in the erroneous belief that his experiment was independent of that of the other observer, would have deduced from relation (3.9b) that the probability of electron arrival at D1 is *zero*.

There is nothing in itself strange about two particles being correlated and flying off in opposite directions; this could occur as well in classical mechanics if an initially stationary object exploded into two pieces. However, once the pieces are separated, the motion of one would not be expected to influence, or be influenced by, subsequent measurements made on the other. What has happened to the "localness" of physical interactions in quantum mechanics?

Perhaps the reader is thinking that the nonlocality manifested by the above two-electron AB *Gedankenexperiment* is an artificial product of the experimental condition whereby if observer 1 receives an electron, then he knows for certain that observer 2 has received an electron. In other words, the experimental configuration is such that one cannot calculate the signal at one detector without it being a joint detection probability for both detectors [relation (3.10c)].

Let us modify the experiment, therefore, so that when observer 1 receives an electron, he *will not know* to which detector the other electron has gone. At the former location of detectors D1 and D2, put two beam splitters, BS1 and BS2, that divide the intensity of an incoming electron beam equally (Figure 3.8b). Thus, an electron incident on BS1

(b)

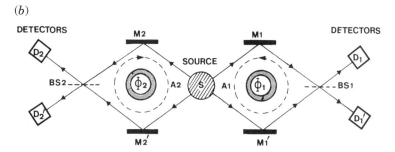

Figure 3.8. (b) Diagram of correlated electron interference experiment with addition of beam splitters BS1 and BS2 to the configuration of Figure 3.8(a) (and corresponding addition of two more detectors). The probability for detecting an electron at a specified detector, irrespective of the fate of the paired electron, is entirely independent of magnetic flux and, in fact, shows no quantum interference effect at all. Interference occurs only in the joint detection of two electrons.

from mirror M1 or M1′ has a 50% chance of reflecting from the surface and a 50% change of being transmitted; likewise for electrons incident on BS2 from mirrors M2 and M2′. Now, instead of having only two detectors as before, let us place *four* detectors (D1, D1′, D2, D2′), one on each side of each beam splitter. With these modifications not every electron incident on the mirrors M1 or M1′ will necessarily go to the detector D1 of observer 1, nor can observer 1 know if the second electron of the correlated pair emitted by the source is received at D2 or D2′.

Alas, the modified two-electron AB experiment does *not* lead to results more compatible with our classical conceptions of locality. If anything, the outcome is stranger than before.

One can readily determine, by extension of the foregoing reasoning leading to the amplitudes (3.10a) and (3.10b), the probability amplitude for each potential pathway of an electron wave from source to mirror to beam splitter to detector.[22] From these amplitudes follows the joint probability for receiving two electrons at any two detectors. For example, the joint probability that one electron arrives at detector D1 and the other electron at detector D2 is

$$P(D1, D2) = \frac{1}{4}\left[1 - \cos\left(\frac{e(\Phi_1 - \Phi_2)}{\hbar c}\right)\right],$$
(3.11a)

whereas the joint probability that one electron arrives at detector D2 and the other electron at detector D2′ is

$$P(D1, D2') = \frac{1}{4}\left[1 + \cos\left(\frac{e(\Phi_1 - \Phi_2)}{\hbar c}\right)\right].$$
(3.11b)

These expressions are similar to relation (3.10c) characterizing the first thought experiment.

However, suppose observer 1 is not interested in where electron 2 goes. In fact, suppose that detectors D2 and D2′ are not even turned on and that *nobody* knows what has happened to electron 2. Observer 1, nonetheless, sits by detector D1 and assiduously notes down the electron counts. What signal does he receive?

The probability that an electron goes to D1 *irrespective* of the detector to which electron 2 goes is simply

$$P(\text{D1}) = P(\text{D1}, \text{D2}) + P(\text{D1}, \text{D2}') = \frac{1}{2}. \tag{3.11c}$$

A constant! The signal that observer 1 receives with his detector alone is completely independent of the magnetic flux of either solenoid. The AB effect seems to have completely disappeared.

In fact, *all* quantum interference has disappeared. Had the geometrical path lengths for the pathways S–M1–D1 and S–M1′–D1 not been equal (as initially specified), then—quite apart from the presence or absence of magnetic flux through any solenoid—there would occur a relative phase shift between the electron waves taking one or the other of these pathways. One would then expect the usual two-slit type of quantum interference to occur. But it does not occur. The phase shift engendered by an optical path-length difference would appear in the argument of the interference term (the cosine function) of both relations (3.11a) and (3.11b). Because these two interference terms have opposite signs, they would again vanish when summed to give $P(\text{D1})$, the probability that observer 1 receives an electron.

The *joint* probabilities, $P(\text{D1}, \text{D2})$ and $P(\text{D1}, \text{D2}')$, *do* show an AB effect. Thus, when either the observer at D2 or the one at D2′ correlates his electron count rate with observer 1, the latter becomes aware of an AB effect. However, if neither the observer at D2 nor the one at D2′ bothers to participate in the experiment, then observer 1 detects no quantum interference effects at all. How can this be? How can observers at one end of the universe destroy the quantum interference of electrons at the other end simply by deciding *not* to observe? Surely this is most odd.

As with other uniquely quantum phenomena, the above *Gedanken-experiment*—which some day will no doubt be performed as a real experiment—has no explanation within classical physics. We cannot satisfactorily account for these results through any imaginable behavior of particles and waves such as we find in the macroscale world.

It should be brought out explicitly at this point that—despite the presence of a vector potential field in the region of space accessible to the electrons—the AB effect even as originally described (with diffraction of a single-electron wave function around one solenoid) is, itself, an intrinsically nonlocal phenomenon. The presence of a vector potential field does not *per se* make the AB effect an *effect* (i.e., something observable) unless the pathways potentially available to the charged particle circumscribe the confined magnetic field. In this sense, the AB effect reflects the topology (global or nonlocal geometrical features) of the multiply connected space through which the particles propagate.

However, quite apart from anything specific to the AB effect, quantum mechanics manifests an essential nonlocality deriving from the wave description of matter. Once linked together in a single quantum system, quantum particles remain a single system even after they have separated sufficiently far that there can be no information exchange or physical interaction between them capable of affecting an experimental outcome. These results *are* strange, even within the framework of quantum mechanics, because most of our past experience with quantum interference phenomena concerned, principally, the self-interference of single particles. The above-described hypothetical system, however, involves inextricably "entangled" quantum states of two particles (to employ terminology first introduced by Schrödinger). No matter how far apart the two particles separate, they constitute a single quantum system manifesting the bizarre effects of nonlocality.

This may seem absurd from the perspective of the familiar experiences that define for most of us the nature of "physical reality." If so, we are in good company, for Einstein himself was sorely plagued by these strange implications of quantum theory.

In 1935, in a paper[23] that has subsequently become a wellspring of voluminous discussions and experimental tests of quantum mechanics, Einstein and his collaborators Boris Podolsky and Nathan Rosen, raised the issue know today as the EPR paradox. Does quantum mechanics provide a complete description of physical reality? EPR posed the question and answered it negatively. What, after all, *is* physical reality? According to EPR,

If, without in any way disturbing a system, we can predict with certainty (i.e., with probability equal to unity) the value of a physical quantity, then there exists an element of physical reality corresponding to this physical quantity.

And, insisted EPR,

... every element of the physical reality must have a counterpart in the physical theory

if a theory is to be regarded as "complete."

How well does quantum mechanics do when judged by these criteria? Not very well, it seems. EPR provided their own example for disqualifying quantum mechanics as a complete theory, but our first two-electron AB experiment will amply serve to illustrate the difficulties. As stated before, without in any way disturbing the electron emitted to the left, observer 1 can determine which path the electron takes to detector D2 by determining which path the electron emitted to the right follows to detector D1. By the EPR criteria, then, these electron paths are elements of physical reality and must have a counterpart in the physical theory. However, we know that if the ensuing quantum interference is not to be destroyed, quantum theory does *not* permit us to know which path either electron has taken. To know this information is equivalent to knowing the transverse components of coordinate and momentum of each electron to a degree of precision higher than that permitted by the uncertainty principle. In quantum theory, these various paths available to the electrons are not elements of an objective physical reality as much as they are elements of a potential reality.

This does not mean, of course, that quantum mechanics is necessarily an incomplete theory. Rather, the EPR definition of reality may not be adequate, an objection that they, themselves, anticipated:

One could object to [our] conclusion on the grounds that our criterion of reality is not sufficiently restrictive. Indeed, one would not arrive at our conclusion if one insisted that two or more physical quantities can be regarded as simultaneous elements of reality *only when they can be simultaneously measured or predicted.* . . . This makes the reality of [such quantities] depend upon the process of measurement. . . . No reasonable definition of reality could be expected to permit this.

Einstein, Podolsky, and Rosen's assertion notwithstanding, physical reality is what it is: the strange reality depicted by the quantum theory. Does an alternative description of physical reality exist? Einstein believed that such a description would some day be found. However, physicists have searched for decades, and are searching still; no alternative theoretical framework has been created, as far as I am aware, that is an successful as quantum theory.

By the time Einstein advanced the views expressed in the EPR paper, he was already largely regarded as out of the mainstream of modern physics. The paper met with a barrage of rebuttals and criticism, although, as Einstein wryly noted, no two critics objected to the same thing. Einstein died four years before the article by Aharonov and Bohm appeared, and it is unlikely, I would surmise, that he ever saw the paper by Ehrenberg and Siday. I have often wondered what Einstein would have said about the AB effect, which so alters our conception of physical reality not only in the domain of mechanics but in electromagnetism as well. As one interested in the foundations of electrodynamics throughout his life, would he have considered the primacy of potentials over electric and magnetic fields a violation of his cherished beliefs, or would he have said—as the young Einstein rashly did upon hearing of Bohr's theory of light production in 1913—"The theory

... must be right?"[24] Would he, the master geometer of physics, have been pleasantly surprised or appalled at a quantum phenomenon dependent on the topology of space? We will never know.

3.4. Heretical Correlations

In addition to electrical charge, the electron is also endowed with an intrinsic angular momentum, or spin, of $\frac{1}{2}\hbar$. Whereas charge is responsible for the interactions leading to the Aharonov–Bohm effect, spin gives rise, at least indirectly, to different quantum interference effects that are in some ways even more remote from our classical expectations. Although the AB effect cannot be accounted for in terms of electric or magnetic forces, it is nevertheless a consequence (albeit a quantum consequence) of classical electromagnetic fields. However, there are consequences of electron spin that have *no* classical roots whatever.

Unlike the (single-electron) AB effect, the electron phenomena to be considered here have not yet been observed in the laboratory, but they must exist if our current understanding of the quantum behavior of matter is correct. These effects are not only impossible to reconcile with the imagery of classical physics, but appear to challenge, as well, the traditional interpretation of particle interference i.e., Dirac's dictum: a particle can interfere only with itself.

It has long been a fundamental tenet of wave theory—pertinent as well to the "wave mechanics" of matter—that wave *amplitudes*, and never intensities (i.e., the squares of amplitudes), interfere. One might well imagine, therefore, that the development of an interferometer, in which the superposition of separate light intensities produced an interference pattern, would be viewed with considerable skepticism. Indeed, that was exactly the response of many physicists to the intensity interferometer of R. Hanbury Brown and R. Q. Twiss (HBT).[25]

Developed in the 1950s for the purpose of measuring stellar diameters by a method less sensitive to mechanical vibrations or atmospheric distortions than suspending a Michelson interferometer[26] at the end of an optical telescope, the HBT instrument functioned as follows (Figure 3.9). Light from a star was received at two spatially separated photodetectors whose electrical outputs were passed through "lowpass" filters. The filters suppressed components of the electric current oscillating at frequencies outside the domain of the radiowave spectrum i.e., outside the range of about 1–100 MHz (1 MHz = 10^6 oscillations per second). The two filtered currents were then multiplied together electronically and averaged over a prescribed time interval. The resulting number, a measure of what is termed the cross-correlation of the incident light, produced an oscillatory curve when

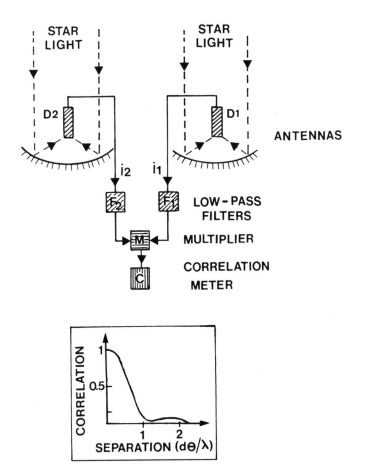

Figure 3.9. Schematic diagram of the Hanbury Brown–Twiss intensity interferometer. Light from an extended source is received by two photodetectors sensitive to the light intensity. The output current of each detector is passed through a filter that admits only components oscillating at low frequencies (corresponding to beat frequencies among the incident optical waves). The two filtered outputs are multiplied electronically, averaged over time, and recorded by the "correlation meter" as a function of receiver separation. Although light intensities are not expected to interfere, the resulting oscillatory correlation curve reveals that some kind of interference has occurred.

plotted as a function of the separation of the two detectors—a clear sign that some kind of interference had occurred.

What is most noteworthy here is that the photodetectors are so-called square-law devices: The electrical output is proportional to the incident light power flux (i.e., intensity). It is the multiplication or correlation of two intensities (not amplitudes) that has produced an interference pattern. How can two intensities interfere?

Actually, the phenomenon does no violence to any known physical principle and can be explained quite simply within the framework of classical wave theory in a way that an electrical engineer would find satisfying and far from surprising. Briefly, each point on the stellar surface gives rise to broad wave fronts at the Earth that illuminate *both* photodetectors of the interferometer. At the surface of each detector, therefore, there occurs a superposition of numerous amplitudes emitted from different locations on the star at different frequencies and with random time-varying phases. What remains from the filtering of the electrical signals are low frequency "beats" produced by the interference of waves at neighboring optical frequencies. For each pair of light-emitting points on the star, the phase associated with a given beat frequency at a given detector consists of two parts: (1) a well-defined component determined by the optical path lengths between the point sources and the particular detector and (2) a random time-varying component resulting from the *difference* in initial random phases of the superposing waves. However—and this is the key point—at a given instant, the random phase difference associated with a particular beat frequency is the *same* for *both* detectors because (in the imagery of classical optics) the same broad wave fronts sweep over both detectors. Thus, multiplication and time averaging of the filtered signals do not lead to the vanishing of all correlations but—for each pair of interfering waves—to the correlation function $c(d)$,

$$c(d) \sim I_1 I_2 \cos\left(\frac{2\pi d\theta}{\lambda}\right), \tag{3.12a}$$

proportional to the mean light intensity (I_1, I_2) at each detector.

The phase of the correlation function varies with detector separation d, angular separation (as seen from the Earth) of the point radiators θ, and mean wavelength λ. Relation (3.12a) must be averaged over all pairs of points on the stellar surface and over all contributing optical frequencies in order to obtain the net signal. Surprisingly, the result is quite simple; suitably normalized, $c(d)$ is effectively the square of the fringe visibility produced by the same light in the Michelson stellar interferometer with mirror separation d. The visibility, or contrast, of the fringe pattern

$$V(d) = \frac{I_{\max} - I_{\min}}{I_{\max} + I_{\min}} \tag{3.12b}$$

is defined as the difference in intensity between neighboring points (near the center of the pattern) of maximum and minimum brightness divided by the sum of these intensities; as a function of mirror separation, $V(d)$ can range between values of zero and unity.[27]

As one may have expected, it is ultimately amplitudes, and not really the intensities, that interfere in an intensity interferometer. Nevertheless, the correlation depends on the product of intensities, rather than the product of wave amplitudes, in marked contrast to our archetypal example of two-slit interference discussed previously. Also, because the highest frequency of the detected beats ($\sim$100 MHz) is roughly a million times lower than the frequencies of the optical "carrier" waves ($\sim$$10^{14}$ Hz), the difference in optical path lengths from the light source to the two detectors need no longer be restricted to values comparable to an optical wavelength ($\sim$$10^{-5}$ cm), as is the case with a Michelson interferometer; light reaching one detector can be retarded with respect to the other by thousands of wavelengths without affecting the correlation [provided the delay is small compared with c/100 MHz = 30 cm].

Technically, the physical quantity actually measured by HBT was not the time-averaged product of the light intensities, but the correlation of the *fluctuations* in intensity at the two detectors. If one represents the instantaneous light intensity received at detector 1 by $I_1(t)$ and the average intensity by $\langle I_1 \rangle$ (which is independent of time for a stable—or so-called stationary—light source), then the fluctuation in light intensity at time t is taken to be $\Delta I_1(t) = I_1(t) - \langle I_1 \rangle$; likewise, the corresponding instantaneous intensity fluctuation at detector 2 is $\Delta I_2(t) = I_2(t) - \langle I_2 \rangle$. HBT measured the time-averaged product of the fluctuations:

$$c(d) = \langle \Delta I_1(t) \Delta I_2(t) \rangle = \langle I_1(t) I_2(t) \rangle - \langle I_1 \rangle \langle I_2 \rangle. \tag{3.12c}$$

This relation differs from that which led to Eq. (3.12a) only by the last term containing the product of the (time-independent) mean intensities.

The fluctuation in the electric current issuing from a photodetector derives principally from two different origins. The major component is the classical *shot noise* associated with the "graininess" of electricity i.e., the fact that charge is transported by discrete units (electrons) rather than by a continuous flow of electrical fluid. The shot noise of one detector is totally independent of the shot noise of the other detector and, consequently, does not contribute to the correlation function $c(d)$ when the detector outputs are multiplied and time-averaged. The smaller noise component, termed *wave noise*, is associated with the incoming light and arises from the myriad random emissions of electromagnetic radiation by the atoms of the hot source. From atom to atom, these emissions vary in amplitude and relative phase at each frequency. Moreover, because the atoms are not all moving with the same velocity relative to the observer, the frequency content of the emissions can also vary from atom to atom as a result of the Doppler effect. The light generated by such a chaotic source (and, indeed, that is the technical term for it: chaotic light) is a superposition of waves of different frequency with amplitudes and relative phase that fluctuate randomly in time, thereby producing the current fluctuations or wave noise at the output of a photodetector. The wave noise at two detectors illuminated by the same source *is* correlated and contributes to $c(d)$.

The preceding heuristic picture of the functioning of the intensity interferometer is rooted in classical physics at least as far as the light is concerned (although the detectors function by means of the photoelectric effect, which is a quantum mechanical process). It did not take long, however, for physicists to wonder about the quantum implications of intensity interferometry and to pose to Brown and Twiss some thorny questions. As a personal observation, I know of many instances where a phenomenon, puzzling from the standpoint of classical physics, received a satisfactory treatment within the framework of quantum physics. In the present case, ironically, a phenomenon happily understandable by means of basic physical optics became a troublesome enigma when examined from the viewpoint of the quantum theory of light.

Although the theory of the intensity interferometer is, in principle, valid for electromagnetic radiation of any wavelength, there are important practical distinctions in the treatment of radio waves and visible light. As summarized by Hanbury Brown[28]:

Radio engineers, before the advent of masers, thought of radio waves as waves and not as a shower of photons. [Because] the energy of the radio photon is so small and there are so many photons, the energy comes smoothly and not in bursts . . . We say that the fluctuations in [the photodetector] output are principally due to "wave noise" and not to "photon noise". By contrast, at optical wavelengths, the energy of the individual photon is much greater and there are relatively few photons, so that we can no longer neglect the fact that the energy comes in bursts. [The] fluctuations . . . are due principally to "photon noise" and not "wave noise".

From the standpoint of quantum optics (i.e., the theory of photons), the correlation of wave noise has a surprising, indeed startling, implication: The photons received at the two detectors are correlated. One might think—and many *did* think—that the random emission of classical waves translates into a quantum imagery of randomly emitted photons, and, as logic would seemingly dictate, randomly emitted photons must arrive randomly at separated detectors. However, this was *not* the case.

Imagine an experiment in which linearly polarized photons emitted from the same source and arriving at two separated detectors are counted, and the experimenters keep track somehow of the number of coincident arrivals i.e., the number of times two photons arrive simultaneously, one photon at each detector, within some specified short time interval. Of course, one expects a certain number of such coincidences to occur accidentally even for completely random arrivals; this can be calculated on the basis of classical statistics. HBT actually did this experiment;[29] they found that the coincident count rate significantly exceeded this background level.

To appreciate just how significant was the observed departure from randomness, it is helpful to consider a superficially different, but conceptually equivalent, experiment and its interpretation. Suppose that instead of measuring the number of coincident linearly polarized photons at two detectors, one measures the time interval (or delay) between consecutive arrivals of photons at *one* detector. For example, one photon arrives and starts a clock; a second photon arrives and stops the clock. The time interval is 8 (in some system of units). The experiment is repeated and the next time interval is 6. After a sufficiently large number of such cycles have been carried out, the experimenter plots the number of recorded events (the consecutive arrivals) corresponding to a particular delay as a function of delay. Perhaps there might have been 1000 pairs of photons with a time delay of 5 units, 950 pairs of photons with a time delay of 10 units, and so on.

The result of such an experiment (Figure 3.10) measures what is termed the conditional probability of receiving a second photon given

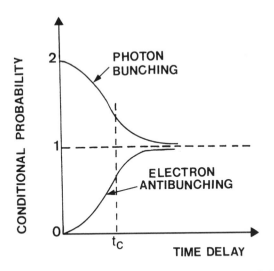

Figure 3.10. The conditional probability of receiving a second thermal photon of specified polarization after detection of a first is higher for a time interval short with respect to t_c (longitudinal coherence time) than for a comparatively long time interval. In fact, the probability of detecting two photons simultaneously (zero time delay) is twice as high. This behavior is illustrative of "photon bunching." Electrons should display "antibunching"; the predicted probability of simultaneous detection of two spin-polarized electrons is zero.

that a first one was already detected; I will call this simply the conditional probability. The outcome, seemingly difficult to reconcile with one's intuitive expectations, appears astonishing. The probability that two photons of the same polarization arrive at the detector *simultaneously* (zero delay) is *twice* the probability that they arrive purely randomly. The random arrivals occur for delays significantly greater than a certain time interval corresponding to the temporal (or longitudinal) coherence time t_c of the light source. As the delay time increases from zero to infinity, the conditional probability falls smoothly from 2 to 1 with t_c as the approximate demarcation between correlation and randomness.

To judge from the written recollection of Hanbury Brown, the correlation of photons wreaked havoc on his tranquillity (not to mention his prospects for external funding). In the words of Brown[30]:

... if one must think of light in terms of photons then ... one must accept that the times of arrival of these photons at the two separated detectors are correlated—they tend to arrive in pairs. Now to a surprising number of people, this idea seemed not only heretical but patently absurd and they told us so in person, by letter, in publications, and by actually doing experiments which claimed to show that we were wrong. At the most basic level they asked *how, if photons are emitted at random in a thermal source, can they appear in pairs at two detectors?* At a more sophisticated level, the enraged physicist would brandish some sacred text ... and point out that ... our analysis was invalidated by the uncertainty relation. ... [Italics added.]

The disturbing question posed to Brown lies at the heart of yet another quantum mystery. I will return to this question later, for it reverberates like an eerie harmony through the phenomena to be discussed shortly. In the parlance of contemporary quantum optics, the pair phenomenon observed by HBT (as well as by others) is termed *photon bunching*. It should be stressed, however, that the graphic imagery of photons grouping together as they propagate through space is misleading. Quantum mechanics does not, in general, permit us to know the path taken by a particle through space, for any intervention by the observer to "see" the particle will disturb its motion. The path of a photon is especially problematical, for, unlike electrons, photons disappear whenever stopped.

The bunching of light (or—depending on how one wants to regard the phenomenon—excess wave noise) is, today, an established and noncontroversial fact. The quantum theory of light accommodates with no difficulty the predictions HBT first made on the basis of classical reasoning. Moreover, the experiments that purportedly proved HBT wrong were eventually recognized as being insufficiently sensitive to detect the light correlations. But what, one might wonder, ought to occur in a HBT-type experiment with *electrons*?

This question occupied my attention during much of my time at the Hitachi Research Laboratory. For one thing, more than three decades after the pioneering studies of Brown and Twiss, no comparable experiments with electrons, as far as I was aware, had been attempted, let alone successfully performed. Indeed, there seemed to be relatively little discussion of the matter at all. Was it possible to observe HBT-type electron correlations with the beam of an electron microscope? What phenomena would result?

Although analogies with light drawn from the classical domain of physical optics can often provide insights into the quantum interference of electrons, the classical explanation of the optical intensity interferometer provides no help whatever in understanding the corresponding electron interferometer. Electrons, as a consequence of their intrinsic spin $\frac{1}{2}\hbar$ are fermions; no more than one electron can occupy a specified quantum state. Electrons, therefore, cannot form classical waves such as light waves, the quantum description of which entails large numbers of photons with identical quantum properties of energy, momentum, and helicity (related to the classical attributes of frequency, wave vector, and polarization). To predict the outcome of an electron HBT-type correlation experiment requires an intrinsically quantum mechanical analysis.

Let us start with a simplified electron correlation *Gedankenexperiment* analogous to the actual photon correlation experiment of HBT. Consider two spatially separated compact sources, S1 and S2, that randomly emit electrons all of which have the same energy and spin component, but whose momentum may vary within a narrow range about the forward direction. Sufficiently far from the sources, so that the electrons can be characterized by plane waves, are two detectors, D1 and D2 (Figure 3.11). What is the joint probability as a function of detector separation that D1 and D2 will each simultaneously receive one particle?

From the standpoint of classical physics, where the particles may be thought of as distinguishable, one can reason that the desired probability is the sum of two contributions: (1) the probability that an electron from source S1 goes to detector D1 and an electron from S2 goes to D2 and (2) the probability of the alternative arrangement whereby an electron from S1 goes to D2 and an electron from S2 goes to D1. (We discount the simultaneous production of two electrons from one source and none from the other.) The two-electron wave functions describing the two configurations are

$$\psi_a(1, 2) = \psi_{S1}(D1)\psi_{S2}(D2), \tag{3.13a}$$

$$\psi_b(1, 2) = \psi_{S1}(D2)\psi_{S2}(D1). \tag{3.13b}$$

For the present discussion, the significant part of each single-electron wave function, represented as a plane wave, is simply a phase factor of the form

SOURCES

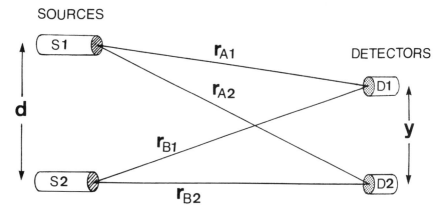

Figure 3.11. Schematic experimental configuration to illustrate electron interference arising for particle indistinguishability. Sources S1 and S2 each emit a spin-polarized electron; detectors D1 and D2 each receive an electron. Because it is not possible to know from which source a detected electron has issued, the amplitudes of the two possible events interfere. The antisymmetry of the electron wave function under particle exchange leads to a joint detection probability $P(D1, D2)$ that vanishes with vanishing source separation d or detector separation y.

$$\psi_S(D) \sim \exp(ikr_{SD}), \qquad (3.13c)$$

where r_{SD} is the path length between one of the sources S and one of the detectors D and $k = 2\pi/\lambda$ is the wave number (or magnitude of the linear momentum in units of $\hbar$) of an electron with wavelength λ. Calculation of the joint probability

$$P(D1, D2) = |\psi_a(1, 2)|^2 + |\psi_b(1, 2)|^2 = \text{constant} \qquad (3.13d)$$

gives a number independent of the relative separation of the detectors. In other words, the randomly emitted particles arrive totally uncorrelated at the two detectors. The result may seen logically satisfying; the electrons begin life independently at two sources and end up randomly at the detectors. Nature, however, does not always favor logical simplicity based on classical reasoning. The conclusion and mode of thinking are again wrong.

I have discussed in the previous section the "ghostly" correlations inherent in the quantum description of the two-electron Aharonov–Bohm effect with widely separated solenoids. These correlations are inexplicable on the basis of classical physics—but at least they evolved from a special initial condition that could very well be understood in classical terms. The electrons in that example, having been produced in a state of zero total linear momentum, were ever afterward (in the absence of external forces) constrained by the law of momentum conservation to propagate with equal and opposite momenta.[31] This law and the corresponding correlation of momenta apply with equal validity to electrons or to pieces of brick. In a manner of speaking, therefore, a correlation was "built in" at the outset by a particular initial condition characterizable in both quantum and classical terms. In the present case, however, electrons emerge randomly from what, to all appearances, are separate sources. What kind of correlation could one possibly expect?

In the quantum mechanical scheme of things, the electrons are indistinguishable particles, and it is not possible, even in principle, to designate from which source a particular electron comes. True, one can always place an electron detector close enough to a source to determine whether it has emitted an electron, but this, of course, is an intervention that alters the motion of the particle (even if the detector was somehow transparent to electrons) and therefore the condition of the originally intended experiment. The arrival of "labeled" electrons at specified detectors represents outcomes that can never actually be distinguished. As is well known in such instances (e.g., in the case of two-slit interference), one must add the amplitudes, and not the probabilities, for each indistinguishable quantum pathway. But in what way are the amplitudes to be added?

It is precisely at this point that the fermionic nature of the electron—an attribute distinguishing it from the photon in as fundamental a way as electrical charge—enters the analysis. The amplitudes for a given process, such as

$$\text{Process a} \begin{cases} \text{particle 1} \rightarrow \text{detector 1} \\ \text{particle 2} \rightarrow \text{detector 2} \end{cases}$$

and the reverse process

$$\text{Process b} \begin{cases} \text{particle 1} \rightarrow \text{detector 2} \\ \text{particle 2} \rightarrow \text{detector 1} \end{cases}$$

must be superposed with *opposite* signs. This is one example of what is usually termed the spin-statistics connection. The exchange of any two identical particles whose spin is an odd half-integer multiple of $\hbar$ follows the above rule; the aggregate behavior of such particles is governed by what is known as Fermi–Dirac statistics. Particles with an even-integer spin are classified as bosons, for they are governed by Bose–Einstein statistics; under particle exchange, the boson wave function incurs no sign change.

Why nature works in this way seems to lie outside the framework of quantum mechanics proper; a satisfactory explanation can be made only in terms of the relativistic invariance and microscopic causality of quantum fields. Once, when asked why spin-$\frac{1}{2}$ particles obey

Fermi–Dirac statistics, Feynman planned to prepare a freshman lecture on it—but failed. "You know, I couldn't do it," he said; "I couldn't reduce it to the freshman level. That means we really don't understand it."[32] Perhaps this overstates the case somewhat, but the principle is nonetheless a deep one.

In the *Gedankenexperiment* under consideration, the two-electron wave function representing the above two indistinguishable processes is therefore

$$\psi(1,2) \sim \psi_a(1,2) - \psi_b(1,2), \tag{3.14a}$$

where the component wave functions are given in relations (3.13a) and (3.13b). Upon substitution of the single-electron amplitudes (3.13c), the appropriately normalized joint detection probability takes the form

$$P(\mathrm{D1}, \mathrm{D2}) = |\psi(1,2)|^2 \sim \frac{1}{2}\left[1 - \cos\left(\frac{2\pi yd}{\lambda r}\right)\right], \tag{3.14b}$$

in which y is the detector separation, d is the source separation, and r is the mean distance of the detectors from the sources (approximately the same for either source to either detector).

The correlation to which the above expression gives rise is very different from that for thermal photons. The joint probability that two electrons arrive at the same location (zero detector separation) is *zero*. This quantum expression of particle avoidance has been termed *antibunching*. The phenomenon of antibunching, which arises from an antisymmetric linear superposition of wave functions as in relation (3.14a), is an example of quantum interference arising, not from space–time differences in alternative geometrical pathways, but from the spin-statistics connection (i.e., from the fact that electrons are fermions).

The joint probability expressed in relation (3.14b) is seen to oscillate repeatedly, giving rise to an infinite number of detector locations at which the electron anticorrelation is perfect [$P(1, 2) = 0$]. This is a consequence of representing the individual electron wave-functions by infinitely extended plane waves instead of by a more realistic wave-packet description. An electron source, such as is found in an electron microscope, produces a beam of electrons whose quantum description would include a distribution of particle numbers, energies, linear momenta, and spin components. Such a source, like a thermal light source, might also be termed chaotic. Despite the added complexity, the essential feature of antibunching (although not necessarily the oscillations) is predicted to persist in the aggregate electron behavior.

One dramatic illustration of the anticorrelation of electrons is the conditional probability of electron arrival at a single detector (Figure 3.10). In contrast to the case of thermal light, the probability of detecting a second electron a short time (compared with the beam coherence time t_c) after receipt of a first is suppressed below that expected for totally random particles. For zero time delay, the probability of two electrons arriving together is predicted to be strictly zero—a quantum consequence of the "minus sign" in electron exchange.

Yet, spin and statistics and minus signs aside, how can one understand in some more tangible way the origin of antibunching? For the case of thermal light, the correlations among photons at least have a classical explanation in terms of fluctuating light waves. However, now, in the absence of a classical explanation, the question to HBT comes back even more forcefully in its fermionic version to haunt us: If spin-polarized electrons are emitted at random in a thermal source, how can they *avoid* arriving in pairs at a detector?

Particle indistinguishability and the uncertainty principle help provide a heuristic answer, but one must be careful not to be trapped by the paradox-laden terminology of classical physics. As posed, the question is not physically meaningful, for its premise cannot be substantiated. *Are* the particles emitted randomly?[33] How would one demonstrate this—other than by inferences based on particle detection? Is there any way to determine the exact instant of emission of each particle without affecting its subsequent motion? Indeed, is there an "instant" of emission?

As pointed out previously, the particles of a beam with an energy uncertainty ΔE do not emerge from their source like mathematical points, but can be represented by wave packets created over a characteristic time interval $t_c = \hbar/\Delta E$, the coherence time. Thus, one could know nothing about the emission time of a hypothetical electron whose wave function is a monochromatic plane wave. If the energy of the beam is not perfectly sharp, but nevertheless defined well enough so that the particles may be assumed to move with speeds close to the mean speed v, an emerging electron will likely be found within a coherence length $\ell_c = v t_c$.

Two particles whose emission events are separated by a time interval long in comparison to the coherence time are characterized by wave packets that effectively do not overlap; there would then be no quantum interference effects engendered by particle exchange and the spin-statistics connection. These particles arrive, therefore, uncorrelated at the two detectors, and one might think of their emissions as random. However, the wave packets of two particles whose emission events occur in a time interval short with respect to the coherence time can overlap, and the subsequent particle motion can manifest, even in the absence of interactions attributable to forces, the "ghostly" correlations of particle exchange.

The above response to the question demanded of Hanbury Brown and Twiss must nevertheless be accompanied by cautionary words, for, like any visualizable explanation, it is couched in the language and imagery of classical physics. Such a description can be grossly misleading. To think of the coherence time as providing a definitive criterion for the overlap or nonoverlap of particle wave packets and, hence, for the occurrence or nonoccurrence of particle correlations is not correct. The particles that pass through the intensity interferometer are all part of a single multiparticle system; the correlations, so to speak, are *always* there.

For example, the coherence time of a 150-kV field-emission electron source, a potentially suitable candidate for an electron HBT experiment, is extremely short, about 10^{-14} s. Suppose one were to attempt to observe electron antibunching by measuring, as described previously the conditional probability that a second electron arrives at a given detector at various delay times after receipt of a first electron. It is a near certainty that any experimental attempt with current technology would have to be made with delay times much longer than the coherence time, perhaps two to five orders of magnitude longer. Yet, quantum theory shows that, even for delay times orders of magnitude longer than the coherence time, the sought-for correlations will not have vanished entirely—they will still be there, albeit weakly, to be disentangled (by statistical analysis of a sufficiently large number of counts) from the random background events.

The subtleties of correlated electron states and the potential pitfalls of adopting too literally the imagery of wave packets show up strikingly in an experimental configuration combining elements of both the Aharonov–Bohm and Hanbury Brown–Twiss experiments. Consider again the diffraction of electrons, produced by a single source, through two narrow apertures between which is placed an AB solenoid with confined magnetic flux (Figure 3.12). There are now, however, *two* detectors whose outputs are correlated so that the joint probability of detection—in essence, the coincident count rate—can be determined. Since the count rate at one detector has been previously shown (in Section 3.2) to vary harmonically with the magnetic flux, one might well expect that the joint count rate at the two detectors must likewise exhibit a flux-dependent quantum interference effect. Yet, surprisingly, if the electrons are correlated, this expectation is not borne out.

Suppose the electron source is again spin-polarized and very nearly monochromatic; the electrons, represented by plane waves of mean wave number k, have a spread in linear momentum about the forward direction. Then, in view of what has been said previously concerning the antisymmetrization of fermion wave functions, the total amplitude for an electron to pass through each slit and arrive simultaneously at each detector takes the form

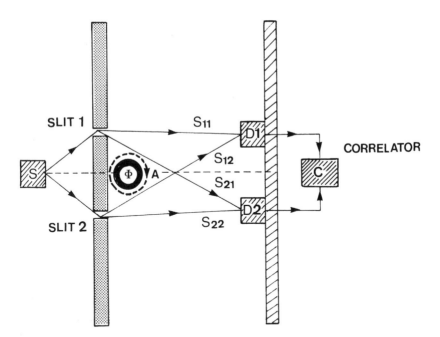

Figure 3.12. Schematic diagram of a hybrid Hanbury Brown–Twiss and Aharonov–Bohm experiment. Electrons emitted from source S pass through slits 1 and 2, around the solenoid (with magnetic flux Φ directed into the page and vector potential field **A** circulating clockwise) and are received at detectors D1 and D2. The correlated output of the detectors shows an interference effect that depends on both the confined magnetic flux and the fermionic nature of the electron.

$$\psi(D1, D2) \sim \exp(iks_{11})\exp(iks_{22})\exp\{i[\alpha(s_{11})+\alpha(s_{22})]\}$$
$$- \exp(iks_{12})\exp(iks_{21})\exp\{i[\alpha(s_{12})+\alpha(s_{21})]\}. \qquad (3.15a)$$

Here, s_{ij} is the distance from the ith slit ($i = 1,2$) to the jth detector ($j = 1,2$). Thus, the first product of amplitudes represents the process for which electrons from slits 1 and 2 respectively arrive at detectors D1 and D2; the second product of amplitudes represents the exchange process. The phase shifts $\alpha(s_{ij})$, incurred by propagation through the vector potential field of the solenoid, are related to the confined magnetic flux Φ by Stokes' law as follows:

$$\alpha(s_{11})-\alpha(s_{21}) = \alpha(s_{12})-\alpha(s_{22}) = \left(\frac{e}{\hbar c}\right)\Phi \equiv \alpha. \qquad (3.15b)$$

It is then a straightforward matter to show that the (normalized) joint probability of electron arrival at D1 and D2, deducible from relations (3.15a,b), is

$$P(D1, D2) = |\psi(D1, D2)|^2 = \frac{1}{2}\{1 - \cos[k(s_{11} - s_{21} + s_{22} - s_{12})]\}. \qquad (3.15c)$$

Although quantum interference occurs, there is no trace of the magnetic flux! The magnetic phase shifts for the direct and exchange processes have cancelled.

It appears, at least at first glance, that the two-slit AB effect with electrons correlated by the spin-statistics connection manifests a curious phenomenological reversal vis-à-vis the AB effect with momentum-correlated electrons discussed in the previous section. In the latter case, the magnetic flux dependence occurs only in the joint detection probability $P(D1, D2)$; the probability of electron arrival at a single detector e.g., $P(D1)$, manifests no AB effect. By contrast, the experimental configuration of Figure 3.12 gives rise to a joint probability $P(D1, D2)$ unaffected by the confined magnetic field, although the probability of electron arrival at each detector individually has been shown earlier to vary harmonically with magnetic flux in the following way:

$$P(D1) = \frac{1}{2}\{1 + \cos[k(s_{11} - s_{21}) + \alpha]\}, \qquad (3.16a)$$

$$P(D2) = \frac{1}{2}\{1 + \cos[k(s_{12} - s_{22}) + \alpha]\}. \qquad (3.16b)$$

If this is the case, then it leads to an extraordinarily puzzling consequence. By arranging experimental conditions so that the geometrical phases are

$$k(s_{11} - s_{21}) = k(s_{22} - s_{12}) = -\frac{\pi}{2}$$

(e.g., symmetrical disposition of D1 and D2 above and below the forward direction) and adjusting the magnetic flux so that $\alpha = \pi/2$, one deduces that $P(D1, D2) = 1$, $P(D1) = 1$, and $P(D2) = 0$. How can it be that there is a 100% coincidence count rate if the individual count rate at one of the detectors is zero!

The origin of the paradox lies in the inconsistent treatment of correlated and uncorrelated electron states. The probabilities $P(D1)$ and $P(D2)$ of Eqs. (3.16a) and (3.16b) were calculated by means of single-particle wave functions and therefore characterize the case of *uncorrelated* electron propagation through the two slits. When only one of the two detectors is registering particles and there is, therefore, no ostensible exchange process, it may seem reasonable to think of the electrons as arriving independently at the detector in one-particle wave packets. This would be incorrect, however. The joint probability $P(D1, D2)$ for uncorrelated electrons is simply the product $P(D1)P(D2)$, and this expression, which is flux dependent, differs markedly from Eq. (3.15c). For correlated electrons pairs, the probability that one electron arrives at a particular detector, let us say D1, irrespective of the fate of the second particle, is obtained by summing (integrating) $P(D1, D2)$ of Eq. (3.15c) over all locations of the other detector, D2. The resulting expression is (as it must be) independent of the magnetic flux and cannot vanish for nonzero $P(D1, D2)$.

Phenomena such as electron antibunching or the AB–HBT (non)effect that involve both quantum interference and particle correlation raise several fundamental questions.

First, it is clear that these phenomena are not interpretable in terms of the self-interference of single particles. Two particles must be present for interference to occur. What is one to make, therefore, of Dirac's dictum that interference between different particles never occurs? Is Dirac wrong?

I do not think so. The critical point is to recognize that, under the specified experimental conditions, there is no way, short of an observer intervention that alters the system and destroys the interference, to determine whether the particle from slit 1 or slit 2 has propagated to a particular detector. There is no distinguishable particle 1 or particle 2; there are only two-particle events. The very act of labeling is a mental construct drawn from classical physics that deceives one into imagining the separate particle trajectories through the interferometer. In reality, all that one knows with assurance is the number of two-particle events recorded in a certain period of time.

Dirac's remark, interpreted more largely, is still valid. When, even in principle, it is impossible to identify the source from which a particle has issued without changing the course of the experiment, it is not possible to say with certainty that the interference is produced by *different* particles.

Second, to what extent does a particular source produce correlated or uncorrelated electrons? The answer to this question gives partial insight into why an electron HBT experiment would be difficult. It is not the coherence time alone, but rather what is termed the beam degeneracy that helps gauge the feasibility of observing quantum mechanical correlations arising from spin and statistics. The degeneracy parameter

$$D = jA_c t_c \tag{3.17}$$

is the product of the particle flux j (number of particles per second through a unit area normal to the beam), the coherence time t_c, and the coherence area A_c, which is approximately the square of the transverse or lateral coherence length, ℓ_s, introduced in Eq. (3.7). Physically, A_c represents the effective surface (perpendicular to the beam) over which an interference pattern can be produced.[34] The degeneracy is a dimensionless statistical parameter indicating the mean number of particles per quantum state. The greater the degeneracy, the greater is the contribution of correlated pairs of particles emerging from the source. For light, there is no limit to the mean number of photons, which are massless bosons, that can occupy a quantum state. For example, D is about 10^{-3} for a mercury arc lamp such as the one employed by HBT in their photon correlation experiment; gas lasers can produce beams with D in excess of 10^{12}. Because there can never be more than one fermion per quantum state, however, the degeneracy of an electron source never exceeds unity.

Unfortunately, the degeneracy of even the most coherent electron beams now available is substantially below unity; the degeneracy of the field-emission source which has been employed in the Hitachi electron self-interference and AB effect experiments is about 10^{-6}. No wonder, then, that even with a flux of some 10^{13} electrons/cm^{2}s, such a beam produces a fringe pattern interpretable entirely as the self-interference of single-particle wave packets. One might think that the more intense beam of an electron accelerator might manifest electron correlations more strongly, but this is not necessarily the case. The lower coherence area attributable to greater beam divergence yields a resulting degeneracy below that of the field-emission source.

The situation is still not hopeless. First, new types of electron sources are being developed with emission tips of nearly atomic size;[35] degeneracies on the order of 10^{-2} have been predicted. Second, with advancing technology, it should be possible to enhance the ratio of signal (the fermionic correlations) to noise (the background random correlations) by use of faster detectors. Other variables being fixed, the signal-to-noise ratio generally increases as the square root of the total length of time the particle count is maintained. During the 1950s, HBT succeeded in observing the correlations of light beams with a degeneracy on the order of 10^{-3} by collecting data for some ten hours. Although the degeneracy of present electron sources is lower, the use of detectors with response times closer to the electron coherence time t_c would reduce the total counting time needed to achieve a desired signal-to-noise level.

Another promising possibility is to use charged particles other than electrons. For example, with a gas source that produces an intense, collimated, nearly monoenergetic beam of helium ions, one could, in principle, study the correlations of either fermions (such as the ^{3}He$^+$ ion) or bosons (like the ^{4}He$^+$ ion) simply by changing the input gas.

Difficult though they may be, these new types of quantum interference experiments will some day be performed, for there are, I believe, few limitations to human ingenuity beyond the laws of physics themselves. And to these rigorous physical laws, I would add another one of a historical nature—namely the development of new experimental methods nearly always leads to significant discoveries.

3.5. HBT Update

In 1986, I first proposed in my lectures "Quantum Optics of Particles: Distinctive Features of a Hanbury Brown–Twiss Experiment with Electrons" (given at the Optical Society of America Annual Meeting) and "New Quantum Effects by Means of Electron Intensity Interferometry" (given at the 2nd International Symposium on the Foundations of Quantum Mechanics in the Light of New Technology) a series of experiments to explore novel quantum interference effects attributable to the fermionic nature and electrical charge of electrons.[36] In the years following, I wrote a number of papers (a few of which are cited in Selected Papers by the Author at the end of this book) investigating in quantitative detail the phenomena to be manifested by such experiments.

The ideas interested researchers in diverse fields of physics who subsequently contacted me for advice on proceeding. Indeed, the first message I ever received after establishing an electronic mail account upon my return to the United States from Japan was a request for information regarding the electron HBT effect. Throughout the 1990s, a number of attempts were made to observe the antibunching of electrons in a coherent electron beam. I began experiments of this kind myself with colleagues at the NSF Center for High Resolution Electron Microscopy at Arizona State University. Over fifteen years have now passed since those two seminal lectures, and, to my knowledge, no experiment employing free-electron beams has yet unambiguously demonstrated the fermionic anticorrelations of electrons.

However, shortly after I began the redaction of this present book, electron antibunching was successfully observed, not in free-electron beams, but in electron currents constrained to flow in two-dimensional "mesoscopic" (i.e., very small, but not atomic sized) semiconductor devices. The decisive feature of these devices is that the relevant electron states are nearly completely degenerate (i.e., have degeneracy parameter close to unity).

In one experiment[37] performed in Switzerland and shown schematically in Figure 3.13, a magnetic field confined mobile electrons to a plane; electrons injected into the device by a voltage source encountered a quantum point contact that served as a tunable beam splitter transmitting a portion of the current (50% in the reported experiment) into a metallic contact and reflecting the rest into another contact. The transmitted (I_t) and reflected (I_r) currents were converted into voltage signals by two 1-kΩ resistors and then amplified. The outputs of the two amplifiers were multiplied electronically, after which a spectrum analyzer determined the correlation $\langle \Delta I_t \Delta I_r \rangle$ of the current fluctuations at a central frequency in the range of 100 kHz to 1 MHz.

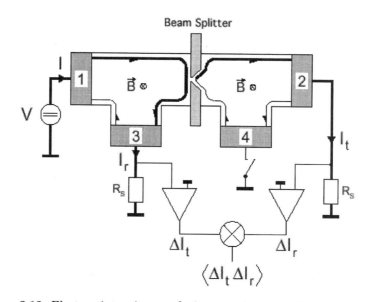

Figure 3.13. Electron intensity correlation experiment performed on electron currents confined to a plane by a magnetic field. Regions numbered 1 through 4 are electron reservoirs; a metallic split gate serves as a beam splitter. Electrons injected by voltage source V move along the upper edge channel to the gate where they are either transmitted into contact 2 or reflected into contact 3. The transmitted and reflected currents are converted by resistors R to voltage signals that are amplified and then correlated. (Adapted from http://haithabu.fy.chalmers.se/abstracts/037.pdf.)

The variation of the measured spectral densities as a function of the incident current I is plotted in Figure 3.14. The upper curve, corresponding to the autocorrelation in the transmitted channel, shows a positive slope as expected for all particles, whether fermions or bosons. The lower curve, corresponding to the cross-correlation between transmitted and reflected currents, shows a negative slope indicative of anticorrelated fluctuations.

An analogous experiment[38] employing a quantum point contact to inject single-mode electrons into a mesoscopic beam splitter at the planar interface between two semiconductor materials (GaAs and AlGaAs) was performed independently by another group of researchers. Figure 3.15 shows a schematic diagram of the interferometer, which is a few hundred nanometers in size; the semiconductor finger constituting the actual beam splitter is only 40 nm wide. Again, the outputs from the beam splitter revealed a negative cross-correlation, signifying that electrons arrived individually (and not in groups) at the beam splitter and were distributed randomly between the transmitted and reflected currents.

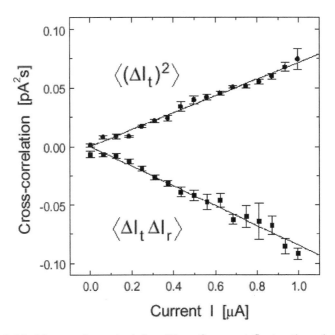

Figure 3.14. Measured spectral densities of current fluctuations (at a temperature 2.5 K) produced by the interferometer of Figure 3.13 with the beam splitter adjusted for 50% transmission. The autocorrelation (upper curve) and cross-correlation (lower curve) vary linearly with input current with positive and negative slopes, respectively. Quantum theory predicts that $\langle \Delta I_r \Delta I_t \rangle / |I| = -2et(1-t) = -0.25(2e)$ for transmission coefficient $t = 0.5$; the experimental result is $-0.26(2e)$. (Adapted from http://haithabu.fy.chalmers.se/abstracts/037.pdf.)

That there exist now experimental methods of performing "intensity interferometry" with electrons is an exciting development, for it is an important first step toward realizing in the laboratory an entirely new class of experiments for probing the collective behavior of quantum particles. Few (if any) physicists who concern themselves with the fundamentals of quantum mechanics would have doubted that electrons in aggregate show antibunching. However, electrons, in fact, should exhibit more varied behavior than that. Theoretical studies[39] have brought to light the existence of new classes of collective electron states that, like thermal photons, give rise to positive cross-correlations or "bunching" even though they are constructed in accordance with Fermi–Dirac statistics. Moreover, certain of these multiparticle states lead to fluctuations in particle number smaller than the fluctuations encountered in coherent electron beams presently employed in interferometry. Use of such states, therefore, could greatly increase the sensitivity of an electron interferometer. Such correlated electron states have yet to be created and explored experimentally—but, as in the

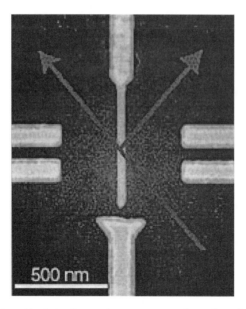

Figure 3.15. Intensity interferometer constructed at the planar interface of GaAs and AlGaAs semiconductors. Electrons are incident upon the beam splitter (40 nm wide) from the lower right and are either transmitted or reflected with equal probability. A negative cross-correlation of the two beams shows that electrons arrive individually at the beam splitter and not, like thermal photons, in bunches. (Adapted from http://www.stanford.edu/dept/news/report/news/april21/antibunch-421.html.)

demonstration of electron antibunching, the appropriate technology will eventually be developed.

*   *   *

Physicists have had some eighty years to adjust to the discovery of the wavelike behavior of matter. Time and familiarity often have a way of dulling astonishment, but neither makes the strange processes of the quantum world more visually accessible today than they were previously. What is one to make of a description of nature that forbids detailed knowledge of motion, manifests force-free interactions between matter and fields, and gives rise to ghostly correlations between arbitrarily separated noninteracting particles? Schrödinger, the person perhaps most responsible for the wave mechanics of matter, wrote in utter frustration

... that a space–time description is impossible, I reject *a limine*. Physics does not consist only of atomic research, science does not consist only of physics, and life does not consist only of science. The aim of atomic research is to fit our empirical knowledge concerning it into our other thinking. All of this other thinking, so far as it concerns the outer world, is active in space and time. If it cannot be fitted into space and time, then it fails in its whole aim and one does not know what purpose it really serves.

Like all humans, scientists have a deep-rooted need for descriptive explanations; mathematical formalism, alone, even if seemingly correct, is somehow insufficient. However, quantum mechanics furnishes predictions, not explanations. Perhaps there will come a time when the mysterious wavelike processes inherent in the structure of quantum theory will be unraveled in a causally explicit way—although I rather doubt it. But neither do I find that doubt disturbing. If not purpose, then surely there is at least great satisfaction in a theory of such broad predictive power that opens up for exploration a world beyond the senses where even the imagination can scarcely follow.

Notes

1. K. Przibram (ed.), *Letters on Wave Mechanics*, Philosophical Library, New York, 1967, p. vii.
2. I have on occasion listened to physicists and teachers expound how the consequences of Einstein's (special) theory of relativity, the ultimate classical theory of space, time, and motion, would seem intuitively obvious if only the basic ideas were presented in elementary or secondary school. I am profoundly incredulous. The intellectual exposure to a set of abstract principles can never make "self-evident" physical phenomena that lie well outside the bounds of a person's familiar experiences.
3. R. P. Feynman, R. B. Leighton, and M. Sands, *The Feynman Lectures on Physics*, Addison-Wesley, Reading, MA, 1965, Vol. III, p. 1–1.
4. A. Tonomura, J. Endo, T. Matsuda, and T. Kawasaki, Demonstration of Single-Electron Buildup of an Interference Pattern, *American Journal of Physics* **57** (1989) 117.
5. I discuss the seemingly impossible (but actually realizable) process of high-resolution imaging *without* lenses in my book *Waves and Grains: Reflections on Light and Learning* (Princeton University Press, Princeton, NJ, 1998, Chapter 7).
6. C. Zimmer, Shell Game, *Discover* **13**(5) (1992) 38–42.
7. G. I. Taylor, Interference Fringes with Feeble Light, *Proceedings of the Cambridge Philosophical Society* **15** (1909) 114.
8. P. A. M. Dirac, *The Principles of Quantum Mechanics*, 4th ed., Oxford University Press, London, 1958.
9. Although few physicists would dispute the mathematical formalism of quantum mechanism, the interpretation of that formalism has given rise to widely differing points of view. One such perspective espoused by Louis de Broglie and David Bohm, based on a transformation of the Schrödinger equation into a form resembling the equation of motion of a fluid, asserts that a particle has a definite position and momentum at all times, although one cannot know what they are. Advocates of this minority point of view argue that "... our *knowledge* of the state of a system should not be confused with what the state *actually* is," or that quantum mechanics

"is constructed so that we cannot 'observe' position and momentum simultaneously but this fact *per se* does not have any bearing on the issue of whether a particle has a well-defined track *in reality*." I find statements like these too metaphysical for my own taste, but readers interested in this point of view may consult the book by P. R. Holland, *The Quantum Theory of Motion* (Cambridge University Press, New York, 1993), from which the quoted passages (with original italicizations) were drawn (p. 77).

10. As used by Feynman, the phrase "the *only* mystery" is somewhat misleading, for it implies that all quantum interference phenomena can be understood on the same basis as the analysis of the two-slit experiment. This is not the case. As one directs attention away from systems of single particles to systems of correlated particles or of particles and fields together, other "mysteries" equally profound arise. These are discussed in the sections to follow, but a more technical elaboration can be found in my book *More Than One Mystery: Explorations in Quantum Interference* (Springer-Verlag, New York, 1995). For the sake of clarity, I wish to emphasize here that neither I nor Feynman ever intended by use of the term "mystery" to imply that the mathematical basis or practical implementation of quantum mechanics is somehow uncertain, unknowable, or arbitrary. Rather, "mystery" is a colorful word conveying that the space–time evolution of particles involved in quantum interference processes cannot be visualized, in contrast to particles in classical physics.

11. R. Lakes, Experimental Limits on the Photon Mass and Cosmic Magnetic Vector Potential, *Physical Review Letters* **80** (1998) 1826.

12. The gradient of a scalar function ϕ expressed in Cartesian coordinates is a vector with components

$$(\text{grad } \phi)_x = \frac{\partial \phi}{\partial x}, \quad (\text{grad } \phi)_y = \frac{\partial \phi}{\partial y}, \quad (\text{grad } \phi)_z = \frac{\partial \phi}{\partial z}.$$

Correspondingly, the curl (or rot) of a vector function $\mathbf{A} = (A_x, A_y, A_z)$ is a vector with components

$$(\text{curl } \mathbf{A})_x = \left(\frac{\partial}{\partial y}\right)A_z - \left(\frac{\partial}{\partial z}\right)A_y,$$

$$(\text{curl } \mathbf{A})_y = \left(\frac{\partial}{\partial z}\right)A_x - \left(\frac{\partial}{\partial x}\right)A_z,$$

$$(\text{curl } \mathbf{A})_z = \left(\frac{\partial}{\partial x}\right)A_y - \left(\frac{\partial}{\partial y}\right)A_x.$$

13. To effect a gauge transformation of a given set of electromagnetic potentials, ϕ and $\mathbf{A}$, into a new set, ϕ' and $\mathbf{A}'$, select an appropriate gauge function Λ and calculate

$$\phi' = \phi - \frac{1}{c}\frac{\partial \Lambda}{\partial t},$$

$$\mathbf{A}' = \mathbf{A} + \text{grad}\Lambda,$$

where c is the speed of light.

14. W. Ehrenberg and R. E. Siday, The Refractive Index in Electron Optics and the Principles of Dynamics, *Proceedings of the Physical Society (London)* **B62** (1949) 8.

15. The magnetic field strength of the Earth is on the order of $1\,\mathrm{G}$ near the Earth's surface. A small household bar magnet can be about 10^2–$10^3\,\mathrm{G}$.

16. Y. Aharonov and D. Bohm, Significance of Electromagnetic Potentials in the Quantum Theory, *Physical Review* **115** (1959) 485.

17. Y. Aharonov and D. Bohm, Further Considerations on Electromagnetic Potentials in the Quantum Theory, *Physical Review* **123** (1961) 1511.

18. R. P. Feynman, R. B. Leighton, and M. Sands, *The Feynman Lectures on Physics*, Addison-Wesley, Reading, MA, 1965, Vol. II, Section 15-5.

19. D. H. Kobe, Aharonov–Bohm Effect Revisited, *Annals of Physics NY* **123** (1979) 381.

20. N. Osakabe, T. Matsuda, T. Kawasaki, J. Endo, A. Tonomura, S. Yano, and H. Yamada, Experimental Confirmation of Aharonov–Bohm Effect Using a Toroidal Magnetic Field Confined by a Superconductor, *Physical Review A* **34** (1986) 815.

21. It has been claimed that a part of the magnetic field of a charged-point particle moving parallel to a conducting surface will penetrate the conductor and thereby provide a classical interaction for interpreting the AB effect. See T. H. Boyer, Understanding The Penetration of Electromagnetic Velocity Fields Into Conductors, *American Journal of Physics* **67** (1999) 1.

22. The amplitude for electron transmission through a (nonabsorbing) beam splitter has a phase shift of 90° relative to the amplitude for reflection. This property, which also characterizes the reflection and transmission of light under comparable circumstances, derives from particle and energy conservation.

23. A. Einstein, B. Podolsky, and N. Rosen, Can Quantum-Mechanical Description of Reality Be Considered Complete? *Physical Review* **47** (1935) 777.

24. Cited in W. Moore, *Schrödinger*, Cambridge University Press, Cambridge, 1989, p. 137.

25. R. Hanbury Brown and R. Q. Twiss, A New Type of Interferometer for Use in Radioastronomy, *Philosophical Magazine* **45** (1954) 663.

26. Conceptually, Michelson's stellar interferometer is another version of the classic two-slit experiment. In lieu of apertures, there are two mirrors a distance d apart, each one reflecting an image of the light source (a star). The two images are made to superpose on a third mirror, thereby giving rise to a pattern of interference fringes similar to that of Figure 3.1.

27. When the mirrors are close together, the superposed amplitudes have a well-defined relative phase; the intensity of the dark fringes approaches zero and the visibility is close to unity. When the mirrors are far apart, the two images constitute essentially independent sources; there is practically no interference and the visibility vanishes.

28. R. Hanbury Brown, *The Intensity Interferometer*, Taylor and Francis, London, 1976, pp. 4–5.

29. R. Hanbury Brown and R. Q. Twiss, Correlation Between Photons in Two Coherent Beams of Light, *Nature* **177** (1956) 27.

30. R. H. Brown, *The Intensity Interferometer*, Taylor and Francis, London, 1976, p. 7.
31. It should be pointed out that the results of the two-electron two-solenoid AB experiments of Section 3.3 would not have been essentially altered if the electron momenta were correlated in some other way than by back-to-back emission. If, for example, the second electron propagated towards mirror M2 (rather than M2′) when the first propagated towards M1, various phase factors would be different, but the nature of the correlation expressed $P(D1, D2)$ in would be the same.
32. D. L. Goodstein, Richard Feynman, Teacher, *Physics Today* **42** (February 1989) 75. The origin of the spin-statistics connection intrigued Feynman, and even near the end of his life, he was searching for a simpler way to understand the matter than that given many years earlier by Pauli. Although he never expounded it at the freshman level, he did give a lucid, physical explanation in his 1986 Dirac Memorial Lecture, "The Reason for Antiparticles," published in R. P. Feynman and S. Weinberg, *Elementary Particles and the Laws of Physics*, Cambridge University Press, Cambridge, 1987.
33. The related question of whether radioactive nuclei decay randomly (with accompanying random emission of alpha, beta, or gamma rays) is taken up in Chapter 8.
34. I discuss degeneracy and coherence parameters more comprehensively in *More Than One Mystery: Explorations in Quantum Interference* (Springer-Verlag, New York, 1995).
35. H.-W. Fink, Point Source for Ions and Electrons, *Physica Scripta* **38** (1988) 260.
36. M. P. Silverman, Quantum Optics of Particles: Distinctive Features of a Hanbury–Brown Twiss Experiment with Electrons, *Optical Society of America 1986 Annual Meeting Technical Digest* (Optical Society of America, Washington DC, 1986, p. 44, and *Optics News* **12**(9) (1986) 123; New Quantum Effects by Means of Electron Intensity Interferometry, *Proceedings of the 2nd International Symposium on the Foundations of Quantum Mechanics in the Light of New Technology*, edited by M. Namiki, Physical Society of Japan, Tokyo, 1986, p. 369.
37. M. Henny, S. Oberholzer, C. Strunk, T. Heinzel, K. Ensslin, M. Holland, and C. Schönenberger, Hanbury-Brown and Twiss Experiment with Fermions, *Science* **284** (1999) 296.
38. W. D. Oliver, J. Kim, R. C. Liu, and Y. Yamamoto, Hanbury Brown and Twiss-Type Experiment with Electrons *Science* **284** (1999) 299.
39. M. P. Silverman, Fermion Ensembles That Show Statistical Bunching, *Physics Letters A* **124** (1987) 27. See also *More Than One Mystery: Explorations in Quantum Interference*, Springer-Verlag, 1995. These correlated electron states resemble in some ways the so-called "squeezed" states of light that have been intensely investigated in recent years in part because they lead to uncertainties in phase or amplitude (for a particular phase component) smaller than that ordinarily expected from the Heisenberg uncertainty principle. This does not violate any physical law, however, because the total uncertainty for both phase components is in accord with the uncertainty principle.

CHAPTER 4

Quantum Beats and Giant Atoms

4.1. The Light from Atomic "Pulsars"

When I was at Harvard University many years ago investigating the structure and interactions of the hydrogen atom, I first learned of a remarkable optical phenomenon. I might well have observed it in my own experiments, had it occurred to me to look. But it did not and I missed seeing at that time a most striking demonstration of the principles of quantum mechanics. By the time I realized this, the apparatus had been "improved" and the effect would not have been produced under the changed conditions. Nevertheless, like a haunting melody, this curious phenomenon known as "quantum beats" has often returned to my thoughts to form a significant part of my scientific interests.[1] Had I looked carefully enough at my hydrogen atoms back in the late 1960s, I would have seen them periodically "winking" at me like the rotating beacon of a lighthouse—like a little atomic pulsar.

Since 1913, when Niels Bohr revealed his semiclassical planetary model of the atom, atomic hydrogen has been a touchstone against which the success of any theory of atomic structure is measured. With but a single bound electron, it is the simplest naturally occurring atom of the periodical table—and even today the only atom for which analytically exact theoretical treatments can be provided. [If exotic combinations of particles are included, then positronium, a bound electron and antielectron (positron) may be considered the simplest atom, for it is a purely electrodynamic system (i.e., not subject to the strong nuclear interactions) containing two apparently structureless particles.] The main legacy of the Bohr theory, retained and refined in the complete quantum mechanics which subsequently followed more than ten years later, is the idea of discrete characteristic states with quantized energies. This means that a bound electron cannot absorb energy in arbitrary amounts but, in marked contrast to classical theories of the atom, only in quantities that take it from one energy eigenstate to another (from the German "eigen" = own or particular). Once the discreteness of atomic states and the quantization of atomic energy were accepted, it followed as a seemingly irrefutable proposition that a free atom, unperturbed by external fields, had at all times to be in one of its eigenstates.

It might be worth noting that there is nothing intrinsically quantum mechanical about the concept of "quantization"—the very feature that gave the "new" mechanics its name. Indeed, the attribute of discreteness of allowed values is encountered in other systems of a purely classical nature, as, for example, the quantized oscillation frequencies of a vibrating string, membrane, or air column. Quantization is a consequence of the imposition of boundary requirements—and this can occur in classical or quantum mechanics. Frequency and energy and closely linked in quantum mechanics by Einstein's relation, $E = h\nu$, in a way without parallel in classical physics. Nevertheless, if I had to give a one-word synopsis of what is phenomenologically unique in quantum mechanics, it would be "interference", not "quantization"; but this is a matter of personal opinion. The phenomenon of quantum beats is the embodiment of interference—the ultimate two-slit (or more) particle interference experiment packed into the diminutive volume of a single atom.

One objective of my Harvard experiments was to probe the structure of the hydrogen atom more thoroughly than had been done previously. Motivating every "hydrogen watcher" is the hope of finding a discrepancy with theory, for, although there is satisfaction in confirming quantum mechanics, it would be far more exciting to disprove it. Bohr's prediction of the spectrum of electronic energy levels

$$E_n = -\frac{Ry}{n^2} \quad (n = 1, 2, 3, ...) \tag{4.1a}$$

for a particle bound in a Coulomb potential [$V(r) = -e^2/r$] was correct as far as it went. Here, the Rydberg constant, Ry, defined by

$$Ry = \frac{2\pi^2 me^4}{h^2} \sim 13.6 \, \text{eV}, \tag{4.1b}$$

in which h is Planck's constant and e and m are the electron charge and mass, respectively, sets the scale of atomic energies. However, the energy level structure of a real hydrogen atom is more complex.

For one thing, the intrinsic spin $\frac{1}{2}\hbar$ (recall $\hbar = h/2\pi$) of the electron gives rise to an electron magnetic dipole moment. The magnetic moment of a classical particle is proportional to the angular momentum of the particle and inversely proportional to the mass. This is also the case for the quantum mechanical electron, although the proportionality constant is a factor of 2 larger than that deduced from classical mechanics. The electron magnetic moment emerges in a natural way from the Dirac relativistic equation of motion.

From the perspective of an observer at rest in the laboratory, the proton in a stationary hydrogen atom is practically at rest. Actually, the electrostatically bound electron and proton orbit in binary-star fashion about their common center of mass. The location of the center of mass is almost coincident with the location of the proton, the more massive particle of the pair by a factor of nearly 1840. As seen from the electron rest frame, however, the proton is in orbit about the electron (just as an Earth-bound observer sees the diurnal passage of the Sun). This picture, of course, is drawn from the imagery of classical physics, which, when pushed too far, can give misleading, if not totally erroneous, results; quantum mechanics does not ordinarily allow us to imagine electron or proton trajectories within an atom. Still, the picture can be useful at times.

The orbiting proton (in the electron reference frame) constitutes an electrical current that produces at the electron site a magnetic field proportional to the electron orbital angular momentum (in the laboratory reference frame). Depending on whether the electron spin and orbital angular momenta are parallel or antiparallel to one another (the only allowed possibilities for a spin-$\frac{1}{2}$ particle), the interaction between the electron magnetic moment and the local magnetic field can slightly augment or diminish the electrostatic (Coulomb) energy of an atomic state.

Each Bohr energy level of given principal quantum number n actually comprises $2n^2$ degenerate states (i.e., states of the same energy) distinguished by quantum numbers designating their orbital angular momentum (L), component of orbital angular momentum along an arbitrarily chosen quantization axis (M_L), and component of spin along that same axis (M_S). Thus, this fine structure or spin–orbit interaction splits the Coulomb energy of states with nonzero angular momentum quantum number L into two close-lying levels. The exact amount of splitting depends on the angular momenta of the states involved, but to a good approximation, it is smaller than the electrostatic energy by the square of the so-called Sommerfeld fine-structure constant, $\alpha_{fs} = e^2/\hbar c \sim 1/137$, and the first power of the principal quantum number, or, more succinctly,

$$\frac{\Delta E_{fs;n}}{\Delta E_n} \sim \frac{\alpha_{fs}^2}{n} \sim \frac{5 \times 10^{-5}}{n}. \qquad (4.1c)$$

The hydrogen fine-structure intervals divided by Planck's constant correspond to Bohr frequencies that generally fall in the microwave or radio-frequency range of the electromagnetic spectrum.

The proton, like the electron, is also a spin-$\frac{1}{2}$ particle with a magnetic moment. However, the relation between the proton magnetic moment and spin is not as simple as for an electron. The reason for this is that the electron, as far as one presently knows, is a true elementary particle with no internal structure, in contrast to the proton, which is thought to be a composite of three more elementary particles known as quarks and has a complex internal structure. The hyperfine, or spin–spin, interaction between the proton and electron magnetic dipole moments splits each fine-structure level further. This hyperfine splitting again depends on the quantum numbers of the states in question, but, to good approximation, is smaller than the fine-structure splitting by the ratio of the electron and proton masses; thus,

$$\frac{\Delta E_{\mathrm{hf};n}}{\Delta E_n} \sim \frac{m}{m_p} \frac{\alpha_{\mathrm{fs}}^2}{n} \sim \frac{3 \times 10^{-8}}{n}. \tag{4.1d}$$

In addition to the Coulombic, spin–orbit, and spin–spin interactions, which have analogs in classical electromagnetism, there are processes that have no direct counterparts in classical physics. These involve the interaction of the bound electron with the "vacuum". Classically, a vacuum is empty space; not so in quantum physics. The quantum electrodynamical vacuum is a roiling sea of ephemeral (or virtual) particles of matter and light (photons) that can affect the properties of real particles although their own existence is so short-lived as to preclude the possibility of direct observation.

One effect of the vacuum on atoms is quite well known, although perhaps not thought about in this context: the spontaneous emission of light. The fluctuating virtual electromagnetic fields of the vacuum stimulate excited atoms to undergo transitions to lower-energy states, thereby emitting real photons. Thus, the interaction of an atom with the vacuum results in a finite lifetime of the excited atomic states. There are also other more exotic processes that affect the atomic energies.

A bound electron, for example, can interact with the vacuum to emit and then immediately reabsorb a photon; this process alters what is known as the electron self-energy. The electron self-energy, the calculation of which yields an infinitely large value, is not measurable. However, the *difference* in self-energy values between a free electron and one bound in a hydrogen atom *is* calculable and measurable. Another such process, referred to as vacuum polarization, involves the emission by the atomic nucleus of an electron–positron pair, the immediate mutual annihilation of this pair to produce a photon, and the absorption of this photon by the bound electron. The net effect of these (and other) virtual processes is to shift the energy levels of different states by different amounts. The most notable shift is between the states designated $n\mathrm{S}_{1/2}$ and $n\mathrm{P}_{1/2}$, which, according to the relativistic quantum theory of Dirac, are predicted to have equal energy in the absence of vacuum processes.[2]

The atomic beam experiments in which I was engaged[3] were undertaken to explore a wide range of hydrogen fine-structure and hyperfine-structure intervals and the shifts induced by vacuum processes. I liked to think of this research as my "Nobel Prize Project"—*not* because the work was destined to win one, but because each principal experimental ingredient of the project had already earned someone else a Nobel Prize. For the development of the "molecular ray method" or beam, Otto Stern receive the Prize in 1943. I. I. Rabi won it in 1944 for his "resonance method," the discovery that one can induce transitions between quantum states of a nucleus by irradiation with a magnetic field oscillating at the appropriate Bohr transition frequency. The procedure, as I employed it, worked just as well with an oscillating electric field applied to electronic fine-structure states of the hydrogen atom. Investigation of the hydrogen fine structure earned Willis Lamb the Prize in 1955; the quantum electrodynamic displacement of S and P states bears his name (Lamb shift). Much later (in 1990), Norman Ramsey, a former Rabi student, was to receive the Prize for a modification of the resonance method whereby *two* spatially separated, but coherently oscillating, radiofrequency fields allowed one to measure nuclear energy level intervals with high precision. The use of this technique, whose theoretical possibilities I studied at great length, significantly improved the precision with which the hydrogen Lamb shift could be determined. In addition, if the creators of the theoretical underpinnings of the experiments were also to be acknowledged, then, of course, the list of Nobel Laureates must include Bohr (1922), for "the investigation of the structure of atoms, and of the radiation emanating from them," Heisenberg (1932), "for the creation of quantum mechanics," and Schrödinger and Dirac (1933), "for the discovery of new productive forms of atomic theory." Newton once remarked that if he saw farther than most, it was because he stood on the shoulders of giants (a comment written during an "unusual fit of modesty" according to one of my historian colleagues). In any event, the predecessors upon whose achievements I relied had no mean stature, either.

Because all hydrogen states, except for the ground state,[4] are unstable and decay radiatively to some lower state(s), one could monitor the effects of external perturbations on them by the corresponding increase or decrease in light emission at the appropriate wavelength. The use of a *fast* atomic beam—a beam in which the atoms move through the apparatus at roughly a hundredth the speed of light— greatly facilitates such spectroscopic measurements. Produced at one location, the atoms rapidly traverse various chambers containing the electromagnetic fields (oscillating at radio or microwave frequencies) for probing desired energy intervals; the atoms continue past a detection window through which the fluorescent photons (the spontaneous decay radiation) can be counted, thereby providing a measure of the number of atoms remaining in the states of interest. When the frequency of an oscillating field equals the Bohr transition frequency for a pair of atomic states,[5] the probability of a transition into or out of the states is greatest, and the photon count rate is maximally affected. (Whether a transition occurs into or out of atomic states coupled by the oscillating fields depends on the relative populations of these states and the time of exposure to the fields.) Thus, use of a fast beam allows for separate regions of creation, spectroscopy, and detection of short-lived states.

One might wonder, however, how a *neutral* hydrogen atom can be brought up to a speed 8000 times that of a passenger jet! After all, unlike the charged particles in high-energy accelerators, a neutral particle cannot be accelerated by electric or magnetic fields. The trick is first to accelerate a beam of *protons* to the desired speed. The protons were produced in an ion source—essentially a cylindrical glass tube, supplied with H_2 gas, inserted through the coil of a powerful radio-frequency oscillator that dissociated the molecular hydrogen into a plasma that gloriously radiated Harvard's crimson color (the Balmer lines of excited H atoms[6]). Extracted from the plasma and then electrostatically accelerated under a potential difference of about 20,000 V, the protons impinged on a thin carbon foil a few hundred atoms thick, capturing electrons as they shot through virtually unaffected in their forward motion. Now, the accelerated proton beam had become a beam of fast-moving hydrogen atoms distributed over a broad range of quantum states.

There were many aspects of the experiment that were challenging, but few more frustrating than these foils which had a tendency to "burn through" just when the collection of data seemed to be going well. Replacing them was a time-consuming affair, for the accelerator had to be shut down and opened to the atmosphere, after which began the tedious task of separating the ultrathin carbon foils from glass microscope slides on which they were mounted by the manufacturer, and of then remounting them (without crumpling or breaking) on a frame to be suspended in the path of the beam. Useful though they had been, I was not unhappy to dispense with the whole business of carbon foils and use an indestructible gas target that accomplished the same task with less aggravation. I had not realized, unfortunately, that what was potentially the most interesting part of the experiment was literally thrown away!

The characteristic feature of the random decay of independent systems—whether alpha-particle decay of atomic nuclei or radiative decay of excited atoms—is the exponential variation in time. This is the inevitable result of a decay process in which the number of particles decaying at any moment is proportional to the number of particles present:

$$\frac{dN}{dt} = -\frac{N}{T}. \qquad (4.2a)$$

Here, $1/T$ is the characteristic decay rate of the particular process; T is said to be the particle lifetime (for the given mode of decay), but this is a statistical quantity referring to the whole ensemble of particles, and not an attribute of an individual particle which may live considerably longer or shorter than an interval T. The differential equation (4.2a) is readily integrated. After a time interval t, the ratio of the number, $N(t)$, of remaining states is to the number, N_0, of initial states is

$$\frac{N(t)}{N_0} = \exp\left(-\frac{t}{T}\right). \qquad (4.2b)$$

The lifetime T, therefore, is the time interval after which the relative population of decaying particles has dropped to $e^{-1} \sim 0.37$.

As I mentioned before, the belief was widespread that an atom unaffected by external perturbations had to be in one of its allowed energy eigenstates. Thus, the beam of fast H atoms emerging from the carbon foil would contain what one could describe as a *mixture* of states. A complete description of such a mixture would entail a tabulation of the statistical frequencies or probabilities with which each hydrogenic state appears; perhaps something like 80% 1S states, 10% 2S states, 5% 2P states, and so forth (if one limits the description to the orbital states of different electronic manifolds). As the beam leaves the foil, the excited states decay in time at rates that depend on the principal and orbital angular momentum quantum numbers. For example, the lifetime of a 2S state is about $\frac{1}{7}$s (effectively infinite on the timescale of atomic processes[7]), whereas that of a 2P state is 1.6 ns ($1\,\mathrm{ns} = 1 \times 10^{-9}\mathrm{s}$). The uniform motion of the beam converts the decay over a time interval to decay over a space interval.

Suppose one examined the light output from the decaying 4S states (lifetime $\sim 230\,\mathrm{ns}$) by placing a filter in front of a photodetector to block all radiation except for the blue Balmer β light (4S to 2P transition) of wavelength about 486 nm. As a function of distance x from the foil, the Balmer β light intensity, proportional to the number of decaying atoms in the 4S state, would be expected to fall off exponentially in accordance with relation (4.2b) as follows:

$$\frac{I(x)}{I(0)} = \exp\left(-\frac{x}{vT}\right), \qquad (4.2c)$$

where the left-hand side is the relative light intensity at x, and v is the beam velocity. To be sure, a photodetector surface is not a mathematical point; the observed signal in an actual experiment comprises photons from decays spanning a range of locations along the beam and striking different points of the surface of the detector. If one is principally interested—as I was at the time—in maximizing the light received, it is advantageous to use a photodetector with a wide window, survey a broad segment of the atomic beam, and count photons of all polarizations. The signal is then given by relation (4.2c) with appropriate averages made over beam length and detector surface. Nevertheless, neither I nor my experimental colleagues had any doubt that the light emitted from a narrow segment of the beam followed the exponential decay law.

Except that it did not![8] Or, rather, it did if one observed all polarizations equally, and it did not, if the light was detected through a polarizer (e.g., a simple sheet of polaroid film). In the latter case, the light intensity oscillated with distance from the foil indicating that the atoms, like miniature beacons, were in some way turning on and off coherently. As the mean distance between atoms in the beam was far greater than a characteristic atomic size, and as the production of atoms by proton impact on carbon apparently took place independently and randomly, there was no reason to believe that different atoms in the beam could in any way cooperate with one another. The oscillating light output reflected, in a profound way, an oscillatory process intrinsic to each atom—but with all the atoms in synchrony. Like the build-up of a pattern of interference fringes by single electrons as described in Chapter 3, the observed intensity oscillations *could* have been produced one atom at a time—provided one had the patience to collect enough photons.

What were those oscillations? Why did they appear only in polarized light? Why did the standard description of radiative decay not work? Ordinarily applicable in all instances of incoherent particle preparation and decay, relations (4.2a) and (4.2b) do not take account of the uncertainty principle.

A proton moving at roughly 10^8 cm/s will pass through a 10^{-6}-cm-thick fixed carbon foil in a time interval of about 10^{-14} s. At some point in that short time interval, a hydrogen atom is created. As I pointed out previously, an uncertainty in the time of production Δt implies an uncertainty in the energy of the system: $\Delta E \sim \hbar/\Delta t$. In the present case, this energy uncertainty is *larger* than all of the hydrogen fine-structure and hyperfine-structure energy splittings! The Bohr frequency for the largest fine-structure splitting (that of 2P states) is about 10^{10} Hz; the largest hyperfine-structure splitting (that of the 1S states) is about 1.4×10^9 Hz. However, the energy uncertainty (expressed in frequency units) of the foil-excited atoms is $\Delta \nu = \Delta E/h = 1/\Delta t \sim 10^{14}$ Hz.

In effect, the experimenter cannot know to what state any atom has been excited. Of course, if he were to intervene in some way to measure the precise energy of each atom in the beam, this energy would turn out to be one of the energy eigenvalues, but then the quantum beats would disappear. If the experimenter does not measure the energy of an atom, then that atom cannot, even in principle, be thought of as being in a well-defined, albeit unknown, quantum state. The appropriate quantum description must entail a linear superposition of all allowable energy states that give rise to the same final condition—i.e., decay to a specified lower state with emission of a photon falling within the passband of the measuring apparatus. (If the passband were sufficiently narrow that detected photons came from one particular state of a linear superposition of excited states, then the energy of the atom would be known and the quantum interference would vanish.)

Let us consider the example of a four-state atom with two close-lying excited states such as that shown in Figure 4.1a. The excitation is assumed for the time being to be nearly instantaneous. The atom goes from the initial state g (assumed here to be the ground state) to some final state f by quantum pathways $g \to e_1 \to f$ or $g \to e_2 \to f$. Since, under the circumstances of the experiment, these paths are indistinguishable, one must add the probability amplitude for each. Suppose that the probability amplitude for transition from the ground state to an excited state e_i is a_i ($i = 1, 2$) and the corresponding amplitude for a transition from the excited state e_i to the final state f is b_i ($i = 1, 2$). Were it possible to "turn off" the interaction of the atom with the vacuum, then, during the time the atom is excited, it would evolve freely in the absence of all external forces and potentials. From the quantum mechanical equation of motion (Schrödinger or Dirac equation), one can readily deduce that the probability amplitude for free evolution in a state of energy E is proportional to the phase factor $\exp(-iEt/\hbar)$. It is the interaction with the vacuum, however, that induces the radiative transition to state f that makes detection of the quantum beat possible. The theoretical effect of the vacuum can be calculated rigorously by means of quantum electrodynamics; the end result—more or less consistent with our intuition—is that, besides the free-evolution phase factor of magnitude unity (no change in number of atoms in the state), there is a decay factor $\exp(-t/2T)$ representing a loss of atoms from the excited state (with a characteristic lifetime of T).[9] Thus, the total probability amplitude for each pathway in expressible as

$$A(g \to e_1 \to f) \sim a_1 b_1 \exp\left(-\frac{iE_1 t}{\hbar}\right)\exp\left(-\frac{t}{2T}\right) \qquad (4.3a)$$

$$A(g \to e_2 \to f) \sim a_2 b_2 \exp\left(-\frac{iE_2 t}{\hbar}\right)\exp\left(-\frac{t}{2T}\right). \qquad (4.3b)$$

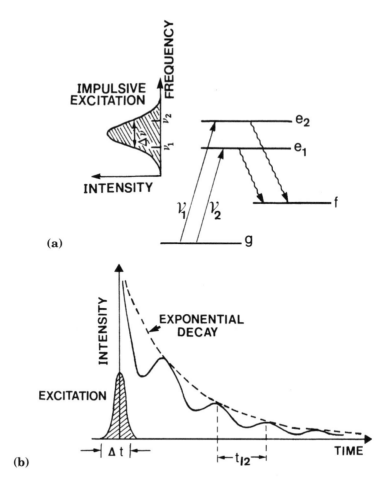

Figure 4.1. (a) Energy level diagram of an atom with ground state g, two close-lying excited states e_1 and e_2, and a lower final state f. An impulsive excitation with frequency components corresponding to the Bohr transition frequencies from g to e_1 and e_2 can drive the atom into a linear superposition of its excited states from which the excited electron radiatively decays to lower state f. (b) The spontaneous emission from atoms prepared as shown in (a) exhibits an oscillatory decay in time with a period t_{12} inversely proportional to the Bohr transition frequency between states e_1 and e_2: $t_{12} = 1/|\nu_2 - \nu_1|$. The spontaneous emission from atoms in an incoherent mixture of the two excited states would decay exponentially in time, as shown by the broken line.

The probability for the transition of the atom out of g and into f with emission of one photon at time t is then

$$P(t) = |A(g \rightarrow e_1 \rightarrow f) + A(g \rightarrow e_2 \rightarrow f)|^2, \qquad (4.3c)$$

which takes the form

$$P(t) = \left[|a_1 b_1|^2 + |a_2 b_2|^2 + 2|a_1 a_2 b_1 b_2| \cos\left(\frac{(E_2 - E_1)t}{\hbar} + \phi \right) \right] \exp\left(-\frac{t}{T} \right). \quad (4.3d)$$

The phenomenon of quantum beats is, in essence, an interference effect in *time* rather than in space; the temporal sequence of transitions between two (or more) sets of internal energy states of the atom is analogous to the spatial pathways through one or the other of two slits in the free-electron "Young's fringes" experiment described earlier. The transition probability—and, therefore, the photon count rate—decays in time as a *modulated* exponential (Figure 4.1b). The "beat" frequency in the quantum interference term corresponds to the Bohr frequency of the excited states; it may thus be seen that the measurement of quantum beats can afford useful spectroscopic information about the energy level structure of an atom or molecules. About this aspect I will have more to say later. The relative phase that appears derives from the phases of the excitation and decay amplitudes, which can be complex numbers.

The system of percussionally excited atoms in a linear superposition of two quantum states with an oscillating relative phase factor $\exp\{-i(E_2 - E_1)t/\hbar\}$ may be likened to a system of synchronously precessing electric dipoles. Although this picture is a classical one, it helps account for some of the features of the quantum beat experiment. The maximum intensity of the radiation from any one dipole sweeps past the detector at the angular frequency $\omega = (E_2 - E_1)/\hbar$. However, for an ensemble of randomly oriented (although synchronously oscillating) dipoles, the net signal is modulated only when light of a particular polarization is observed; the precession can alter the distribution of the radiation, but not the total amount generated (which decreases exponentially in time). This is reflected as well in the quantum mechanical expression (4.3d), which characterizes a transition induced by a particular component of the electric dipole of the atom (a vectorial quantity) and, therefore, the emission of a photon of specified polarization. If one adds together comparable expressions for *all* possible electric dipole transitions from the given excited states to *all* substates (if there are more than one) of the final level f—which is tantamount to observing light of all polarizations—the quantum beat term vanishes, a consequence of the quantum rule known as the Wigner–Eckart theorem.

The picture of precessing dipoles also helps one visualize what should transpire as one increases the time interval over which the excited states are prepared. If the duration of the excitation T_e is short compared to the period of dipole precession $2\pi/\omega$, the dipoles precess together and emit radiation in phase. As T_e lengthens, however, different dipoles are set precessing at increasingly later times and emit radiation increasingly out of phase with that emitted by dipoles established earlier. Once T_e is comparable with, or longer than, $2\pi/\omega$, the precession of the dipoles is no longer synchronized, the radiation is no longer in phase, and the quantum beats are washed out. As a quantitative criterion, the range ωT_e of precession angles spanned by a system of dipoles created over a period T_e must be small compared with unity (1 radian) if the dipoles are to precess synchronously. Thus, for the beats to be observable, one would expect the inequality

$$\omega T_e < 1 \qquad (4.4a)$$

to hold. This criterion also follows readily from the quantum mechanical analysis. With $\omega = (E_2 - E_1)/\hbar$ and $T_e \sim \hbar/\Delta E$, one recognizes in relation (4.4a) a consequence of the uncertainty principle:

$$(E_2 - E_1) < \Delta E. \qquad (4.4b)$$

The uncertainty in the energy of the bound electron must be greater than the energy interval separating the excited states; otherwise, the atom would be in a definite excited state and beats would not occur.

Excitation of atoms by a carbon foil is not the only means by which quantum beats can be produced. Indeed, in many respects it is advantageous to excite the atoms optically, for example with a pulsed laser. Whereas electron capture from a carbon foil gives rise to many excited states simultaneously, excitation with a tunable laser permits one to select specific excited states of interest. Another commendable feature about optical excitation is that the excitation amplitudes (the a_i) can be determined precisely; the theory of the interaction of atoms with light is, if not simple, at least well understood. By contrast, the capture of electrons by proton impact on a solid target of multielectron atoms is a more difficult process to treat theoretically.

The theory of quantum beats produced by light pulses predicts some unusual optical effects that, as far as I know, have yet to be demonstrated experimentally. They must exist, however, if our understanding of the interaction of atoms with light is correct.

Before the development of powerful pulsed lasers, the light sources used to excite atoms were weak in the sense that the majority of exposed atoms remained in their ground state; that is, the probability of a transition was low and the lifetime of the ground state was long (in principle, infinitely long in the absence of radiation). An atom that absorbs a photon from a weak light pulse undergoes effectively one transition to the excited states and subsequently—after passage of the pulse—decays by spontaneous emission to lower states, including the ground state. If the light pulse, like the carbon foil excitation, is short compared to the dipole precession time $(2\pi/\omega)$, then, as one would expect, the probability of spontaneous emission is negligibly small throughout passage of the pulse.

If a light pulse is sufficiently intense, however, it can stimulate atoms to *absorb and emit* photons during its passage. Indeed, many such cycles of excitation by light absorption followed by stimulated emission to the ground state could occur over the duration of one pulse. Thus, as a consequence of strong light excitation, the atomic ground state acquires a finite lifetime T_p inversely proportional to the light intensity at the transition frequency. T_p is the inverse of the excitation rate; it is the "pumping" time, the mean time between the successive absorption and stimulated emission of a photon. When the pulse has passed, the excited atoms again decay exclusively by spontaneous emission.

What effect should all of this cycling back and forth between ground and excited states have on the quantum beats? Very little, one might imagine. After all, the observed beats occur in the *spontaneous* emission from *freely* precessing atomic dipoles *after* passage of the light pulse. This is certainly true if the light pulse is short compared with the precession time. However, a theoretical study of the effect of increasing the pulse duration produced a most surprising result.

One might expect, in view of the reasoning behind relation (4.4a), that the contrast[10] of quantum beats should diminish and ultimately vanish as the duration of the light pulse, T_e, exceeds the precession time characteristic of the excited states. This was indeed the case for light pulses of weak to moderate intensity. However, when the intensity of a long ($T_e > 1/\omega$) light pulse was increased sufficiently so that the ground-state lifetime was short ($T_p \ll 1/\omega$), the quantum beats reappeared strongly! How was it possible—to refer again to the classical analogy—for apparently randomly phased dipoles to emit light synchronously?

The explanation of this baffling phenomenon turned out to be simple, but subtle. Because of the frequent cycles of excitation and stimulated emission, the precession of the dipoles is interrupted so often that their overall dispersion in phase angle remains small. The system of randomly excited and de-excited atoms resembles somewhat the "random walk" of a drunkard through the woods: He bumps into trees, falls down, gets up and starts off again—sometimes in the original direction, sometimes in the opposite direction. At the end of a certain time, he has progressed in a random direction from his point of origin by a distance that varies as the square root of the number of steps.

During the passage of the light pulse in a time interval T_e, the number of successive absorption and stimulated emission processes that occur is approximately $N = T_e/T_p$. Each time an atom is re-excited from the ground state, the corresponding dipole can precess either in the original sense or in the opposite sense. Over the time T_p that the atom is excited (before another stimulated emission to the ground state occurs), the corresponding dipole precesses through an angle $\theta \sim \omega T_p$. The dispersion in phase over the whole system of atoms, like the mean displacement in a random walk problem, is $\Delta\theta = N^{1/2}\theta$. The criterion for the appearance of quantum beats, $\Delta\theta < 1$ radian, is then expressible as

$$\frac{T_e}{T_p} > (\omega T_e)^2. \tag{4.4c}$$

Relation (4.4c) is equivalent to that of relation (4.4a) when the excitation is weak and the ground-state lifetime is long, $T_p > T_e$. However, even when $\omega T_e > 1$, a pulse of long duration should still lead to quantum beats if the pumping time (i.e., ground-state lifetime) is made short enough by an intense illumination. This would indeed be an interesting phenomenon to observe.

*   *   *

By the time my atomic beam experiments were completed, the results did not show any discrepancy with quantum mechanics, and I thought I knew all I ever wanted to know about hydrogen—at least for a while. I soon realized, however, that atoms similar in electronic structure to hydrogen have an intrinsic interest all their own, especially when they are so large that *one* such atom could accommodate some 50,000,000 "ordinary-sized" atoms in its volume!

Here was a whole new domain of atomic physics to explore—through a portal opened by pulsed lasers and quantum beats.

4.2. Anomalous Reversals

The atomic hypothesis has been around for some two millennia. Despite compelling evidence provided by the study of chemistry and the kinetic properties of gases, acceptance of the actual existence of atoms was strongly resisted by a number of renowned scientists (e.g., Wilhelm Ostwald and Ernst Mach), even as late as the first decade of the 20th century.

One problem, of course, is that atoms are ordinarily much smaller than the least object that could be seen through a microscope. Physical optics teaches us that one cannot resolve objects of a size inferior to the wavelength of the light used for viewing.[11] The characteristic diameter of an atom is some three orders of magnitude smaller than the wavelength of visible light. As pointed out previously, the wavelength of electrons in an electron microscope can be a fraction of an atomic diameter; with such a microscope, one can (in a manner of speaking) "see" structures interpretable as an aggregate of atoms.

However, this is a recent development. No one at the turn of the 20th century could have conceived of seeing an atom.

The scale of molecular size was already roughly known in the 19th century by means of chemical experiments or kinetic experiments to determine Avogadro's number from macroscopic quantities of matter. I can recall, as a student, having to estimate the length of some kind of oleic molecule from the amount of substance required to form a monomolecular film of an aqueous substrate. (How the instructor could be certain that the film was monomolecular was never made clear to me!) By knowing the molecular formula, I could then estimate the size of a carbon atom. No theory based on classical physics, however, was able to predict the characteristic size of an atom or molecule. The reason, in short, is that Planck's constant was not known.

Bohr's semiclassical theory of the atom in 1913 was the first to provide a natural scale of atomic size, the Bohr radius a_0,

$$a_0 = \frac{h^2}{4\pi^2 m e^2} \sim 5 \times 10^{-9}\, \text{cm}, \tag{4.5a}$$

in terms of Planck's constant and the electron charge and mass. The Bohr theory showed that the characteristic size of the orbit of an atomic electron in an energy state of principal quantum number n is

$$r_n = n^2 a_0. \tag{4.5b}$$

Before h entered the physicist's lexicon of physical constants, the only natural length scale that could be constructed from known particle attributes and universal constants was the so-called "classical electron radius,"

$$r_0 = \frac{e^2}{mc^2} \sim 3 \times 10^{-13}\, \text{cm}, \tag{4.5c}$$

which was orders of magnitude smaller than the size of atoms inferred from experiment; it is more characteristic of the size of the atomic nucleus.

The n^2 dependence of atomic size implies, however, that highly excited atoms are *not* necessarily small—that, in fact, they are larger than some of the observable and manipulatable objects still adequately treated by the laws of classical physics. Were it possible to raise an electron to the $n = 100$ level, the orbital radius would be $10^4\, a_0$, or about $0.5\,\mu$m (recall: $1\,\mu$m $= 10^{-4}$ cm), which is already on the order of the size of some bacteria. Pulsed lasers indeed make such excitations possible; for example, under laboratory conditions, barium atoms[12] have been excited to electronic levels in the vicinity of $n = 500$ with a corresponding Bohr radius of $12.5\,\mu$m. Atoms in comparable states of excitation also occur naturally in the interstellar medium. Note that human red blood cells have diameters of 6–8 μm, and most other human cells fall in the range of about 5–20 μm.

There is something fascinating about an atom that one should be able to "see"! Unfortunately, it is not possible to do anything of the kind. To see the atom requires that one illuminate it and that the atom scatter the light to a detector. However, a highly excited atom—generally termed a Rydberg atom—is markedly sensitive to its environment; the least perturbation will likely de-excite or ionize it. The interaction of an atom with an electric field, for example, ordinarily depends on the atomic polarizability, which is a measure of the extent to which the field can displace electric charge from its equilibrium position. Polarizability has the dimension of volume, and one might expect that the atomic polarizability would scale as the cube of the orbital radius, or as n^6. This is not strictly the case—the scaling goes approximately as n^7—but it provides a good indication of the difficulty faced by someone wanting to probe, but not destroy, a Rydberg atom.[13] The polarizability of a hydrogen atom in the level $n = 100$ would be over 1 million million times greater than that of the atom in its ground state.

What makes highly excited atoms particularly interesting to study is, among other things, that they are systems at the threshold between the quantum world and the classical world. This is the "antitwilight zone," so to speak, where quantum strangeness is expected to merge into classical familiarity by means of the correspondence principle. Since the electrostatic force that binds the electron to the nucleus has the same inverse-square distance dependence as the gravitational force that binds the planets to the Sun, one might think of an atom with the outer valence electron excited into a Rydberg state as a miniature planetary system with the electron orbiting a central core (nucleus plus unexcited electrons) of unit net positive charge. For such a system, the characteristics of the quantum states should be reasonably well described by Kepler's laws.

Kepler's first law, for example, states that the orbit of a planet about the Sun is an ellipse, with the Sun at one focus. The electron orbits are also elliptical, although, for simplicity, only circular Bohr orbits are usually discussed in elementary textbooks. In the "old" quantum mechanics [i.e., the Bohr theory and its various elaborations (principally by Arnold Sommerfeld) predating the creation of a consistent quantum theory in 1925], the atomic orbits were classified as "penetrating" or "nonpenetrating." The penetrating orbits are highly elliptical (like the orbits of comets) and take the electron near or through the core; the nonpenetrating orbits are more nearly circular (like the planetary orbits of the solar system) and widely circumnavigate the core (Figure 4.2).

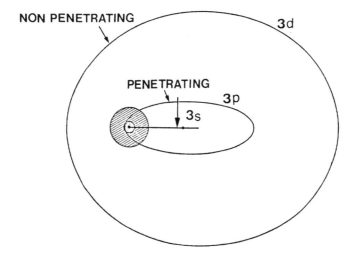

NON PENETRATING

3d

PENETRATING

3p

3s

Figure 4.2. The three lowest valence electron orbits according to the classical model of the sodium atom. The shaded circle represents the core of filled electronic shells and the atomic nucleus. Orbits are designated penetrating or nonpenetrating according to whether or not they pass through the core. (Adapted from H. E. White, *Introduction to Atomic Spectra*, McGraw-Hill, New York, 1934, p. 103.)

To those whose have studied the modern quantum theory of the atom before (if ever) encountering the old quantum theory of electron orbits, the correlation between quantum states and Bohr orbits may at first be a little surprising. For example, the electron probability distribution in an S state, a quantum state of zero angular momentum, is spherically symmetric; textbook pictures often represent the S-state electron distribution as a fuzzy ball. Classically, however, the state with zero angular momentum is the ultimate penetrating orbit where the ellipse has degenerated into a straight line right through the core. The higher the angular momentum, the more nonpenetrating is the orbit, and the less spherically symmetric is the probability distribution. There is no contradiction here, for what is being pictured are two entirely different things. An orbit represents the sequential motion of the electron in time; a stationary-state probability distribution does not represent motion and has no causal implications at all. However, the fact that the S-state probability distribution is nonvanishing at the nucleus is consistent with the classical linear trajectory. The quantum wave functions for all other angular momentum states have a node or zero amplitude at the origin.

One might also be surprised to learn how low the angular momentum of a nonpenetrating orbit can be. Consider the orbits corresponding to the lowest three states of the sodium atom, which, like hydrogen, has a single outer valence electron (Figure 4.2). Although the $3s$ and $3p$ orbits are clearly penetrating, the $3d$ orbit remains well outside the core.[14] (Nevertheless, beware of classical pictures! I shall return to this point shortly.)

Kepler's third law states that the square of the orbital period of a planet is proportional to the cube of the semimajor axis (i.e., one-half the long axis) of the ellipse. Thus, it follows for the nearly circular orbits of high angular momentum that the orbital period T_n varies as the $\frac{3}{2}$ power of the radius or, from relation (4.5b), as the cube of the principal quantum number:

$$T_n \propto n^3 a_0^{3/2}. \tag{4.6a}$$

According to classical electrodynamics, which should adequately account for radiation production in the domain to which the correspondence principle applies, an oscillating or rotating charged particle should emit electromagnetic waves of the same frequency as the frequency of periodic motion. The Keplerian electron in level n should, therefore, emit light at a frequency

$$f_n \sim \frac{1}{T_n} \propto n^{-3} \tag{4.6b}$$

that varies as the inverse third power of the principal quantum number as it continuously spirals inward to a lower energy orbit corresponding to level $n - 1$. The above relation readily follows from the quantum mechanical formula, expression (4.1a), for the hydrogen atom energy spectrum; because E_n varies as n^{-2}, the energy interval between levels n and $n - 1$, and, therefore, the radiation frequency, varies as n^{-3} in the limit of large n.

The classical picture of a gentle transition between close-lying non-penetrating orbits with emission of low-energy radiation is substantiated quantum mechanically by means of the selection rules governing transitions between angular momentum states. The largest value of angular momentum (in units of $\hbar$) that an electron in level n may have is $n - 1$, and, as pointed out earlier, the emission of a photon carries away one unit of angular momentum. Since a state with angular momentum $n - 2$ can occur only in the manifold of states of principal quantum number $n - 1$, the emitting electron undergoes a transition from the level n to the level $n - 1$ in accord with the classical picture. For electrons with large, but not necessarily maximal, angular momentum, there is a range of lower levels that can be reached from a given Rydberg level n. Nevertheless, for n large enough, the energy intervals—and therefore the radiation frequencies—still vary essentially as n^{-3}.

Classical reasoning also allows us to draw an important conclusion concerning the lifetime of the Rydberg states corresponding to nonpenetrating orbits. The total power radiated by a nonrelativistic accelerated charged particle was first shown by the English physicist J. J. Larmor to vary as the square of the acceleration a as follows:

$$\text{Radiated power} = \frac{2e^2a^2}{3c^3}. \tag{4.7a}$$

This relation is known as the Larmor formula. By Newton's second law of motion, the acceleration of the Rydberg electron is proportional to the (inverse square) electrostatic force keeping it in orbit. Thus, the radiated power

$$\text{Radiated power} \propto a^2 \propto r^{-4} \propto n^{-8} \tag{4.7b}$$

varies as the inverse eighth power of the principal quantum number. Because the quantity of energy carried away by each photon varies as n^{-3} [from relation (4.6b)], the time spent in level n,

$$t_n \sim \frac{\text{Radiated energy}}{\text{Radiated power}} \propto n^5, \tag{4.7c}$$

should vary as the fifth power of n. Higher-angular-momentum Rydberg states, then, are predicted to be *very* long-lived. This prediction is not inconsistent with the previous statement that such states are extremely sensitive to environmental perturbations; Rydberg states are long-lived when they are left alone.

Unlike the case of a nearly circular orbit, the acceleration of an electron in a penetrating elliptical orbit depends on the electron location. The force—and therefore the acceleration and rate of light emission— are greatest, however, in the vicinity of the pericenter, the point of the orbit closest to the focus where the core is located. The distance to the pericenter is largely independent of the energy, and therefore of the principal quantum number, of the orbiting particle. Because an electron emits light significantly only when passing through the pericenter, the time spent in orbit is just proportional to the orbital period. Thus, from relation (4.6b),

$$t_n \sim T_n \propto n^3; \tag{4.7d}$$

the radiative decay lifetime of the low-angular-momentum Rydberg states should vary as the cube of the principal quantum number. These states, too, are long-lived.

According to the Larmor formula, an electron in a penetrating orbit should radiate energy at a greater rate than an electron in a nearly circular nonpenetrating orbit within the same electronic manifold; the acceleration near the pericenter, a distance on the order of a few Bohr radii from the core, is much greater than acceleration at a distance of $n^2 a_0$ from the core. This is again substantiated by quantum mechanical selection rules. Low-angular-momentum states are found in electronic manifolds of both large and low n (provided only that the angular momentum quantum number L does not exceed $n - 1$). Since the probability for an electric dipole transition between two states varies as the third power of the frequency of emitted radiation,[15] other things being equal, quantum mechanics favors a large "quantum jump" (i.e., a transition from a state n,L to a lower energy state n', $L - 1$ with $n' \ll n$). Thus, radiative decay of low angular momentum Rydberg states should lead to photons of higher energy than radiative decay from high angular momentum states of corresponding principal quantum number.

The above properties of Rydberg states are reasonably well confirmed experimentally, and one might be tempted to suppose that the simple picture of a distant outer electron orbiting a central nucleus and inner electron core with little mutual interaction is an adequate model for a highly excited atom—at least for the classically nonpenetrating orbits (which exclude the S and P states). It would then follow that singly excited Rydberg atoms, regardless of the distinguishing properties of the parent ground-state atoms, should exhibit essentially hydrogenic behavior. In many ways, this expectation is realized. With regard to binding energies, polarizabilities, lifetimes—and indeed every atomic property of which the calculation involves the radial coordinate r to a non-negative power—a Rydberg atom increasingly resembles a hydrogen atom the larger n becomes.

Nevertheless, upon closer scrutiny, this comfortable agreement crumbles in some rather curious ways, as in the case of the anomalous fine structure of the sodium atom. Because of the relative simplicity of its electronic configurations, the ease with which one can work with it experimentally and the convenient region (that of visible light) into which many of its spectral lines fall, the sodium atom makes an excellent system for the investigation of Rydberg states. With a single-valence electron outside an inert gas (neon) core, sodium Rydberg states may be expected to resemble closely the excited states of hydrogen. Thus, the observation and interpretation of marked nonhydrogenic behavior of some property of sodium would be of considerable theoretical interest in atomic physics.

To understand what is anomalous about some sodium fine-structure levels, let us first reconsider the fine structure of hydrogen, as this is the model for normal structure. I explained previously that the fine-structure splitting of the Bohr energy levels originates in the spin–orbit interaction (i.e., the interaction between the electron magnetic dipole moment and the local magnetic field produced by the apparently circulating proton). Since the charge of the electron is negative, the orientation of the electron magnetic moment (proportional to the electron charge) is opposite that of the electron spin. One con-

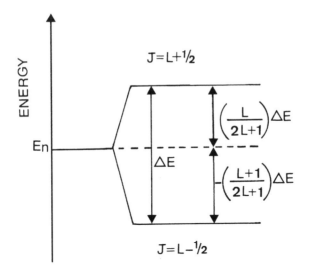

Figure 4.3. Normal (hydrogenic) fine-structure level ordering. In the presence of the spin–orbit interaction, an energy level E_n $(n > 1)$ splits into two levels; the one with electron total angular momentum quantum number $J = L + \frac{1}{2}$ lies above the other with $J = L - \frac{1}{2}$.

sequence of this is that the energy of an electron is lowered when the electron magnetic moment is aligned parallel to the local magnetic field; this is the state for which the electron spin and orbital angular momenta are *anti*parallel. Conversely, the electron energy is raised in the opposite configuration where the two momenta are parallel. Thus, the fine structure is considered normal (or hydrogenic) when the state with electron total angular momentum quantum number $J = L + \frac{1}{2}$ lies higher (less tightly bound) than the state with $J = L - \frac{1}{2}$ (Figure 4.3). The energy interval, derived from the Dirac theory of the hydrogen atom, is

$$\Delta E = E_{n,L,J=L+1/2} - E_{n,L,J=L-1/2} = \frac{\alpha_{fs}^2 Ry}{n^3 L(L+1)}, \tag{4.8}$$

where α_{fs} is the fine-structure constant $e^2/\hbar c \sim 1/137$.

For the D states of excited sodium, not only is the magnitude of the fine-structure splitting not accurately given by relation (4.8), but the ordering of the levels is reversed, the $J = \frac{5}{2}$ lying lower than the $J = \frac{3}{2}$ states, thereby giving a ΔE of opposite sign. That the sodium D states of the ground $(n = 3)$ level are inverted has been known at least since the 1930s by means of optical spectroscopy with high-quality interference gratings. From measurements of the light-absorption spectrum corresponding to transitions from 3P to 4D and from 3D to 4P states, experimenters were able to infer the 3D fine-structure splitting. Subsequent measurements of the same kind, and by the more recent experimental method known as two-photon spectroscopy,[16] showed that the inversion of the sodium fine structure persisted in levels $n = 4, 5,$ and 6.

Why is the fine structure in these low-lying levels anomalous? The explanation is not a simple one; there may, in fact, still be no consensus among theoreticians as to the relative importance of various proposed mechanisms. Speaking generally, however, the excited valence electron does not experience a purely central potential ($V \propto 1/r$) as a result of the presence of the other ten electrons that (together with the nucleus) comprise the core. The interaction between the excited electron and the core is referred to as "core polarization." In the classical picture, the orbits of the core electrons are perturbed by the penetration of the valence electron; this distorts the potential in which the valence electron finds itself and ultimately changes the energies of the two fine-structure states from what they would be in hydrogen where there is no subsystem of core electrons. A classical picture, however, can be misleading. Not all sodium fine-structure levels are inverted; the $P_{3/2}$ and $P_{1/2}$ levels are normally ordered even though P states correspond to highly penetrating classical orbits.

In accordance with the correspondence principle, one might expect that, beyond some threshold value of the principal quantum number, the fine-structure ordering must reverse and the energy splitting become progressively more hydrogenic as n increases. The spectroscopic method of quantum beats allows this supposition to be tested without at the same time perturbing the states by probing.

Because the experimental task in question requires the measurement of small energy intervals, it is a significant advantage that quantum beat frequencies are insensitive to atomic motion. As is well known from classical physics, the frequency of a light wave (in fact, any kind of wave) emitted by a source moving with respect to an observer is perceived to be shifted, either higher or lower depending on the direction of relative motion, in comparison with the frequency emitted by a stationary source. Known as the Doppler effect, this frequency shift has been the nemesis of many a spectroscopic investigation. Were all atoms to move at the same speed in the same direction, the simple displacement of a spectral line could be taken into account easily. The net effect, however, of a large number of atoms moving in different directions with a wide spread of speeds is to produce Doppler-broadened spectral lines whose overlap could obscure fine details of atomic energy-level structure.

Since a quantum beat results from the interference between photons that could be emitted from any two of a set of superposed states of the same atom, these photons are Doppler shifted to the same extent. Consequently, their *difference* frequency, which is the frequency of the quantum beat, is largely independent of the dispersion in atomic velocities.[17] It must be said, of course, that when the atomic transition actually occurs, a single atom emits but one photon and does *not* produce a beat in the photodetector output. The observed beat is the product of many such individual emissions—yet the phenomenon does *not* originate in the interference of photons from different emission events (i.e., from different atoms). As in the previously discussed case of interference with free electrons, the *capacity* for quantum interference is intrinsic to each atom. Anyone who finds it hard to visualize just exactly how this occurs is not alone, for this process is again one of the central mysteries of quantum mechanics.

The experiments were performed in the early 1970s at the Ecole Normale Supérieure (ENS) in Paris where I was a guest scientist at the spectroscopy laboratory founded by Alfred Kastler and Jean Brossel. To my good fortune, there had just returned to the ENS a former student, Serge Haroche, who, during a postdoctoral stay at Stanford University in California, had also become interested in quantum beats and was in the process of starting up research in this area at the ENS. We joined forces.

To generate a linear superposition of excited sodium D states directly from the ground state would have required a tunable pulsed laser in the ultraviolet; such a light source did not exist. The problem was solved by exciting the atom in two stages. First, the yellow light from a pulsed dye laser was used to "pump" sodium atoms from the 3S ground state to the $3P_{3/2}$ state, and then, before the 3P states could decay, the atoms were irradiated with the blue light from another pulsed dye laser to bring them into the desired linear superposition of nD states. Both lasers were tunable; by adjusting the wavelength of the second laser, one could select electronic manifolds of different principal quantum number n.

Since anomalous fine structure had already been observed for levels 3–6, we looked for quantum beats in the light issuing from 7D states. There were no beats! Our first thought was that we might have discovered straightaway the "cross-over" level in which the $D_{5/2}$ and $D_{3/2}$ states are almost degenerate (giving rise to "beats" of zero frequency). However, the application of a small magnetic field, which reduced the energy interval (and Bohr frequency) between certain $D_{5/2}$ and $D_{3/2}$ states, did generate beats, thereby suggesting that the beat frequency in the absence of a magnetic field was not zero, but instead too large to be produced by our pulsed laser or to be measured by our photodetector. (By the uncertainty principle, the frequency spread of the second light pulse must be greater than the Bohr frequency associated with the two 7D fine-structure levels if a quantum beat is to be produced. Also, the response time of the detector must be shorter than the beat period if the beat is to be detected.) We estimated, then, that the 7D fine-structure interval must have exceeded about 150 MHz. This turned out to be the case for 8D states too. Starting with 9D, however, our apparatus began to register field-free quantum beats.

Despite the theoretical simplicity of the experimental procedure, it is worth noting that the experiment had not been an easy one. The two tunable dye lasers, relatively compact and uncomplicated affairs, were themselves pumped by the ultraviolet (UV) radiation from a third laser, a huge and powerful apparatus that shot massive electrical discharges through a chamber of nitrogen gas. The subsequent de-excitation of the nitrogen molecules gave rise to about one million watts of UV radiation. Not only was the electrical noise from these discharges "deafening" to the rest of the electronic apparatus, but the switching device (or thyrotron), which triggered the release of the large amount of electrical energy stored in an extensive bank of capacitors, often failed to work. Worse still than an electrically noisy laser was a malfunctioning one that sat unproductively quiet. Many laboratory hours were spent in tedious searches through the morass of cables filling the power cabinet of the laser in the (usually vain) hope that the device could be started up again without intervention of the manufacturer's repairman (who was generally servicing another laser somewhere else in Europe).

Nevertheless, with perseverance, the experiment was eventually brought to the state where fine-structure quantum beats in a succession of increasingly high Rydberg states could be measured. At that point, unfortunately, my time was up and I had to leave France to meet other commitments. Continuation of the work after my departure led to the surprising result that the quantum beats in levels 9 to 16 showed *no* tendency at all to become hydrogenic. The energy splittings, deducible directly from the beat frequencies, continued to depart from the hydrogenic interval of relation (4.8). And the level ordering remained inverted.[18]

The level ordering, it should be noted, is not deducible from the measurement of field-free quantum beats, because the latter provides information only on the magnitude of the energy interval, not on its sign. The order can be determined, however, by a judicious application of the Stark effect, the shifting of atomic energy levels by a static electric field. In the presence of an electric field, all of the nD states become more tightly bound, i.e., shift downward on an energy diagram. However, the $D_{5/2}$ substates shift downward to a greater extent than the $D_{3/2}$ substates. Thus, if the $J = \frac{5}{2}$ states lie below the $J = \frac{3}{2}$ states (anomalous ordering), the energy intervals—and, consequently, the quantum beat frequencies—increase with increasing electric field strength. The combined results of previous research and the quantum beat experiments showed that all measured intervals from $n = 3$ to $n = 16$ were anomalous!

Now, a principal quantum number of 16 is not exactly close to infinity, which is ideally the limit at which quantum and classical mechanics are assumed to give equivalent descriptions of a physical system. It is not even close to 500, which represented, more or less, the upper limit of atomic excitation achieved in a terrestrial laboratory when I began work on *And Yet It Moves*. Nevertheless, a sodium atom in the $n = 16$ level is a highly excited atom; it is large enough (classically speaking) to contain over 4000 ground-state hydrogen atoms. Moreover, the energy of the excited electron is about 96% of the energy required for ionization out of the ground state. It was the pattern of measurements, however, that was most significant; the quantum beat frequencies all fell on a smooth empirical curve (Figure 4.4), the

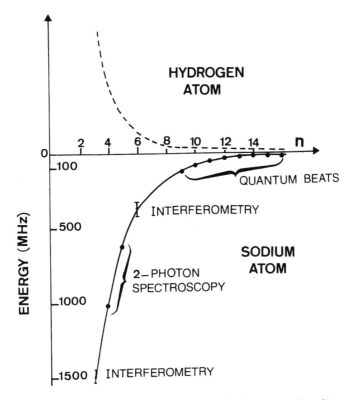

Figure 4.4. nD fine-structure intervals of the hydrogen and sodium atoms. Extrapolation of the curves to high n suggests that the sodium fine structure remains anomalous irrespective of the degree of excitation. [Adapted from C. Fabre *et al.*, *Optics Communications* **13** (1975) 393.]

extrapolation of which did *not* lead to hydrogenic behavior as n approached infinity.

Why not? Is the correspondence principle violated? Understanding this puzzling experimental result taught me several lessons. The first was not to underestimate what can be learned from the classical model of the atom. Clearly, the anomalous behavior persists because of the interaction of the outer electron with the core. It had been implicitly assumed at the outset that the higher the state of excitation, the less penetrating would be the orbit and the weaker would be the interaction with the core. We would have done well to think more carefully about the theory of classical orbits. The shape of an elliptical orbit can be quantified by the eccentricity of the ellipse, which is the ratio of the distance of one focus from the center to the length of the semimajor axis. For example, for a circle, both foci coincide at the center and the eccentricity is zero. The eccentricity of the orbit of an electron in a level n with angular momentum quantum number ℓ subject to an inverse square force can be shown to be

$$\varepsilon = \sqrt{1 - \frac{\ell(\ell+1)}{n^2}}. \tag{4.9}$$

For a state of maximum angular momentum, $\ell = n - 1$, the eccentricity, $\varepsilon = n^{-1/2}$, approaches zero as n approaches infinity, as expected for a circular orbit. However, for the d states ($\ell = 2$), the expression for the eccentricity, $\varepsilon = \sqrt{1 - 6/n^2}$, shows that as n increases, the orbit becomes more elongated, and not necessarily less penetrating. In fact, as illustrated in Figure 4.5 for a series of d orbits, at the pericenter the penetration is about the same irrespective of the principal quantum number. Thus, a $16d$ state might well be expected to interact with the core as much as would a $3d$ state.

The second lesson was not to overestimate what can be learned from the classical model. The classical $3d$ orbit was considered to be nonpenetrating and should *not* have interacted significantly with the core. From the perspective of quantum mechanics, however, the radial portion of the nd wave function is *always* penetrating—for any n. The "loops" of the wave function extend into the core, falling monotonically to zero within a distance of about five Bohr radii from the center. The spin–orbit interaction depends on the expectation (or mean) value of the inverse third power of the electron radial coordinate $\langle r^{-3} \rangle$; thus, the behavior of the wave function close to the nucleus and electron core can be significant even if the mean radial distance, $\langle r \rangle \sim n^2 a_0$, is very large.

What kinds of interactions specifically occur between the outer electron and the core to invert the fine-structure order? Many studies have been undertaken with varying degrees of success to answer that question. Regrettably, it would appear that the more quantitatively suc-

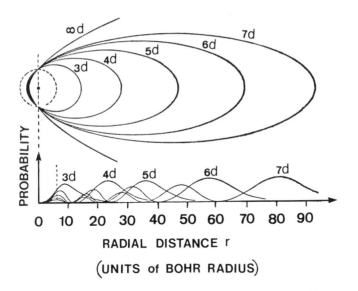

Figure 4.5. Comparison of classical orbits and corresponding quantum mechanical radial probability distributions of the nd electron. Although the mean orbital size increases with principal quantum number n, the orbital properties near the core are largely insensitive to the degree of excitation. [Adapted from H. E. White, *Introduction to Atomic Spectra*, McGraw-Hill, New York, 1934, p. 113.]

cessful the analysis, the less insightful is its underlying basis in terms of a visualizable mechanism. One of the earliest proposals is that the inversion results from "configuration mixing." The actual state of the excited sodium atom is not exclusively the state to which the valence electron has been nominally excited (e.g., the nd states), but is, in fact, a linear superposition of other states of excitation, or configurations, brought about by the electrostatic interactions among the electrons. One such configuration might include the excitation of the valence electron to the nd state and simultaneous promotion of a core electron to the $3p$ state. These "virtual" configurations with two excited electrons cannot be detected directly, but they are believed to influence the relative ordering of the nD fine-structure levels.

If further experiment and theoretical analysis sustain this picture, then a highly excited atom is, indeed, a marvelous structure. Nearly macroscopic in size—from the perspective of what can be resolved by a light microscope—it has many of the attributes of a miniature planetary system subject to Kepler's and Newton's laws, while preserving in the fine details of its energy-level structure the effects of strange quantum processes without parallel in the macroscopic world of classical physics.

Postscript: Apply the
Correspondence Principle with Caution!

Atoms with one highly excited electron can differ in significant ways from the hydrogen atom because, no matter how high the state of excitation, the wave function of that outermost valence electron penetrates (and is influenced by) the inner electron core. In leading to this "lesson," I had pointed out previously how the properties of highly excited hydrogen are well accounted for by application of the Bohr correspondence principle, which is the assertion that the predictions of quantum theory must agree with those of classical theory in the appropriate classical limit. Although there are several ways in which this classical limit can be formulated (e.g., by letting Planck's constant h approach zero), the operational limit for atoms has long meant high excitation, i.e., letting the principal quantum number n become large without bound. This was the manner in which Bohr first applied the principle in his classic 1913 paper on the hydrogen spectrum.[19]

Surprisingly, studies published while this redaction was under way have claimed that the Coulomb potential (which diminishes with distance as $1/r$) is one of relatively few cases in which Bohr's original correspondence principle actually works.[20] It is alleged to fail, for example, in systems subject to long-range potentials that vary as $1/r^k$ with k greater than 2 (which includes the important case of the van der Waals interaction with $k = 3$ and 6). In these cases, contrary to what physicists might have expected, the use of classical reasoning (as implemented in so-called semiclassical approximations) gives increasingly poor results for the most highly excited states.

Although a full explanation of this counterintuitive discovery lies in mathematical details that would be all but opaque to a casual reader, the following observation is not too wide of the mark. A state with large quantum numbers is not necessarily more "classical" than a state with smaller quantum numbers. Rather, the condition for validity of a semiclassical approximation is that the change in momentum of an electron (or any quantum particle) over the distance of its de Broglie wavelength be small in comparison to the momentum itself. This requires that the potential through which the particle moves not vary too rapidly.[21]

For quantum particles subject to long-range interactions, it is possible for the semiclassical condition of validity to be violated within the classically allowed region of the potential—that is, within a region where the total energy of the particle does not exceed the potential energy. Moreover, the spatial extent of this violation can increase with higher levels of excitation. When this occurs, the Bohr correspondence principle breaks down and the quantum world becomes, to physicist and layman alike, a correspondingly stranger place.

4.3. Quantum Implications of Traveling in Circles

Looking out over the countryside from the hills of the Hainberg Wald, I could see the red tile roofs of Göttingen rise above the surrounding verdant forest like clusters of mushrooms. Although outwardly similar to other medieval German towns of Lower Saxony, Göttingen was different. At the entrance to the *Rathskeller*, or Town Hall Cellars, the old proverb *"Extra Gottingam non est vita,"* may once have depicted the world of physics with only mild exaggeration. I went to Göttingen in the late 1970s as a guest professor at the Physikalisches Institut, the institute founded in 1921 by Max Born and James Franck.

To walk through this charming former Hanseatic town largely spared the ravages of time and war is to walk back in history. All around the town center are the beautiful old Renaissance frame houses, the Fachwerkhäuser, with white plaster and brown beams, each succeeding level overhanging the street a little further than the previous one. I presume, although I could be mistaken, that this construction afforded the ancient tenants the best configuration for jettisoning their refuse onto the streets below. However, the Göttingen I saw was clean and bright.

On Market-place, close by the *alte Rathaus*, or Old Town Hall, stood the *Gänselieselbrunnen*, the Goose-girl Fountain, which was something of a town symbol. Tradition required that male doctoral candidates of the university, dressed in tailcoat and top hat, climb the pedestal and kiss the bronze Goose-girl after passing their examinations. Unfortunately, the *Gänseliesel* had to be removed for repairs, but that minor inconvenience, I soon discovered, did not cause the innovative Göttinger to break with the past. Walking near the fountain one day, I encountered a mule-drawn wagon filled with top-hatted *jungen Doktoranden* bringing with them their own quite lively "Gänseliesel" who took her place on the fountain pedestal. Tradition was preserved!

In the 1920s and 1930s, Göttingen was one of the world's centers of physics and mathematics. It was said to be the "Mecca of Physics"; David Hilbert, Max Born, and James Franck were its "prophets," and researchers, students, and visitors made the pilgrimage there from all parts of the globe. Although this golden age had long passed, I was glad to make my own pilgrimage and draw inspiration from the physical reminders of a period of scientific creativity that was not likely to occur again. I missed that age, having been born too late. To assuage this sense of loss, I often strolled through the quiet *Stadtfriedhof* (cemetery) west of town and looked at the inscriptions on the gravestones which recalled people and events closely associated with the physics and mathematics that intensely interested me. There were Carl Friedrich Gauss, Max Planck, Max von Laue, Max Born, David Hilbert, Otto Hahn, and others less well known.

The appointment of Gauss in 1807 as professor of mathematics and director of the *Sternwarte* (Observatory) marked the initial point of ascendancy of Göttingen as a center of scientific excellence. Gauss was probably the greatest mathematician of his time, but he also employed his extraordinary talents on physically important problems in astronomy, geodesy, and electromagnetism. His studies of spatial curvature and non-Euclidian geometry led to mathematical advances that would one day serve Einstein in the creation of general relativity. The tradition of contributing both to pure mathematics and fundamental science passed from Gauss to his former student, Bernhard Riemann, and ultimately to Hermann Minkowski and to David Hilbert, all professors of mathematics at Göttingen. Hilbert's writings on differential equations and eigenvalue problems provided exactly the mathematical foundations needed by the Göttingen quantum physicists—who, to their misfortune, did not pay close enough attention until after Schrödinger "scooped" them in the discovery of the nonrelativistic quantum mechanical wave equation.

Recorded on the gravestones of a number of those luminaries buried in the *Stadtfriedhof* were words or symbols that distilled from a lifetime of work the core of their goals and achievements. On Hilbert's stone, for example, could be read the words, "Wir müssen wissen. Wir werden wissen" ("We must know; we will know"), epitomizing, I presume, his struggle to provide a unified logical foundation to all of mathematics. This dream was effectively shattered by the epoch-making paper of Kurt Gödel on formally undecidable propositions. Max Born's stone bore the basic commutation relation between momentum and coordinate, $pq - qp = (h/2\pi i)\mathbf{1}$, that represented a fundamental distinction between quantum and classical mechanics. He had first written that expression down when he converted into matrix notation[22] a perplexing calculation left with him by his young assistant, Werner Heisenberg. That calculation marked the genesis of a true quantum mechanics. On the tombstone of Otto Hahn was displayed the uranium fission reaction that astounded his contemporaries when they learned that atomic nuclei could be split in half.

Of greater significance to me personally than even the tangible links to the past found in the *Stadtfriedhof* was the intangible, almost spiritual, tie of the Hainberg Wald. How often must those woods have served as the backdrop for intense discussions on quantum physics between the Göttingen physicists and their visitors—between Born, Bohr, Einstein, Heisenberg, Pauli, and many others. Heisenberg recalled of his first meeting with Bohr in the summer of 1922:

... Bohr came to me and suggested that we go for a walk together on the Hainberg outside Göttingen. Of course, I was very willing. That discussion, which took us back and forth over Hainberg's wooden heights, was the first thorough discussion I can remember on the fundamental physical and philosophical problems of modern atomic theory, and it has certainly had a decisive influence on my later career.[23]

How I would have liked to hear those seminal conversations whose echoes faded long ago but reverberate still in the writings of the creators of quantum physics. In an indirect and less momentous way, the Hainberg was to influence my own research.

My wife and I lived on the periphery of Göttingen right across the road from the paths that led into the Hainberg woods. In the mornings, I arose before sunrise and worked intensely for hours on a variety of quantum mechanical problems. I would then stop around noontime and go for a run through the woods by a long circuitous loop that took me along tranquil leaf-strewn pathways among the tall trees, past fields and farmland and the exercise stations of an *Erholungsgebiet*, and eventually back home. Ironically, as I made my way around the Hainberg, I began to think about the curious behavior of quantum systems that, in a manner of speaking, travel in circles.

The questions that aroused my curiosity at the time concerned the experimental distinguishability of different ensembles of quantum systems. The nature of the problem is subtle; there is no direct parallel in classical physics. If one wants to know whether one collection of macroscopic objects is different from another, he can, in principle, look at the two collections, count their constituents, probe them, smell them, taste them, or whatever. The issue, however, is not so clear when treating a collection of objects whose behavior is quantum mechanical. According to orthodox quantum theory, specification of the wave function of a system provides all the information about the system that is allowable by physical law. If the system comprises a statistical mixture of subsystems in different quantum states, then the maximum information is contained in the so-called density matrix, effectively a tabulation of the fractional composition of each subset of objects characterized by the same wave function. But is it always clear when two seemingly equivalent wave functions are actually different, or when two seemingly different wave functions are physically the same?

A wave function ψ is not itself an experimentally observable quantity but always enters in bilinear combination (i.e., in pairs of ψ and the complex conjugate function ψ^*) the mathematical expressions describing quantities that are observable. Consequently, the wave function of a quantum system is not unique; for example, the same quantum state can be represented by an infinite number of wave functions differing only by a phase factor of the form $e^{i\phi}$. In my investigation of the information content and experimental implications of different wave functions, I wondered whether it was truly the case that such a phase factor has no physical consequences at all.

One way by which the wave function can incur a phase factor $e^{i\phi}$ is through a rotational transformation. The rotational properties of wave functions play an important role in quantum theory; these properties are determined by the angular momentum of the particles whose quantum behavior the wave functions are presumed to describe. A particle like the electron or neutron, which has an intrinsic spin angular momentum of $\frac{1}{2}\hbar$, is characterized by a *spinor* wave function. A spinor is a mathematical object with two components—like a matrix with two rows and one column. In the context of quantum physics, the upper component gives the "spin-up" contribution to the wave function, whereas the lower component gives the "spin-down" contribution. Because the spin attribute of being "up" or "down" is defined with respect to an arbitrary quantization axis, the spinorial components will be modified if a new quantization axis is chosen. The new components of a spinor are determined from the old components by a rotational transformation.

Suppose the new axis is inclined at an angle θ to the old axis. The rotational transformation peculiar to spinors involves the sines and cosines of $\theta/2$. Now, this leads to a curious result, for one can imagine a new axis inclined at 2π radians or 360° to the original axis, which, according to our familiar notions of classical reasoning, is no new axis at all; it is the same axis as the original one. Nevertheless, the algebraic prescription governing the rotation of spinors gives rise to the transformation

$$\text{New spinor} = e^{i\pi}(\text{Old spinor}) = -\text{Old spinor};$$

that is, each component of the new spinor is the *negative* of the corresponding component of the old spinor. The negative sign can be regarded as a phase shift of π radians, since $e^{i\pi} = -1$ (an expression that was, itself, at one time rather puzzling[24]). A rotation by 360° does not reproduce the same spinor wave function.

In so far as one is discussing abstract mathematical objects and theoretical changes of coordinate axes, the above rotational property of spinors need not be disturbing. Purely mathematical relations do not have to satisfy criteria imposed by the real world. But physics obviously must. According to quantum mechanics, the "passive" view of a rotation as a reorientation of the coordinate axes with the wave function (e.g., a spinor) held fixed is equivalent to the "active" view that the wave function itself is rotated with respect to a fixed coordinate system. If spinors are to be suitable representations of actual fermionic particles, it is then a legitimate question to ask what, if any, are the observable consequences of physically rotating by 360° a system characterized by a spinor.

The inferences to be drawn from quantum mechanics texts and monographs seemed to indicate that *no* consequences would result. The "holy P. A. M. himself," as Schrödinger referred to Dirac, asserted in his classic work, *The Principles of Quantum Mechanics*[25]:

We thus get the general result, *the application of one revolution about any axis leaves a ket unchanged or changes in sign according to whether it belongs to eigenvalues. . . which are integral or half odd integral multiples of* $\hbar$. A state, of course, is always unaffected by a change of sign of the ket corresponding to it. [Italics included in the original text.]

[The idea of bras and kets, it should be noted, was created by Dirac to represent the state of a quantum system in a totally general way; when a ket is combined in a specified way with a coordinate bra, the resulting bracket (bra-ket) yields the wave function.] Dirac's general result has been frequently cited in the pedagogical literature of quantum mechanics; because the results of all measurements are representable by expectation values (in effect, integrals) bilinear in the wave function, two wave functions differing only in overall sign cannot lead to different physical predictions.

Although the above conclusion is not incorrect, it, nevertheless, seemed to me that its application to particle rotation required further thought. For one thing, I discerned a distinction between "mathematical" rotations (whether viewed passively or actively) that involved the reorientation of a coordinate system or mathematical function like the spinor wave function, and a "physical" rotation whereby some force of interaction was required to cause a real particle such as an electron to depart from rectilinear motion. The former case, whereby an isolated system (measuring apparatus included) is rotated, has no experimental counterpart, and the global phase factor to which it gives rise is, as Dirac states, not observable. However, this is not what is ordinarily meant by a rotation. It is the latter case, where only a part of the system is rotated relative to a fixed part that provides a reference against which the rotation is measured. The 2π phase change of a spinor-characterized portion with respect to a fixed portion of a larger encompassing system *does* have physical implications.

There is an interesting example drawn from classical physical optics that illustrates in an analogous way the issues underlying the observability of spinor phase. In the theory of scalar diffraction—which ignores the electromagnetic nature of light and the corresponding property of light polarization—the amplitude of diffracted light at some point P is given by the so-called Helmholtz–Kirchhoff integral. The precise specification of the light amplitude $\psi(P)$, which can be found in almost any optics textbook, is not needed here, but it will be instructive to note that it takes the form

$$\psi(P) = -i \times \{\text{integral over diffracting surface}\}.$$

Textbooks discuss nearly all aspects of this amplitude that derive from the surface integral: the dependence on wavelength, the angle of diffraction, the approximations that lead from Fresnel (near-field) to Fraunhofer (far-field) diffraction, and the application of the integral to various obstacles and apertures. Yet, I am aware of no standard text that considers the experimental consequences of the factor $-i$. Although it corresponds to a phase shift of 90° or $\pi/2$ radians [$e^{-i\pi/2} = -i$] between the incident and the diffracted light, a reader may well be left with the feeling that it is an artifact of the mathematical approximations underlying the derivation and therefore of no physical significance. But this is not so. By superposing, as in holography, a coherent background of incident light on the radiation diffracted by an object, one can observe in the resulting interference pattern the effect of the $\pi/2$ phase shift.[26]

The above parallel with light suggests that the 360° rotation of an object characterized by a spinor wave function should be observable as well by means of some type of *quantum* interference. And this is indeed the case. Split-beam interference experiments with neutrons,[27] in which the spin of a neutron in one component of the beam was made to precess in a magnetic field relative to the spins in the field-free component, showed that the intensity of the recombined beam oscillated as a function of the magnetic field with a periodicity indicative of a spin rotation of 720° (or 4π radians) rather than 360°.

The interpretation of the double-beam interference experiments on neutrons, however, is not without its difficulties. At the core of the problem is the Heisenberg uncertainty principle: one can never know whether a particular neutron followed a path through the magnetic field or through the field-free region. Although, in principle, the relative spin rotation of neutrons following two classical paths can be measured, this measurement would destroy the interference pattern. Because simultaneous observations of neutron spin rotation and of the interference pattern are incompatible, the notion of relative rotation ceases to have a meaning, as it corresponds to nothing that is measurable—at least in the case of fermions.

The problem does not arise, however, for massive bosons[28] because there is always an additional quantum state (the M = 0 state) that is insensitive to the presence of the magnetic field. (The photon, though often said to be a spin-1 particle, is excluded, for it has only two spin components.[29]) The theoretical expression for the interference pattern may then be decomposed into an incoherent sum of two terms: one characterizing the initial quantization axis of the particles, the other representative of a rotation with respect to that axis. For bosons, therefore, the concept of relative spin rotation retains a meaning in the setting of a split-beam interference experiment. However, when a boson wave function is rotated by 360°, nothing "interesting" happens, for it leads to the same wave function. Hence, there is not much point in studying rotated bosons.[30]

To avoid ambiguities, either semantic or otherwise, associated with the question of spinor rotations, I thought about the physical implications of wave-function transformations in a more general context. The idea of whether *cyclic* quantum transitions were detectable first occurred to me in the form of a mental game as I jogged over the Hainberg. I imagined looking at a friend who is sitting in a chair facing me. I turn my back to him and he can either do nothing or get up, leave the room, and return to his seat. When I turn around, he is in the chair facing me as before. Can I tell by looking at him whether or not he had left his seat?

In the quantum world of atoms, this question has a strange and interesting counterpart. Is there a physically observable difference, for example, between an atom that has undergone a transition to a different state and then returned to its original state, and an atom that has never left its original state at all?

According to quantum mechanics, the properties of an atom ought to depend only on its current state, not on its past history. If an atom is in a 3S state, then it manifests all the properties expected of the 3S state, irrespective of how it happened to get there—i.e., whether it was produced by radiative decay from a 3P state, decay from a 5F state, or absorption from a 2P state. In general, all atoms in a particular quantum state are indistinguishable. Nevertheless, the idea of a cyclic transition has experimental implications. One day, while running and thinking about this question, there occurred to me the possibility of an experimental demonstration by means of quantum beats.

Let us consider an atom (practically speaking, a collection of many identical atoms) with *three* nondegenerate excited states; a pulse of light drives the atom into a linear superposition of the lower two (Figure 4.6). Observed as a function of time following the excitation, the fluorescent light intensity of a specific polarisation will oscillate in the familiar way at an angular frequency ω_{12} corresponding to the energy interval of the two superposed states:

$$I_0(t) = A + B\cos(\omega_{12}t + \phi). \qquad (4.10a)$$

Here, A, B, and ϕ are again constants that depend on the electric dipole matrix elements for excitation and spontaneous emission. (I ignore, as inessential in the present discussion, the exponential decay factor.) Suppose, however, that to this standard quantum beat experiment one adds a radiofrequency (rf) electric field that can induce transitions between the excited atomic states 2 and 3. In general, after exposure to the rf field, the atom is in a linear superposition of all three excited states.

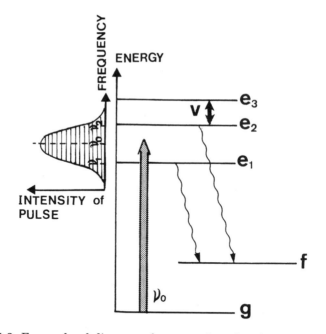

Figure 4.6. Energy-level diagram of an atom impulsively excited by a laser pulse (of mean frequency ν_0) from its ground state g into a linear superposition of states e_1 and e_2 by an external perturbation V. Whether or not the cyclic transition had occurred can be inferred from the ensuing quantum beat signal in the spontaneous emission to lower state f.

When the condition of resonance is met, whereby the radiofrequency exactly matches the Bohr frequency ω_{23}, the theoretical description of the effect of the rf field conveniently simplifies. It resembles precisely, in fact, the mathematical expression for rotation of a spinor; the rotation angle corresponds to twice the product of the rf transition matrix element and the time of exposure to the field. Setting the rf field strength or exposure time so that the associated rotation angle is π radians results in all the atoms in state 2 being driven into state 3. A "rotation" of 2π results in all atoms in state 3 being driven back again to state 2. However, like a 360° spinor rotation, in the process of leaving and returning to state 2, the quantum mechanical amplitude incurs a phase shift of π radians; it is the negative of what it was before the transition.

The occurrence of the minus sign would not ordinarily be observed in the spontaneous emission from an atomic state since the transition probability depends on the square of the absolute magnitude of the amplitude. It is to be recalled, however, that the transitions engendered by the rf field have not, under the conditions of the experiment, affected the amplitude for remaining in excited state 1. The amplitude for excited state 1 is analogous to the coherent background radiation in a holographic demonstration of the $-i$ phase factor. Thus, the negative sign of the amplitude for state 2 represents not a global phase factor, but a 180° phase shift relative to the amplitude of state 1. An experimental consequence of this phase shift shows up as a reversal in sign of the oscillatory component of the fluorescent light intensity:

$$I_{2\pi}(t) = A + B\cos(\omega_{12}t + \phi). \tag{4.10b}$$

One can enhance the effect by measuring the difference signal

$$I_0(t) - I_{2\pi}(t) = 2B\cos(\omega_{12}t + \phi) \tag{4.10c}$$

and thereby eliminate the constant term. Without the spinorial sign change, the difference signal would be identically zero.

For an atom in a linear superposition of eigenstates, there *can* be an experimentally observable distinction between "doing nothing" and undergoing a cyclic transition from one of the component states. This does not contradict Dirac's assertion regarding the nonobservability of global phases, but rather serves to emphasize that, whether a phase is global or not, depends on what one does to, or with, the system. The $-i$ phase factor, for example, in the diffraction of light is an unimportant global phase factor if one simply measures the intensity of the diffracted radiation; it becomes a relative phase, however, in the holographic recording of this scattered light. The history of a system does matter.

*   *   *

The strange nature of cyclic transformations in quantum physics impressed itself upon me once again not long afterward when I was studying the interaction of charged particles with inaccessible magnetic fields, such as occur in the Aharonov–Bohm (AB) effect. The various experimental configurations I described in Chapter 3 all involve the propagation of unbound charged particles, i.e. particles that leave their source, diffract around a current-carrying solenoid (or similar structure), and are detected. My focus on atoms made me wonder about an entirely different type of AB configuration—one like a giant planetary atom in which an electron orbited, not a nucleus, but a long solenoid confining a magnetic field. What effect would the magnetic field have on the orbital properties of the electron?

Like other problems relating to the AB effect, this one, too, has its subtleties. The wave function $\psi_M(2\pi)$ of a particle with well-defined angular momentum M (in units of $\hbar$) that has wound once around the cylinder is simply related to the initial wave function $\psi_M(0)$ by a phase factor as follows:

$$\psi_M(2\pi) = e^{2\pi i M} \, \psi_M(0). \tag{4.11a}$$

To determine whether or not the "rotated" wave function is equivalent to the initial wave function requires that one know the values that the number M is allowed to take.

In the absence of the inaccessible magnetic field, the problem reduces to the quantum mechanical two-dimensional, or planar, rotator, a familiar system whose properties are well understood.[31] The spectrum of angular momentum eigenvalues is the set of all integers, $M = 0, \pm 1, \pm 2$, and so on, where states with angular momentum quantum numbers differing only in sign correspond to circulations about the origin in opposite directions. As in the case of the corresponding classical system, the kinetic energy of rotation is proportional to the square of the angular momentum. Thus, pairs of states with angular momenta $\pm |M| \hbar$ are degenerate—as one would infer from the symmetry of the system; there is no reason to expect that, in the absence of external forces, a clockwise rotating particle should have a different energy than one rotating counterclockwise.

With the confined magnetic field present, however, analysis of the planar rotator leads to a curious ambiguity, for two entirely different solutions to the equation of motion (Schrödinger equation) emerge. According to one solution, the magnetic field has no effect on the energy of the system, but leads to angular momentum eigenvalues that depend on the magnetic flux Φ,

$$M = M_0 + \frac{\Phi}{\Phi_0}, \tag{4.11b}$$

where M_0 is an integer (one of the eigenvalues of the field-free planar rotator) and Φ_0 is the value of the fluxon,

$$\Phi_0 = \frac{hc}{e}. \tag{4.11c}$$

According to the other solution, however, the magnetic field has no effect on the angular momentum eigenvalues, but leads instead to system energies that depend on the magnetic flux. Which solution, then, gives the "right" answer?

It should be noted here, because the distinction is now important, that the angular momentum that enters the phase factor of relation (4.11a) is the *canonical* angular momentum. This is the dynamical variable that determines (through the commutation relations of its components[32]) the behavior of a physical system under rotation. It is *not* necessarily the same thing as the "quantity of rotational motion," which, for the circular trajectory of a point particle, is familiarly given by (mass) × (speed) × (orbital radius). This latter dynamical variable is the *kinetic* angular momentum; it is always an observable quantity, whereas the canonical angular momentum need not be. For the field-free planar rotator, there is no difference between the kinetic and canonical angular momenta. When the rotator is in the presence of a vector potential field, however, these two dynamical quantities are no longer the same.

Careful examination of the origin of the two solutions shows that one is not "more correct" than the other, but that they refer to physically different systems, and, as in the case of cyclic atomic transitions, the history of the system plays a significant role. The quantum states of the second solution, in which the energy is flux dependent, characterize a particle orbiting a confined magnetic field that was "turned on" at some indefinite time in the past. By Faraday's law of induction (one of the Maxwell equations of classical electromagnetism), an electric field is produced throughout the time interval that the magnetic field is growing from its initially null value to the final constant value it will subsequently maintain. This electric field exerts a torque on the particle, thereby doing work and changing the initial kinetic angular momentum and kinetic energy of the particle to the values that characterize the second solution. By contrast, the solution with energy independent of flux represents a system in which the particle orbits a region containing an already existing uniform magnetic field. Such a field does no work on a charged particle (even if the particle were immersed in the field[33]) and, therefore, cannot alter the particle energy and kinetic angular momentum. Nevertheless, as we have seen before, a constant magnetic field can have quantum mechanical consequences with no counterpart in classical physics. In the present case, the bound-state AB effect modifies the spectrum of canonical angular momentum eigenvalues, and *this* has implications for cyclic transformations.

From relations (4.11a)–(4.11c), it is seen that the wave function of a particle that has undergone N revolutions about an inaccessible constant magnetic field is given by

$$\psi(2\pi N) = e^{2\pi i N \Phi / \Phi_0} \psi_M(0). \tag{4.11d}$$

Is there any physical distinction between orbiting the magnetic field once and making multiple passages around the field? If the magnetic flux is an integral multiple of the fluxon, the cyclical rotations will have no effect on the particle wave function because any phase factor of the form $e^{(2\pi i \times \text{integer})}$ is unity. There is no general reason, however, for the flux to be quantized, in which case the ratio Φ / Φ_0 can be arbitrary. If a value of Φ is chosen such that the phase in relation (4.11d) is π radians for one rotation, then the wave function changes sign for an odd number of rotations, just like a spinor. This is very interesting because nothing has heretofore been said about the spin of the orbiting particle. It could, in fact, be a spinless particle—a charged boson—which, as a result of the presence of the magnetic flux, behaves like a fermion under rotation! For arbitrary values of Φ/Φ_0, the wave function characterizes a particle that behaves under rotation neither like a fermion nor like a boson. Some physicists call it an "anyon."

How might one demonstrate such strange behavior? One possibility is to employ, again, a split-beam quantum interference experiment. Imagine dividing a collimated beam of charged particles coherently into two components, one of which is made to circulate N times in a clockwise sense about an AB solenoid bearing flux Φ_1, while the other circulates an equal number of times in a counterclockwise sense about a second AB solenoid bearing flux Φ_2 (Figure 4.7). Suppose that the

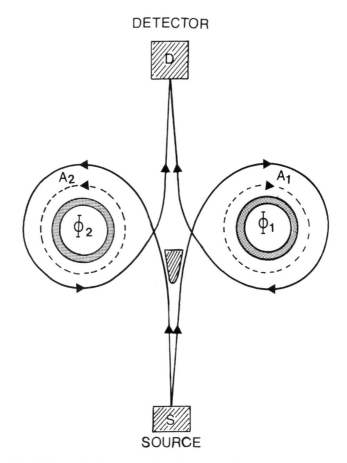

Figure 4.7. Schematic diagram of a split-beam electron interference experiment whereby an electron, issuing from source S, makes an integral number of revolutions about one or the other flux-bearing solenoid. Interference in the forward intensity of the recombined beam reveals the number of "windings." For appropriate settings of the magnetic flux, charged bosons can be made to behave like fermions under rotation.

magnetic fields within the two solenoids are oppositely directed. Suppose, too, that the "bending" magnetic fields outside the cylinders are uniform and equal in magnitude so that the radius of the particle orbit around each solenoid is the same. With this configuration, there is no net contribution to the relative rotational phase shift from the external magnetic fields or from the spin angular momentum of the particle. Upon recombination of the components at a distant detector, the forward beam intensity can be shown to vary with the magnetic flux within the solenoids according to

$$I(2\pi N) \sim I_0 \cos^2\left[\left(\frac{N(\Phi_1 - \Phi_2)}{\Phi_0}\right)\right], \tag{4.12}$$

where I_0 is the incident beam intensity.

The magnetic flux dependence of the signal reveals the topological parameter N, known as the winding number. In a two-dimensional space, there is a topological distinction between closed paths that make an unequal number of turns about the symmetry axis. Two such paths cannot be converted into each other by a continuous deformation without being cut. In the split-beam AB experiments described in Chapter 3, the charged particles propagate from source to detector by two types of topologically different paths: those that pass once to the left side of the solenoid, and those that pass once to the right. Some theoretical work suggests that a complete description of the AB effect should take account of contributions from paths of all possible winding numbers connecting the source and the detector. Thus, in addition to the configuration of two standard classical paths, one would need to include configurations where either the left or the right or both paths make one or more full loops around the solenoid before extending to the detector. On the basis of the experiments that have been done so far, it would seem that such a description may not be needed; the observed fringe patterns can be adequately accounted for by assuming that only the two classical paths contribute. The proposed experiment, were it to confirm the prediction of relation (4.12), would provide unambiguous evidence of the influence of winding numbers in the Aharonov–Bohm effect.

*   *   *

In the years following my stay in Göttingen, quantum systems made to undergo some kind of cyclic process have become of widespread interest in physics. Pursuant to the work of M. V. Berry[34] in particular, deep and beautiful connections have been shown between such seemingly disparate concepts as quantum phase and spatial curvature, and among phenomena as diverse as spinor rotation, optical activity (the rotation of the electromagnetic field of linearly polarized light), the Aharonov–Bohm effect, and superconductivity.

Like all cyclic paths, my own in Göttingen eventually drew toward a close as the day came for me to return home. Setting out for the train station very early on a cold, wintry morning, I took one last look at the countryside from which I derived so much pleasure. It was snowing, and in the bright moonlight, the Hainberg was radiant with a thick, soft, white mantle of snow. Tall fir trees, their pinnacles lost in the blackness of the sky and their heavily laden branches slung low, lined the path from my front door to the roadway like sentinels before a magical forest. Elves could have walked out of the woods at that moment and I would not have been surprised, so vividly did the scene recall the enchanting landscapes of the old German *Märchen*. As the sight of woods gave way to the sight of houses, the sense of ancient mystery faded and an unforgettable experience came to an end—as had that marvelous period over seventy-five years ago when quantum mechanics was created in Göttingen.

4.4. Long-Distance Beats

As I have already explained, quantum beats are produced by the radiative decay of *individual* atoms (or molecules) excited into a linear superposition of nondegenerate energy eigenstates. The fact that each atom may be in a superposition state does not, in itself, guarantee that the ensemble as a whole will manifest quantum interference effects because a large dispersion in the relative phases of the wave function from atom to atom will lead to unsynchronized emissions and, hence, to no net modulation of the light intensity. In some seemingly paradoxical cases—e.g., the "restoration" of beats by a sufficiently intense broad-band laser pulse of *long* duration—the unsynchronized excitation of atoms need not, in fact, lead to a dispersion in phase large enough to destroy the beats. Nevertheless, if there is one thing that might seem certain—for a science whose foundation (metaphorically speaking) rests on the uncertainty principle—it should be this: An atom that is *not* in a linear superposition of its energy eigenstates cannot give rise to quantum beats. Right?

Well, not exactly . . . even though to believe otherwise may seem patently absurd and a violation of fundamental quantum mechanical principles. Indeed, so counterintuitive was the experimental possibility that presented itself to me one day when I was thinking about the Einstein–Podolsky–Rosen paper that, despite long acclimatization to the intricacies of quantum physics, I was myself startled by its strange implications. I think of this effect, which accentuates the intrinsically nonlocal features of quantum mechanics, as "long-distance beats."

It should be mentioned at the outset that the production of quantum beats is not restricted to single-atom systems. Indeed, under appropriate circumstances, the modulation of atomic fluorescence can also occur as a result of the linear superposition of states of a multiatom system. Consider, for example, two identical atoms, each with a single ground state g and two close-lying excited states e_1 and e_2. If the atoms are near enough to one another—i.e., separated by a distance shorter than an optical wavelength—an incoming photon could be absorbed by either atom and raise that atom to one or the other of its excited states. Now, if the excited atom radiatively decays back to its ground state, the final condition of the system is simply two ground state atoms and an emitted photon.

This may be summarized in the following way:

Process I

$$\text{Atom A:} \quad g \to e_1 \to g + \text{photon } \omega_1,$$
$$\text{Atom B:} \quad g \to g;$$

Process II

$$\text{Atom A:} \quad g \to g,$$
$$\text{Atom B:} \quad g \to e_2 \to g + \text{photon } \omega_2, \qquad (4.13a)$$

where each photon is designated by its angular frequency (a measure of its energy). The frequency of the emitted photon depends on the excited state from which emission occurs, but if the detection process does not discriminate between photons, then there is no way to tell which atom had been excited. Consequently, Processes I and II are indistinguishable: To determine the net probability of photon emission, one must add the *amplitudes* of the two processes. The result is that the light intensity, to which photons emitted by one or the other of the paired atoms contribute, is modulated at the Bohr frequency corresponding to the energy interval of the excited states, $\omega_{12} = \omega_2 - \omega_1$.

It is important to note that the excited atom *must* decay back to its ground state if quantum beats are to occur. Were it to decay to some other low-lying state—call it f, for example—then the above two processes (with g + photon replaced by f + photon) would be distinguishable, because one could, in principle, search out the atom in state f and thereby determine which atom had been excited. For processes with distinguishable outcomes, one adds probabilities, not amplitudes; no quantum interference then occurs.

Because the two atoms of the pair may have different velocities relative to a stationary observer, the photon emitted from one atom can be Doppler shifted to an extent different from that of the photon emissible by the other atom. (Remember that only *one* photon is actually emitted by a pair of atoms. The beat arises not from the interference of two simultaneously present photons, but from the interference of probability amplitudes describing the two radiative processes that could potentially occur.) Depending on the distribution of atomic velocities, the spread in photon frequency can be much greater than the Bohr frequency, with the result that quantum beats from different atoms would be out of phase; no net modulation of the atomic fluorescence would be observable. As pointed out previously, the quantum beats produced from single-atom systems are not sensitive to the Doppler effect.

An interesting alternative to the sequence of events (4.13a) is one in which either atom is brought into the *same* excited state, for example e_1. In fact, one could dispense entirely with the need for two excited states, and quantum beats could still occur. Experimental evidence for just such an effect has been provided by the photodissociation of diatomic calcium molecules.[35] An incoming photon dissociates the diatomic molecule into two atoms, either of which could be raised to an excited state and subsequently decay, emitting a photon different in frequency from the one that was absorbed. The two interfering pathways may be outlined as follows:

Process I: Incoming photon + $M_2 \rightarrow M + M^* \rightarrow 2M$ + photon ω_0

Process II: Incoming photon + $M_2 \rightarrow M^* + M \rightarrow 2M$ + photon ω_0,

$$(4.13b)$$

where M_2 represents the diatomic molecule, M a ground state atom, M^* an excited atom, and ω_0 the photon angular frequency in the rest frame of the emitting atom. If there is no excited-state energy interval, then what determines the beat frequency? Upon dissociation, the two atoms recoil with equal, but oppositely directed, velocities of magnitude v. Thus, with respect to a stationary observer in the laboratory, the frequency of the photon emitted in Process I is Doppler shifted in the opposite direction to that of the photon emitted in Process II. The quantum amplitudes for emission of the differentially Doppler-shifted photons interfere, giving rise to beats at the frequency $2(v/c)\omega_0 \cos\theta$, where θ is the inclination of the axis of the dissociating molecule to the observation direction. The Doppler effect may make the light beats from processes (4.13a) impossible to observe, but it is essential for the production of quantum beats by processes (4.13b).

One conceptually important feature of the single-atom quantum beat phenomenon is that the beat frequencies, according to standard quantum theory, always correspond to level splittings of the emitting upper states and *never* to level splittings in the final lower states. Indeed, this feature has served as one of the tests distinguishing quantum electrodynamics from competing theories of radiative phenomena based on semiclassical considerations. The reason that beats at the Bohr frequencies of the final states cannot occur is that the decay pathways to alternative final states are distinguishable and, therefore, the amplitudes for these processes cannot interfere with one another.

Interestingly, in a two-atom system the quantum beats *can* occur at frequencies corresponding to *final* state splittings. Consider, for example, an ensemble of atoms, each with two nondegenerate ground states g_1 and g_2, and one excited state e. Suppose that two atoms, one in state g_1 and the other in state g_2, are irradiated with light of spectral width greater than the ground-state Bohr frequency. An incoming photon, then, could excite either one atom or the other to the state e, from which the atom subsequently decays by emission of a photon. If it is once more required that each atom radiatively decays back to its original state, then the following two processes are indistinguishable

Process I

$$\text{Atom A:} \quad g_1 \rightarrow e \rightarrow g_1 + \text{photon } \omega_1$$
$$\text{Atom B:} \quad g_2 \rightarrow g_2;$$

Process II

$$\text{Atom A:} \quad g_1 \rightarrow g_1$$
$$\text{Atom B:} \quad g_2 \rightarrow e \rightarrow g_2 + \text{photon } \omega_2 \qquad (4.14)$$

to the extent that the detector again does not discriminate between the energy of the emitted photons. Interference between the amplitudes for these indistinguishable processes leads to a quantum beat at the frequency $|\omega_2 - \omega_1|$ corresponding to the ground-state Bohr frequency.

It is implicit in all of the foregoing discussion that in order for quantum beats to be observable in either the initial upper or final lower states of a radiative transition, the two atoms must be within an optical wavelength of each other. One might think that this limitation on atomic separation is attributable exclusively to the excitation process; that is, if the atoms were further apart than an optical wavelength, then an incoming photon could not "reach" both atoms simultaneously, with the result that there would then not be two indistinguishable excitation pathways. This is not strictly the case, for the spatial extension of a wave packet—which provides a more appropriate description of a photon than does a plane wave—can be much longer than the mean wavelength (as discussed in Chapter 3 in the context of the wavelike properties of free electrons). The limitation is actually connected with the decay process.

A photon is not detected instantly after emission, but only after the so-called retarded time interval r/c, where r is the distance between the de-excited atom and the detector. The amplitude for the emission and detection of a photon of angular frequency ω (wavelength $\lambda = 2\pi c/\omega$) contains the phase factor $e^{i\omega r/c}$. Ordinarily, this phase factor does not play a significant role, for it vanishes in the calculation of the corresponding probability. However, when the photon can be emitted by either of two atoms, A or B, then a phase factor containing the appropriate atom–detector distance r_A or r_B appears in the amplitude for each indistinguishable pathway. The emission probability will then vary harmonically (i.e., in an oscillatory way) with the phase $(2\pi/\lambda)|r_A - r_B|$. If the separation between the atoms is much larger than a wavelength, the preceding phase will vary rapidly with the point on the detecting surface at which the photon is received, and the quantum beat signal will be effectively averaged away.

Although the foregoing two-atom quantum beat phenomena have their points of interest, they do not, I think, present conceptual difficulties beyond those already intrinsic to the interpretation of quantum beats from single-atom systems. Two closely separated atoms may be regarded more or less as a kind of "bondless" molecule, and, after all, the quantum interference of molecular states is no more problematical than the interference of atomic states. To put the matter a little differently, an experimentalist who has assembled the apparatus necessary for coherently exciting individual atoms would not be too surprised to find that he has also coherently excited pairs of atoms if the atomic density were high enough (on the order of one atom per cubic wavelength). All of the atoms are still close together in one small container and subjected to the same beam of light.

However, is it conceivable that two atoms—one in London and the other in New York, for example—can be coherently excited and yet the local experimentalist making observations in each respective laboratory would not even know? Let us return to the problem that intrigued me.

Imagine a transparent container (a resonance cell) filled with an atomic vapor excited by pulses of light. Each atom has a ground state g, nondegenerate excited states e_1 and e_2, and some lower state f (not necessarily the ground state) to which the excited states can radiatively decay. Disregarding for the moment the precise nature of the light source, I will simply say that photons arrive regularly at time intervals longer than the excited-state lifetimes (so that the interaction of an *excited* atom with an incoming photon and the possibility of stimulated emission can be ignored) and are distributed randomly over two possible frequencies. If an arriving photon has frequency ω_1, an absorbing atom is raised from the ground state to excited state e_1. Similarly, the absorption of a photon of frequency ω_2 raises the atom to excited state e_2. Note, however, that the two kinds of photons are each sufficiently sharply defined in frequency that *no* one photon can raise an atom into a linear superposition of excited states e_1 and e_2. This is important because, although a local observer may not know what kind of photon is going to arrive, he does know—or at least he thinks he knows!—that each atom is excited into a well-defined energy state and not a superposition of states. The seminal condition (4.4b) for the occurrence of quantum beats is not met, and, consequently, the fluorescent light emission following each excitation should simply decay exponentially in time.[36]

In fact, no measurement—light intensity or otherwise—made only on the sample of atoms in the cell would cause the observer to think that the atoms were in some way coherently excited and capable of exhibiting quantum interference. This is rigorously demonstrable by examining the quantum mechanical density matrix for the atoms, i.e., the mathematical construction that provides, in principle, a complete theoretical description of the states of the atoms. For example, if the atoms in the cell were excited into linear superpositions of states ψ_1 and ψ_2 representable by a wave function of the form

$$\psi = a_1\psi_1 + a_2\psi_2, \tag{4.15a}$$

then the density matrix ρ would take the form

$$\rho = \begin{pmatrix} \langle |a_1|^2 \rangle & \langle a_1 a_2^* \rangle \\ \langle a_1^* a_2 \rangle & \langle |a_2|^2 \rangle \end{pmatrix}; \tag{4.15b}$$

where the brackets $\langle\,\rangle$ imply an average of the enclosed quantity over all the atoms of the ensemble. For the conditions of the experiment described above the density matrix would simply comprise a tabulation of the probabilities of finding an atom in each of its states; that is, only the diagonal elements containing the absolute magnitude squared of the various expansion coefficients (such as $\langle |a_1|^2 \rangle$) would appear. The off-diagonal elements (such as $\langle a_1 a_2^* \rangle$) involving products of different coefficients designate the extent to which the system of atoms is coherently excited and, therefore, capable of manifesting quantum interference effects. *And,* if the vapor is sufficiently rarefied so the mean separation between atoms in the cell is much larger than an optical wavelength, the previously described type of quantum interference between the states of two (or more) atoms cannot occur either.

Surprisingly, the atoms can still give rise to a strange nonlocal type of quantum interference. Suppose that the first experiment were in New York. Imagine an identical experiment, the "mirror-image" of the first, set up in London; photons arrive regularly with random distribution over the same two frequencies at this station, too, and excite the same types of atoms into one or the other of its excited states. The London observer measures the fluorescent light intensity following each pulse and deduces, like the observer in New York, that the atoms

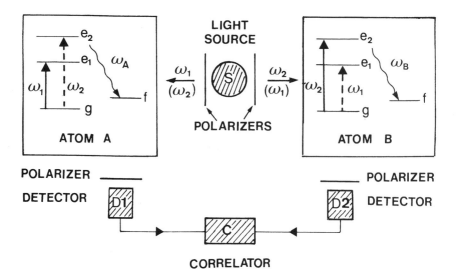

Figure 4.8. Schematic configuration of a "long-distance" quantum beat experiment. Pairs of photons of angular frequency ω_1 and ω_2 are emitted "back to back" by source S and excite widely separated absorbing atoms A and B into one or the other of two excited states e_1, e_2. The spontaneous emission from each atom separately recorded at detectors D1 and D2 manifests the simple exponential decay in time characteristic of incoherently excited systems. The joint detection probability, however, oscillates in time at the Bohr transition frequency for the two excited states.

are incoherently excited. However, if those two separated observers were to compare their results, they might notice something remarkable.

What links the two separated experiments is that the arriving photons are produced "back to back" (i.e., with correlated linear momenta) by the same source which shall be located, let us say, midway between each station (Figure 4.8). (The location of the laboratories in London and New York is solely for the purpose of dramatizing an unusual quantum effect. I am not concerned with problems attendant to placing a light source somewhere over the Atlantic Ocean, or with the fact that the surface of the Earth is curved!) Although the direction of emission of a photon of either frequency is random, there is one requirement that must be met: If a photon of frequency ω_1 is emitted in a certain direction, then a photon of frequency ω_2 is emitted in the opposite direction. Thus, possibly unknown to the two observers, each excitation of an atom into state e_1 in New York is accompanied by an excitation of an atom into state e_2 in London, and vice versa.

As far as each local observer is concerned, the atoms are incoherently excited into energy eigenstates from which they should decay exponentially in time. Looked at globally, however, there are again two indistinguishable quantum pathways:

Process I

$$\text{Atom A:} \quad g + \text{photon } \omega_1 \rightarrow e_1 \rightarrow f + \text{photon } \omega_{A1}$$

$$\text{Atom B:} \quad g + \text{photon } \omega_2 \rightarrow e_2 \rightarrow f + \text{photon } \omega_{B2}$$

Process II

$$\text{Atom A:} \quad g + \text{photon } \omega_2 \rightarrow e_2 \rightarrow f + \text{photon } \omega_{A2}$$

$$\text{Atom B:} \quad g + \text{photon } \omega_1 \rightarrow e_1 \rightarrow f + \text{photon } \omega_{B1} \qquad (4.16)$$

The emitted photons are labeled by the emitting level *and* atom since different atoms can have different velocities relative to the stationary observers. Note, again, that the final atomic state f does not, as before, have to be the initial state, since both atoms undergo excitation and decay.

To realize that these two processes are occurring, however, the two observers would need to compare, or correlate, the optical signals that each measures locally. Suppose the New York observer at detector D1 counts a certain number of fluorescent photons emitted within a time interval Δt_1 about the time t_1; likewise, the London observer at detector D2 counts photons in the time interval Δt_2 about time t_2. By multiplying these counts together (electronically), averaging over repeated trials, and varying the detection times, the two observers can construct the joint probability $P(t_1, t_2)$ for receipt of two photons separated by the time interval $|t_1 - t_2|$. The theoretical treatment of such an experiment involves not the individual density matrices of the two separated ensembles of atoms, but the density matrix for the *entire* system of atoms. However far apart the two collections of atoms may be and however independently they may go about their business after an excitation, they still constitute a single quantum system. The density matrix for this total system contains both diagonal elements (characterizing the populations of the various states) and off-diagonal elements (signifying coherence terms that can give rise to quantum interference phenomena).

The amplitudes for the two processes in expressions (4.16) interfere with the consequence that the jointly detected fluorescence signal, which takes the general form

$$I(t_1, t_2) \sim A(t_1, t_2) + B(t_1, t_2)\cos\left[\omega_{12}\left((t_2 - t_1) - \frac{(r_B - r_A)}{c}\right)\right], \qquad (4.17)$$

manifests a quantum beat at the Bohr frequency ω_{12} as a function of the delay between photon detections. Here, the term $A(t_1, t_2)$ contains the matrix elements for the independent excitation and exponential decay of the two atoms to final state f; the factor $B(t_1, t_2)$ contains the matrix elements for the correlated excitation and decay of the two atoms.

The long-distance two-atom beats are insensitive to the motion of the atoms, as is seen from the fact that the atomic velocities do not appear in relation (4.17). Why is it that the quantum beats produced by the local two-atom system are strongly affected by the Doppler effect, whereas beats issuing from the correlated excitation of two distant atoms are not? In the first case (local two-atom system), the beat originates from the interference of photons potentially emissible by *different* atoms; the beat frequency is then given by the difference of two optical frequencies, each of which can independently span the full Doppler width produced by the distribution of atomic velocities. In the second case (nonlocal two-atom system), however, a beat arises from interference between the atom A amplitudes of Processes I and II in the sequence of events (4.16); corresponding interference between the atom B amplitudes also leads to this same beat frequency. Because the interferences always involve amplitudes for photon emissions by the *same* atom, the resulting quantum beat is largely unaffected by Doppler shifts.

A significant feature of the oscillatory quantum interference term in relation (4.17) is that both the time delay, $\Delta t = (t_2 - t_1)$, and the difference in retardation times, $\Delta t_r = (r_B - r_A)/c$, are multiplied *not* by the optical frequencies of the emitted photons, but by the much smaller beat frequency ω_{12}. For the quantum beat to persist when the signal is averaged over the finite surfaces of the two separated detectors, the phase $\omega_{12}\Delta t_r$ must be small compared with about 1 radian as the optical path lengths from all A atoms (in New York) to all points of detector D1, and from all B atoms (in London) to all points of detector D2, vary. This, however, does not pose a severe experimental restriction at all. For a beat frequency ω_{12} of about 10^8 per second, for example, one can have $|r_B - r_A| < {\sim}300\,\text{cm}$, which is much greater than a wavelength (${\sim}10^{-5}\,\text{cm}$) and is quite easily attainable in the laboratory. As long as the distribution of retardation times satisfies the criterion

$$\omega_{12}\frac{|r_B - r_A|}{c} < 1, \tag{4.18}$$

it is immaterial how far apart the two collections of atoms may be; because the two atoms of the correlated pair interact with distinctly different photons, the "size" of the photon is inconsequential.

It will be recognized by now that the phenomenon of long-distance beats illustrates in the context of *bound* electron states correlations of a nature similar to those manifested by *free* electrons in the "quantum interference disappearing act" described in Chapter 3. In the terminology invented by Schrödinger, the atoms of the two separated ensembles are in "entangled" quantum states—multiparticle states that cannot be expressed as a product of the states of single particles. How atoms that undergo random excitation and exponential decay when observed locally can correlate their activity with other atoms engaged in correspondingly random behavior an arbitrary distance away is not explicable in terms of any classical mechanism. Entangled states give rise to quantum phenomena not only unaccountable within the framework of classical physics but that sometimes seem bizarre even by the expectations of quantum physics when those expectations are based on the study of single-particle or uncorrelated multiparticle systems.

To Schrödinger, the feature of entanglement was the most characteristic property of the wave mechanics he created, but it was not a feature which pleased him. "Measurement on separated systems," he wrote, "cannot directly influence each other—that would be magic".[37] The mystery of the quantum world, metaphorically speaking, is that it gives us magic without magic!

Notes

1. I have written in more technical detail of these interests in *More Than One Mystery: Explorations in Quantum Interference* (Springer-Verlag, New York, 1995) and in *Probing the Atom: Interactions of Coupled States, Fast Beams, and Loose Electrons* (Princeton University Press, Princeton, NJ, 2000).

2. The fine-structure states of atoms are labeled by (1) the principal quantum number n specifying the electronic manifold, (2) a letter indicative of the orbital angular momentum L in units of $\hbar$ (S = 0, P = 1, D = 2, F = 3, G = 4, and so on in alphabetic sequence), and (3) a numerical subscript giving the total (i.e., orbital + spin) electron angular momentum J in units of $\hbar$.

3. I discuss these experiments and the underlying theory of electric resonance spectroscopy in the book *Probing the Atom: Interactions of Coupled States, Fast Beams, and Loose Electrons* (Princeton University Press, Princeton NJ, 2000).

4. The $1S_{1/2}$ ground level actually comprises four states distributed within two hyperfine components designated by the total (electron orbital + electron spin + nuclear spin) quantum number F. The energy interval between the lower, or more tightly bound, $F = 0$ state (the true ground state) and the three degenerate $F = 1$ states (with magnetic quantum numbers $M_F = -1, 0, +1$) corresponds to a frequency lying in the microwave region of the spectrum—the 1420-MHz line of great importance in radioastronomy and astrophysics.

5. The Bohr frequency ν_{12} for a quantum transition between a state with energy E_1 and a state with higher energy E_2 is given by $h\nu_{12} = E_2 - E_1$.

6. The crimson color of excited atomic hydrogen arises from the spontaneous emission of red photons (656.3 nm) in the transition $n = 3 \to n = 2$ and blue photons (486.1 nm) in the transition $n = 4 \to n = 2$.

7. Ordinarily, a single photon is emitted when the bound electron undergoes a transition to a lower energy state. However, an electron in a 2S state can decay only to a 1S state, a process that involves no change in orbital angular momentum. Because an emitted photon would have to carry away one unit of angular momentum, a one-photon 2S → 1S transition is forbidden by angular momentum conservation. The 2S state can decay by emission of two photons; this process has a low probability of occurrence, and the 2S lifetime is correspondingly long.

8. J. Macek, Interference Between Coherent Emissions in the Measurement of Atomic Lifetimes, *Physical Review Letters* **23** (1969) 1.

9. The *amplitude* contains a factor $\exp(-t/2T)$ in order that the *probability* (proportional to the square of the amplitude) diminish as $\exp(-t/T)$.

10. The contrast of the beats is analogous to the visibility of the fringes of an interference pattern. In an expression such as Eq. (4.3d), it is the ratio of the coefficient of the time-dependent quantum interference term to the sum of the two time-independent terms representing spontaneous emission from the individual excited states.

11. This is a consequence of light diffraction, the light usually being observed far (i.e., many wavelengths) from the diffracting object. Less well known, however, is the fact that structures smaller than a wavelength can be resolved when observed very close to the diffracting object (under so-called "near-field" conditions).

12. J. Neukammer *et al.*, Spectroscopy of Rydberg Atoms at $n \sim 500$, *Physical Review Letters* **59** (1987) 2947.

13. The quantum mechanical expression for the polarisability contains terms involving the product of two radial matrix elements divided by an energy interval. Since a radial matrix element increases as the square of n and the energy interval decreases as the cube of n, the polarizability increases as the seventh power of n.

14. It is conventional to employ lowercase letters (s, p, d, etc.) for single-electron orbits and uppercase letters (S, P, D, etc.) for the overall quantum state of a multielectron atom. (For the hydrogen atom there is no distinction.) A good discussion of the orbits of the old quantum theory is given by H. E. White (*Introduction to Atomic Spectra*, McGraw-Hill, New York, 1934, Chapter 7).

15. The emitted intensity, proportional to the product of the transition probability and the light frequency, therefore varies as the fourth power of the light frequency in accord with what one would deduce from the classical Larmor formula, relation (4.7a), for an orbiting charged particle.

16. Under appropriate conditions, a bound electron can absorb two photons, each of about one-half the energy required to effect the desired transition. By angular momentum conservation, the allowed transition is governed by the selection rule, $\Delta L = 0$ or 2 (depending on photon polarization). Two-photon transitions can occur between the sodium 3S and nD states. I discuss multiphoton transitions quantitatively in my book *Probing The Atom: Interactions of Coupled States, Fast Beams, and Loose Electrons* (Princeton University Press, Princeton, NJ, 2000).

17. The frequency of light emitted by an atom moving at a nonrelativistic speed v toward a stationary observer is of the form $\omega = \omega_0(1 + v/c)$, where ω_0 is the

corresponding frequency in the atomic rest frame and c is the speed of light. The difference in frequency between two waves emitted by the same source is then $\Delta\omega = \Delta\omega_0(1 + v/c)$. Because the difference frequency is many orders of magnitude smaller than the optical frequency of either wave, $\Delta\omega_0 \ll \omega_0$, and since $v/c \ll 1$, the dispersion in atomic velocities does not result in any significant broadening of the quantum beat signal.

18. C. Fabre, M. Gross, and S. Haroche, Determination by Quantum Beat Spectroscopy of Fine-Structure Intervals in a Series of Highly Excited Sodium D States, *Optics Communications* **13** (1975) 393.

19. N. Bohr, On the Constitution of Atoms and Molecules, *Philosophical Magazine Series 6* **26**(151) (1913), 1–73. This paper is reprinted in the 50th anniversary volume: N. Bohr, *On the Constitution of Atoms and Molecules* (Benjamin, New York, 1963).

20. B. Gao, Breakdown of Bohr's Correspondence Principle, *Physical Review Letters* **83** (1999) 4225. The argument made by Gao has been contested, however, by others who believe that he has interpreted the Bohr correspondence principle incorrectly. See the comments by C. Eltschka *et al.*, [*Physical Review Letters* **86** (2001) 2693] and C. Boisseau *et al.*, [*Physical Review Letters* **86** (2001) 2694].

21. Expressed mathematically, the semiclassical condition of validity is $\lambda |dp/dr| \ll |p|$, where λ is the de Broglie wavelength [see Eq. (3.2a)] and p is the linear momentum.

22. The coordinate q and linear momentum p are mathematical operators expressible, among other ways, as infinite-dimensional matrices. It is from the noncommutability of operator multiplication that the Heisenberg uncertainty relations formally arise.

23. B. L. van der Waerden, *Sources of Quantum Mechanics*, North-Holland, Amsterdam, 1967, p. 22.

24. In general, $e^{i\phi} = \cos\phi + i\sin\phi$. The identity, $e^{i\pi} + 1 = 0$, discovered by Leonhard Euler (1707–1783), contains the most important symbols of modern mathematics and has been regarded in the past as a sort of "mystic union" in which 0 and 1 connoted arithmetic, π stood for geometry, $i = \sqrt{-1}$ designated algebra, and the transcendental number e represented analysis. It seems fitting, somehow, that this identity be the point of origin of another of the "mysteries" of quantum mechanics.

25. P. A. M. Dirac, *The Principles of Quantum Mechanics,* 4th ed., Oxford University Press, London, 1958, p. 148

26. I first learned of such an experiment from Brian Thompson of the University of Rochester, New York, while participating in his 1977 National Science Foundation Short Course on Coherent Optics. Following publication of *And Yet It Moves,* Michael Berry of the University of Bristol, UK, brought to my attention that L. G. Gouy (~1890) observed the $\pi/2$ phase advance of a wave passing through a focus.

27. H. Rauch *et al.*, Verification of Coherent Spinor Rotation of Fermions, *Physics Letters* **54A** (1975) 425; S. A. Werner *et al.*, Observation of the Phase Shift of a Neutron Due to Precession in a Magnetic Field, *Physical Review Letters* **35** (1975) 1053.

28. J. Byrne, Young's Double Beam Interference Experiment with Spinor and Vector Waves, *Nature* **275** (1978) 188.

29. The spin components of the photon are oriented either parallel or antiparallel to the direction of propagation; a transverse (M = 0) substate is disallowed as a result of the vanishing photon rest mass. This profound point is discussed by E. P. Wigner, Relativistic Invariance and Quantum Phenomena, in *Symmetries and Reflections*, Indiana University Press, Bloomington, 1967, Chapter 5. It is more accurate to attribute to the photon a "helicity" (projection of spin onto linear momentum) of $1\,\hbar$.

30. Actually, the study of rotating bosons can be quite interesting, although for reasons far removed from the principal themes discussed in this chapter. In Chapter 9, however, I discuss a form of matter known as a Bose–Einstein condensate (BEC), which is a coherent system of identical bosons. Rotation of a BEC can give rise to an array of quantized vortices.

31. I discuss comprehensively the physics of the two-dimensional rotator in field-free space and in the presence of electric, magnetic, and vector potential fields in the book *More Than One Mystery: Explorations in Quantum Interference* (Springer-Verlag, New York, 1995).

32. The commutation relations of the canonical angular momentum take the form: $(L_x L_y - L_y L_x) = i\,\hbar L_z$ with corresponding forms for even permutations of the coordinates x, y, and z.

33. A uniform time-independent magnetic field produces a local Lorentz force that acts at right angles to the direction of particle motion. Since force is perpendicular to displacement, the force can do no work on the particle.

34. M. V. Berry, Quantal Phase Factors Accompanying Adiabatic Changes, *Proceedings of the Royal Society London* **A392** (1984) 45.

35. P. Grangier, A. Aspect, and J. Vigue, Quantum Interference Effect for Two Atoms Radiating a Single Photon, *Physical Review Letters* **54** (1985) 418.

36. If the two excited states have the same lifetime, then the characteristic decay rate is the inverse of that lifetime. If the lifetimes of the two states are different, then the fluorescence is describable as a sum of two exponentially decaying curves.

37. W. Moore, *Schrödinger*, Cambridge University Press, Cambridge, 1989, p. 310.

CHAPTER 5

And Yet It Moves: Exotic Atoms and the Invariance of Charge

5.1. A Commotion About Motion

If I had to describe in a word what physics is all about, I am tempted to say "motion"—as construed, of course, in a suitably broad sense to include not only the movement of particles but also such phenomena as the flow of fluids, the propagation of waves, the conversion of heat to work, and the transitions between quantum states. The word "physics," according to my unabridged English dictionary, traces its origin to a Greek term meaning "natural things," which seems appropriate enough, as far as it goes. Actually, in some ways it goes too far, for one can accommodate all of the life sciences in that definition—and, in fact, the term "physic" once meant medicine or the art of healing. The Japanese refer to physics as "butsurigaku," derived from "study of transformation." That also has some good points, but reminds me too much of chemistry. I will stay with "motion."

The concept of motion is fundamental in ways that go well beyond its role in physics. It underlies our very perception of reality. When Zeno of Elea propounded his paradoxes[1] in the 4th century BC, it was not merely to induce mental paralysis in his fellow mathematicians, but to spread the doctrine of his mentor, the mystic philosopher Parmenides, who stressed that the world of sense is nothing but illusion. Thus thought Zeno:

If motion, which pervades everything, can be shown to be self-contradictory, and hence unreal, then everything else must assume the same unreal quality. By convincing people of the unreality of motion, I can ... successfully discredit the world of the senses.[2]

Today, the fall of an apple or the movement of a planet, although probably still a mystery to the majority of laymen, does not ordinarily provoke deep philosophical discussion, much less a conceptual crisis, among physicists. Although not necessarily easy to describe mathematically, the motion of a macroscopic object is at least in principle a decidable proposition. One can tell whether it occurs or not. Look up at the Moon for a while on a clear night; its position in the sky changes. It moves around the Earth. Or does the Earth move around the Moon? In any event, something is moving, and that is a fact, Zeno notwithstanding.

Zeno was born some 2500 years too soon, for, if he enjoyed paradoxes, he would have loved quantum mechanics! Among other things, quantum mechanics disabuses us of the certainty of motion—of the perception of motion through our senses. In contrast to the observable motion of macroscopic-sized objects, one cannot "see" an elementary particle move. For an object to be seen, it must either emit light or be illuminated. Whereas the reflection of light from the Moon or the emission of light from a firefly will scarcely alter the object's subsequent motion, the interaction of light with an elementary particle can change the behavior drastically. One can, of course, illuminate an elementary particle with "softer" (i.e., less energetic) photons that do not substantially change its momentum. The energy of a photon, however, is inversely related to its wavelength, and the illuminated particle cannot be localized to within a region more sharply defined than the wavelength of the light used for viewing. The detailed movement of an elementary particle that has emitted or scattered soft photons would be lost in an indistinct blur. At the quantum level, motion cannot be seen directly; it must be inferred from physical theory.

Heisenberg, a seminal contributor to the creation of quantum mechanics who had given much thought to the nature of physical theory, was fairly definite about what should *not* go into one[3]:

... it seems necessary to demand that no concept enter a theory which has not been experimentally verified at least to the same degree of accuracy as the experiments to be explained by the theory.

Unfortunately, as Heisenberg himself recognized:

... it is quite impossible to fulfill this requirement, since the commonest ideas and words would often be excluded.

Words like orbit, trajectory, velocity—familiar terms in the Newtonian lexicon defining the classical conception of motion.

There would be no problem, Heisenberg contended, if physicists would only remain content, for example, with the images of particle tracks on photographic plates, such as those made in 1911 by C. T. R. Wilson,[4] and not attempt to "classify and synthesize" the results or to "establish a relation of cause and effect between them." Without a causal description, however, how could one ever know if something was moving? Each image of an object, like the arrow in one of Zeno's paradoxes, would lie suspended and temporally disconnected from the images that came before and after.

Figure 5.1. Tracks of alpha particles formed in a Wilson cloud chamber. (Adapted from W. Heisenberg, *The Physical Principles of the Quantum Theory*, Dover, New York, 1930, p. 5.)

But surely an elementary particle moves. If not, then what is one really to make of the Wilson photographs (Figure 5.1)? These photographs show the explicit tracks of alpha rays (helium nuclei) and beta rays (energetic electrons) that passed through the supersaturated water vapor of a cloud chamber. The trails were formed from minute droplets of water that have condensed about molecules ionized by collisions with the energetic charged particles. Is not each track the trajectory of some particle that has made its way through the chamber?

Technically no—not a trajectory. The continuity of the tracks produced by numerous, but discrete, particle collisions and ionizations is illusory. Furthermore, with each random collision, the subsequent motion of the particle is changed in an unpredictable way. A definite location and velocity at each instant of time cannot be assigned to the ionizing particle. Nevertheless, quantum mechanical analysis does establish a picture of events at the microscopic level that conforms to expectations based on classical mechanics to within limits set by the Heisenberg uncertainty relations. From the wave function of the ionizing particle, the most probable value of the particle location at any instant can be inferred, and, in the absence of subsequent disturbances, this locus of points traces out the classically predicted straight-line path. Each successive collision with a water molecule, however, modifies the wave function and increases the uncertainties with which the location and linear momentum of the particle can be known. The angular deviations of orbital segments between successive collisions depend on the relative linear momenta of the ionizing particle and the atomic electrons with which it interacts. For sufficiently energetic and massive particles, like the alpha particles in Wilson's experiments, the ionized molecules will lie effectively on straight lines; the paths of the much less massive beta particles are irregularly curved.

For all practical purposes, the Wilson tracks furnish an adequate record of particle motion even if one cannot assign to each point an instantaneous coordinate and velocity. No harm is done in believing that somewhere within each track a particle passed by. But what about the electrons *within* an atom; can one tell if a bound electron is moving?

From the semiclassical point of view embodied in the Bohr model of the hydrogen atom, an electron bound to a nuclear center of charge $+Ze$ moves about the nucleus in a circular orbit of principal quantum number n with a speed

$$\frac{v_n}{c} = \frac{Ze^2}{n\hbar c}. \tag{5.1}$$

The dimensionless combination of constants $e^2/\hbar c \sim 1/137$ will be recognized as the Sommerfeld fine-structure constant α_{fs}, which sets the scale for the interaction of charged particles with electromagnetic fields. Although nonrelativistic ($v/c \sim 0.007$), the speed of the electron in a ground-state hydrogen atom ($Z = n = 1$) is predicted by the Bohr model to be some 2 million meters per second, by no means a trivial quantity when judged by ordinary experience.

In the quantum description of the hydrogen atom, however, the situation is not so simple. For one thing, corresponding to the uncertainty relation between coordinate and linear momentum is a rotational counterpart between angular momentum and angular location. Dynamical quantities such as linear momentum, angular momentum, energy, and the like, which in classical mechanics can be regarded as properties of an individual particle, have a twofold significance in quantum mechanics. On the one hand, they represent mathematical expressions—or operators—that obey well-defined algebraic relations independent of any particular physical system. On the other hand, they refer to the mean values to which these operators lead when applied to the wave function of specific quantum systems. Whether the wave function and the associated dynamical variables actually characterize one particle or a large collection (an "ensemble") of identical particles ideally prepared all in the same way is a matter of debate among physicists concerned with the foundations and interpretation of quantum theory.

In any event, an uncertainty relation can be established between any two measurable quantities for which the corresponding quantum mechanical operators do not commute. For example, the familiar Heisenberg uncertainty principle relating the uncertanties (technically, the statistical variances) in coordinate and linear momentum is readily derived from the noncommutability of the coordinate and linear momentum operators (the equation inscribed on Max Born's tombstone).[5] Strictly speaking, there is no quantum mechanical operator for angle, as there is for linear coordinate, and the establishment of a corresponding uncertain relation between angle and angular momentum has been somewhat problematical. Nevertheless, several such relations have been proposed,[6] all leading to the following consequence: Because the wave function of a hydrogenic electron—in contrast to that of a free electron—is characterized by a sharp (definite) angular momentum, the electron angular coordinate is totally delocalized about the nucleus. It would seem, therefore, that the concept of an orbit or trajectory within the atom is of little, if any, significance.

A second feature of the quantum mechanical hydrogen atom contrasting starkly with the corresponding Bohr planetary model is that the angular momentum of the 1S ground state is zero. Classically, a particle with zero angular momentum either passes through the axis of rotation or has zero velocity. The quantum calculation of the mean electron orbital radius (for a state with principal quantum number n and atomic number Z) yields in accord with the Bohr model:

$$r(n, Z) = \frac{n^2 a_0}{Z},\qquad(5.2a)$$

where

$$a_0 = \frac{\hbar^2}{me^2}\qquad(5.2b)$$

is the Bohr radius, about 5×10^{-9} cm for an electron of mass $m = 9 \times 10^{-28}$ g and charge $e = -4.8 \times 10^{-10}$ esu. Is the electron moving?

The calculation of the electron velocity is in some ways an undefined problem, for it is not velocity, but momentum, that plays a key role in the formulation of quantum mechanics. If one utilizes the nonrelativistic relation between velocity and linear momentum of a particle with mass m,

$$\boldsymbol{v} = \frac{\boldsymbol{p}}{m},\qquad(5.3a)$$

the quantum calculation leads to a null velocity. However, in contrast to classical mechanics, this does not mean that the electron *speed* is necessarily zero. Calculation of the kinetic energy

$$K = \frac{1}{2}mv^2 = \frac{p^2}{2m} \tag{5.3b}$$

of the ground-state electron results in a root-mean-square speed the same as that of relation (5.1). Matters become yet more confusing when one employs the relativistic electron theory of Dirac rather than the nonrelativistic Schrödinger theory. The velocity operator in the Dirac theory, defined as the time rate of change of the particle coordinate operator[7]

$$v = \frac{dx}{dt}, \tag{5.3c}$$

has but two allowable eigenvalues: $\pm c$. This would seem to indicate that the electron can move only at the speed of light! Yet, the relativistic calculation of the kinetic energy again leads (in the nonrelativistic limit) to a particle speed effectively equivalent to relation (5.1). What is to be made of all this?

Following Heisenberg's warning, one is forced to renounce all hope of visualizing the motion of an electron in the energy eigenstates of an atom. Although such states—referred to as stationary states because their properties do not vary in time—may have nonvanishing expectation values of kinetic energy, angular momentum, and other dynamical variables, there is still *no* sequential connection between neighboring points in the resulting electron probability distribution.

Recognizing this helps to eliminate a number of potentially paradoxical situations. Except for the 1S ground state, the radial distribution of the electron in every other hydrogenic stationary state has nodes; that is, for certain calculable distances from the nuclear center, the probability of finding the orbiting electron is exactly zero. Since the probability of finding the electron at all other points is nonvanishing, the probability distribution consists of disconnected regions. For example, the radial distribution of an electron in the 2S state consists of two disconnected regions separated by a spherical surface of radius $2a_0$ (twice the Bohr radius) on which the electron probability is zero. Given that there is but one bound electron, one might be tempted to inquire how the electron can move from the inner to the outer region if it can *never* be found at a point exactly $2a_0$ from the nucleus!

The answer is simply what has been stated before: The properties of a stationary state do not correspond to a causal description of particle motion. Because two points, however close, of a stationary-state probability distribution do not represent points on a particle trajectory and provide no information at all as to how an electron may have passed from one point to the other, the presence of nodes leads to no paradoxical behavior. According to an ensemble interpretation, we may regard the stationary states as representative of the statistical properties of a large number of similarly prepared atoms. Imagine (never mind how!) photographing the instantaneous location of the single 1S electron in one million hydrogen atoms produced by the same source. The composite radial distribution of the electrons would resemble the stationary-state radial distribution predicted by quantum mechanics for the 1S state. Moreover, because the act of photographing the atoms perturbs the subsequent motion of the electrons, one would not be able to predict from the photograph of a particular electron where it will be an instant later.

The statistical interpretation enables us to understand as well how the 1S electron can have zero angular momentum and a non-vanishing mean speed and orbital radius. This is again what one might expect of the average angular momentum of a great number of atoms with electrons moving in randomly oriented planar orbits, clockwise or counterclockwise in equal measure. None of these properties characterizes the sequential movements of an individual electron.

In marked contrast to the alpha and beta particles moving through a Wilson cloud chamber, the bound electrons in an atom leave no tracks by which to infer movement. Does a bound electron actually move? Does this even matter?

5.2. The Electric Charge of a Moving Electron

One of the attributes of particles that is in some ways both familiar and mysterious is that of electric charge. The theory of quantum electrodynamics provides a comprehensive and (as far as experiment has been able to confirm) correct description of the interaction of charged matter with electromagnetic fields. Yet, curiously enough, we do not know exactly what charge is, only what it does. Or, equally significantly, what it does not do.

Electric charge does not, on balance, change. The conservation of electric charge is one of the most strictly observed conservation laws of physics. To my knowledge, no reproducibly documented violations have ever been reported. Moreover, it is *not* simply a question of global charge balance, as, for example, in a process by which an electron is created at one end of the laboratory and a positron at the other end. Charge conservation is local; there is to be no violation in any space–time region within limits set by quantum mechanical uncertainties. Conceptually, the conservation of electric charge can be understood as arising from a special kind of symmetry in the laws of electromagnetism—the requirement that the basic equations of motion be unaffected by a phase transformation of the fundamental fields.[8]

There is another, perhaps even more profound, sense in which electric charge does not change: It is independent of its velocity. Although often taken for granted, this is a rather remarkable fact of nature, for many physical properties *do* depend on velocity. The apparent mass or inertia of a particle, for example, increases with particle speed; as the speed of the particle approaches that of light, an increasingly greater force is required to effect a given incremental increase in speed. To accelerate a massive particle to the speed of light would require an infinitely large force—and so cannot be done. The charge of a particle, however, does not change at all; it is said to be a Lorentz invariant.

The significance of the Lorentz transformation, named for the Dutch physicist H. A. Lorentz who discovered it in the course of developing a classical theory of the electron, was first recognized by Einstein, who derived the relations in a much more fundamental way—independent of any theory of matter—through consideration of measurements of space and time. The Lorentz transformations enable two observers moving relative to one another to compare their measurements of spatial and temporal intervals and, consequently, to relate all other dynamical quantities (acceleration, force, energy, momentum, etc.) that are based on, or in some way connected to, space–time measurements. The importance of these relations goes well beyond the domain of mechanics. To be considered viable, a physical theory must at the least be compatible with the special theory of relativity and expressible in a form that remains invariant under a Lorentz transformation. The theory is then said to be Lorentz covariant. This means that there is no preferred reference frame—no special state of motion of an observer—for which the theory is valid. If the theory is valid, it must be recognized as such by all observers in uniform motion with respect to one another.

The space–time properties of the electromagnetic field are largely determined by the requirement of covariance *and* the Lorentz invariance of electric charge. The invariance of charge, however, unlike charge conservation, is not at present known to follow from any deeper principle, but must be taken at the outset as an experimental fact. It is conceivable, if certain types of particle decays are allowed, that charge invariance may be connected with charge conservation. For example, to maintain exact charge conservation in the disintegration of a proton into a positron plus neutral particles, the proton and positron must have equal charges irrespective of their speeds. I know of no reproducible observation of proton decay. However, relativistic quantum theory requires that particles and antiparticles have charges of exactly the same magnitude. Hence, if protons decayed into positrons, then the charge of the proton and the charge of the electron would also have to be equal in magnitude. If they were not, then the beta decay of a neutron (into a proton, electron, and antineutrino) would violate the conservation of charge.

Because charge invariance is one of the conceptual pillars of electromagnetism as we know it, the solidity of the empirical foundation upon which it rests is no trivial matter. Just how do physicists know that electric charge is independent of velocity?

One of the strongest arguments advanced for the Lorentz invariance of charge is the electrical neutrality of atoms and molecules. The essence of the argument is as follows. Suppose the charges of the electron and proton at rest to be equal in magnitude. Then, were the Lorentz invariance of charge *not* valid, there would be a charge imbalance for the bound system within which these particles are in relative motion. If motion were to have an effect on the magnitude of charge, one could not expect exact cancellation of the nuclear and electronic charge in composite systems as different as the hydrogen atom, the helium atom, and the hydrogen molecule (H_2).

In a stationary hydrogen atom, as discussed above, the speed of the 1S electron relative to the proton is calculated to be $v \sim 0.007c$. From relation (5.1), one would infer that the ground-state (1S) electrons in a helium atom ($Z = 2$) should move twice as fast as the electron in hydrogen. Actually, of even more interest is the speed of a proton in a helium nucleus composed of two protons and two neutrons. A simple heuristic argument shows that the protons in helium are moving with relativistic speeds within the potential well that binds them by means of the strong nuclear interaction. The uncertainty in the linear momentum of a particle confined to a spatial region of size r is about h/r (where h is Planck's constant). The characteristic size r of the helium nucleus is on the order of 10^{-13} cm. Taking h/r as an estimate of the maximum value of linear momentum of a proton of mass $M = 1.67 \times 10^{-24}$ g and equating it to the classical relativistic expression

$$p = Mv\gamma, \qquad (5.4a)$$

where

$$\gamma = \frac{1}{\sqrt{1-(v/c)^2}}, \qquad (5.4b)$$

leads to a proton speed of about $0.8c$, over 50 times greater than the electron speed in helium. (If the nonrelativistic expression $p = Mv$ were used, the calculated speed of the proton would exceed the speed of light.) The electrons in a ground-state hydrogen molecule are in "molecular orbitals" formed from the atomic 1S states and have a speed comparable to that of the electron in a hydrogen atom. The maximum speed of the two protons, oscillating about their equilibrium separation of nearly 0.1 nm at a frequency on the order of 10^{13} Hz, is approximately 6×10^{5} cm/s or $2 \times 10^{-5}c$—considerably less than that of the two protons in the helium nucleus.

In a set of experiments performed over forty years ago, J. G. King[9] established that the fractional difference in charge between electrons and protons in helium atoms and in hydrogen molecules is zero to within a few parts in 10^{20} or 100 billion billion! The experimental technique is as ingenious as it is simple. In brief, the gas whose charge was to be measured was allowed to escape from an electrically insulated metal container attached to an electrometer; a "deionizer" swept out of the gas stream any ions or free electrons. If there were a charge imbalance in the remaining gas molecules, then the outflow of gas would result in a current flow to the container that would be registered by the electrometer. To within an experimental uncertainty leading to the above awe-inspiring limit on electron–proton charge equality, no such current was found. It would seem that the Lorentz invariance of charge rests on indisputably firm ground.

But does it really? Although the demonstration of atomic and molecular neutrality are convincing, there is still one conceptually untidy step in the chain of reasoning to charge invariance. How do we know that the experiments actually examine the variation in charge with particle speed? How do we know, that is, that within the atom (or within the nucleus) the particles are truly moving? Call it a matter of professional ethics, if you will, but can one in good faith claim that an electron in a stationary state is moving when he wants to justify the Lorentz invariance of charge, and then forego a space–time description of this motion in order to avoid quantum mechanical paradoxes? I wonder what Heisenberg would say.

Although the visualization of motion is by no means necessary or even relevant to the consistency of quantum mechanics, it does play a certain important role in special relativity. In the latter, one must be able to imagine the placement of clocks and meter sticks in different inertial reference frames for the purpose of performing space–time measurements. If the Lorentz invariance of charge is to be inferred from the state of motion of bound elementary particles, then— although the equation of motion must necessarily be quantum mechanical—it must still be meaningful to conceive of a Lorentz transformation relating the rest frame of a particle to the rest frame of another particle or of a stationary observer. However, given that a bound electron cannot even in principle be located without alteration of its state of motion, is such a conception meaningful?

The question of electron motion within an atom first started gnawing at me when I was an undergraduate student a few years after the King experiments were performed. I wondered whether there was any phenomenon at all exhibited by a bound electron that could manifest directly some element of the kinematics of special relativity. It seemed to me that, despite the fact that one could not picture the motion of a bound elementary particle, somehow the question ought to be answerable in the affirmative. The assertion that a particle is in motion relative to a stationary observer has an observable physical consequence: A clock moving *with* the particle must exhibit the relativistic effect of time dilation (also called time dilatation). In the words of the mnemonic familiar to those who study special relativity: "Moving clocks run slow." A time interval determined from a moving clock would appear *shorter* than the same interval measured with a system of clocks relatively at rest.[10]

If the electron were a classical charged particle orbiting a center of force, this effect would, in principle, be manifested in the Doppler shift of radiation emitted from different points in the electron trajectory. As in a binary-star system, there would occur both a blue and a red shift for any angle of observation other than at 90° to the plane of orbital motion. The electron, of course, is not a classical particle—and the immediate proof of that is the very existence of atoms. If the electron radiated as described, it would spiral into the nucleus and, as may be deduced from the Larmor formula (4.7a), the atom would collapse in about 10^{-11} seconds! In the quantum mechanical atom, radiation is not continuous, but occurs only when the electron undergoes a transition between states; the radiation frequency, unrelated in general to the calculable orbital frequency, is not Doppler shifted unless the entire atom is moving.

There is, however, an alternative and peculiarly quantum mechanical clock associated with an elementary particle, namely its natural lifetime. However, the electron, as far as one knows, is a stable particle; there is no other negatively charged particle of lower mass to which it can decay. The electron lifetime should be infinite, and no effect of motion on it would be observable. But suppose the electron *could* decay. Would its lifetime be lengthened if it were bound in an atom?

5.3. The Exotic Atom

At first acquaintance, a so-called exotic atom may seem like a chimera of atomic physics—but it is, in fact, quite real. Since an inevitable comparison with a planetary system is made whenever atoms are discussed, imagine, again, looking at the Moon one night and seeing it vanish in a burst of light. Something similar can happen in an exotic atom—except that the burst might well be one of electrons and neutrinos (ghostly spin-$\frac{1}{2}$ particles with no charge and unknown, if any,[11] mass that interact extremely weakly with all matter). Or imagine the Solar System with one of the planets orbiting *inside* the Sun. That, too, can happen in an exotic atom. An exotic atom is formed when one of the electrons of an ordinary atom is replaced by another more massive (and unstable) negatively charged elementary particle.

The story is oft repeated that I. I. Rabi, when first apprised of the existence of the muon in the 1930s, grumbled, "Who ordered *that?*," in displeasure at the increasing complexity of nature.[12] The number of "elementary" particles known today is so large that most can scarcely be considered elementary, but the muon still is. With a mass of about 207 times the electron mass, the muon is described by particle physicists as essentially a heavy electron.[13] There is one critical difference, however: the muon has a finite lifetime. In its rest frame, the muon lasts about $2.2\,\mu s$ before transforming via a weak interaction (like the β decay of a nucleus) into an electron, an electron antineutrino, and a muon neutrino. Like the electron, the negative muon can bind to a positive atomic nucleus to form an atom—one of the exotic atoms. (To the positive electron, or positron, there also corresponds a positive muon.)

A few microseconds may seem like a minuscule amount of time for a particle, and the atom it forms, to stay around. In comparison to initial expectations, however, the muon lifetime was anomalously long. In 1935, based on analogy with the transmission of the electromagnetic force by photons, Hideki Yukawa predicted the existence of a carrier of the strong nuclear force. The electromagnetic force is known to be of infinite range as a consequence of the zero rest mass of the photon. Knowing that the strong force is of short range (about the size of a nucleon, $10^{-13}\,\mathrm{cm}$), Yukawa was able to predict that the mass of the sought-for nuclear particle should be about two hundred times the electron mass. Not long afterward, a particle of that approximate mass was observed in the cosmic ray showers that reached the Earth's surface. However, *this* particle did not interact strongly with atomic nuclei; if it had, its existence would have been a fleeting $10^{-23}\,\mathrm{s}$, and it would never have passed through the Earth's atmosphere to be detected at ground level. The discovery was actually the muon.[14]

According to relations (5.2a) and (5.2b), the ground-state orbital radius of a negative muon bound to a positive nucleus of charge $+Ze$ should be smaller than that of the corresponding electron orbit by the ratio of the electron and muon masses:

$$r_\mu(Z) = \frac{(m/m_\mu)a_0}{Z} \sim \frac{a_0}{207Z}. \qquad (5.5)$$

The orbital speed, however, depends only on particle charge and should be the same, $v/c \sim Z\alpha_{fs}$, for both the muon and the electron. Hence, a bound muon with lifetime t_0 can complete about $vt_0/(2\pi r_\mu) \sim 3 \times 10^{12}Z^2$ revolutions before undergoing weak decay. In terrestrial terms, this is the equivalent of over 1000 billion years, many times longer than the current age of the Earth (or of the Universe for that matter). The muonic atom would therefore seem to be a reasonably stable system with a well-defined ground state.

The existence of exotic atoms was inferred by Enrico Fermi and Edward Teller in 1947 from the different behavior of positive and negative muons coming to rest in iron or graphite.[15] Positive muons, repelled by the positive atomic nuclei, were expected to decay naturally at their characteristic rate producing "disintegration electrons"; negative muons—at a time when the muon and the predicted Yukawa particle (pion) were not yet recognized as distinctly different particles—were expected to be captured by the atomic nuclei and not give rise to electrons. This expectation was fulfilled for iron. In graphite, however, disintegration electrons emerged equally abundantly from both positive and negative muon decays in sharp disagreement with contemporary expectations. To explain this anomaly, Fermi and Teller assumed that the negative muon was captured into a Bohr orbit from which it, too, decayed naturally rather than by nuclear capture. They showed that a negative muon initially captured into a high-lying atomic state cascaded down into the ground level—generally referred to as the K shell—in a time interval on the order of 10^{-12} s for atoms in condensed matter and 10^{-9} s for atoms in a gas—i.e., in an interval short compared with the muon lifetime. Fewer than about 1% of the muons decay from states other than the ground state.

Once the muon reaches the K shell, it is in an orbit some 200 times smaller than that of a K-shell electron; largely unaffected by the surrounding shells of atomic electrons, the muon can be treated to good approximation as if it were the only bound particle. The radius of a nucleus of atomic number Z and mass number A (number of protons and neutrons) is usually represented by the formula

$$R = A^{1/3}r_0, \tag{5.6}$$

where $r_0 = 1.3 \times 10^{-13}$ cm is about one-half the so-called classical electron radius, e^2/mc^2. From relations (5.5) and (5.6), it is seen that for muonic silver ($Z = 47$, $A = 108$), the muon 1S orbital radius of about 5×10^{-13} cm lies at the periphery of the nuclear surface. In muonic uranium ($Z = 92$, $A = 238$), the ground-state muon orbit of radius about 3×10^{-13} cm lies well within the uranium nucleus (radius $\sim 8 \times 10^{-13}$ cm).

When captured by a nucleus, the negative muon interacts with a proton to give rise to a neutron and muon neutrino. The probability of this process increases with nuclear charge because the degree of overlap of the muon and nuclear wave functions is correspondingly greater as the orbital radius is smaller. For $Z = 11$ (muonic sodium), the rates of weak decay and nuclear capture are approximately equal; for heavy nuclei, the mean lifetime of a bound muon is essentially determined by the nuclear capture process. Nevertheless, because the end products of the two processes are different, one can experimentally distinguish them and study the natural decay of bound muons by monitoring the rate of production of decay electrons.

The natural lifetime of the muon constitutes an ideal clock for testing the time dilation effect of special relativity. Indeed, such a test was first made over half a century ago in a now classic experiment by Bruno Rossi and David Hall[16] whereby the decay rate of cosmic ray muons was shown to depend on particle speed in accordance with the relativistic relation

$$t_\mu = \gamma(\beta)t_0. \tag{5.7}$$

Here, t_μ is the observed lifetime (reciprocal of the decay rate) of muons moving relative to a stationary observer with speed parameter $\beta = v/c$, t_0 is the lifetime in the muon rest frame, and $\gamma(\beta)$ is the dilation factor given by relation (5.4b).

I recall learning of the time dilation of muons in an educational film[17] depicting a modified repetition of the Rossi–Hall experiment some twenty years later, at about the time the question of electron motion first occurred to me. In the film, the lifetime of energetic muons ($v/c \sim 0.995$) deduced from muon fluxes measured at the top of Mount Washington in New Hampshire and at sea level in Cambridge, Massachusetts was found to be lengthened by nearly a factor of nine. I wondered: Would the lifetime of a muon in the K shell of a muonic atom also be lengthened?

A quick estimate of the dilation factor of the bound muon lifetime can be made simply by substituting the Bohr speed $v/c = Z\alpha_{fs}$ into relation (5.4b). This would suggest, for example, that in muonic uranium, where $v/c \sim 0.41$, the decay rate of a K-shell muon should be only about 0.77 that of a free muon at rest. This is not, however, a rigorous way to deduce the properties of a quantum system, which should in principle, be determined by means of the expectation value of an appropriate operator. But what operator would correspond to time dilation? Since the relativistic expression for the total (mass + kinetic) energy of a free particle of mass M is

$$K = Mc^2\gamma \tag{5.8a}$$

and since the Hamiltonian operator H that governs the time evolution of a quantum system is the sum of K and the potential energy V, it seemed to me reasonable to define a relativistic time-dilation operator by

$$\gamma = \frac{H - V}{Mc^2}.$$ (5.8b)

The expectation value of γ for the $1S_{1/2}$ ground state of a point Coulomb potential $V = Ze^2/r$, evaluated by means of the exact relativistic Dirac wave functions, led to the satisfying result

$$\gamma(Z) = \frac{1}{\sqrt{1 - (Z\alpha_{fs})^2}}$$ (5.8c)

in agreement with the semiclassical relativistic theory of the Bohr atom. For intermediate and heavy muonic atoms in which the muon orbit lies closer to, or within, the nuclear surface, the potential seen by the muon can be decidedly different from that of a Coulomb potential. In fact, a muon orbiting deep inside a nucleus with uniform distribution of positive charge will experience a potential similar to that of a harmonic oscillator. Nevertheless, once the potential V is known, relation (5.8c) can be used to determine the relativistic dilation of a bound-state lifetime.

Coincidentally, although I was not to realize this until some years afterward, accurate determinations of the lifetime of a number of moderate and heavy muonic atoms[18] had been published shortly after King's experimental test of electron and proton charge equality. In experiments performed with the 200-MeV synchrocyclotron at the University of Liverpool, negative muons traveling at a speed of about $0.94c$ were brought to rest in various targets and the resulting time distribution of decay electrons was monitored. This distribution is governed by the familiar exponential decay law (there are no quantum beats here!)

$$N(t) = N_0 e^{-t/t_\mu},$$ (5.9)

where $N(t)/N_0$ is the fraction of undecayed muons at time t. Before each selected target was used, an experimental run was made with a carbon target that furnished an effective measurement of the muon decay rate under circumstances comparable to free-muon decay (since $Z = 6$ leads to $\gamma \sim 1.00$).

So . . . do bound muons move? Summarized in Table 5.1 is the experimentally observed ratio $R = t_0/t_\mu$ of the bound- and free-muon decay rates. Also shown is the theoretically expected ratio $R = 1/\gamma(Z)$, where, for ease of calculation, the point Coulomb potential was assumed. In the case of heavy muonic atoms, where the time-dilation effect is greatest, a more appropriate model of the electrostatic

Table 5.1. Experimental and Theoretical values of $R(Z)$

Element	Z	$R(Z)_{\text{theo}}$	$R(Z)_{\text{expt}}$
V	23	0.99	1.00 ± 0.04
Fe	26	0.98	1.00 ± 0.04
Ni	28	0.98	0.96 ± 0.04
Zn	30	0.98	0.93 ± 0.04
Sn	50	0.93	0.87 ± 0.04
W	74	0.84	0.78 ± 0.04
Pb	82	0.80	0.86 ± 0.04

potential should, in principle, be employed. Nevertheless, it is clear that the overall trend shows basic agreement between experiment and theory.

Let us not reach too hasty a conclusion, however. Actually, to be completely above board, it must be noted that the muon decay rate is influenced not only by relativistic time dilation but also by dynamical and statistical effects incurred through the binding. The two principal effects, which modify the decay rate in opposite ways, are termed the electron Coulomb field effect and the phase-space effect. In the first, the positively charged nucleus attracts the electron emitted in the muon decay and thereby produces a greater overlap of the muon and electron wave functions near the nucleus than would otherwise be the case for a freely propagating electron (described by a plane wave). The electron Coulomb field effect thereby enhances the muon decay rate. From the perspective of classical physics, it may seem strange that the end product (an electron) formed *after* the decay has occurred can have an antecedent effect on the rate of occurrence of that very process. However, the weak decay of a particle with concomitant production of new particles is not a classically explicable process. In the phase-space effect, the volume of phase space—in essence the number of quantum mechanical states—accessible to the decay products of a bound muon is restricted by the electrostatic binding force in comparison to that of a free muon. Hence, the phase-space effect reduces the decay rate.

Detailed calculations[19] of the decay rate that take account of all contributing processes suggest that at least in the light- and medium-mass elements there is a fortuitous cancellation between the Coulomb field and phase-space effects; the observed difference in lifetime between bound and free muons could then be principally attributed to the kinematic effects of special relativity. For the heavy elements as well, the accord between theory and measurement would be impaired were the effect of time dilation not included. It does seem that bound muons are in some kind of motion after all.

As for the electron—well, it is just a light muon, and my doubts have been dispelled. Who can even begin to imagine what strange scenes an electron encounters in its endless unfathomable trek around, near, and possibly through the nucleus? Trajectory and velocity has it none, and yet it moves.

5.4. The Planetary Atom

It is difficult to imagine anyone brought up in the 20th century not being familiar with the ubiquitous atomic energy symbol, the small central sunlike disk surrounded by three noncoplanar electron orbits. The symbol, which calls to mind nuclear power plants and nuclear submarines and has decorated a vast number of technical and nontechnical documents of the former U.S. Atomic Energy Commision, has come to represent the atomic (or, more accurately, nuclear) era. Physicists have long known that such a graphic depiction is a gross idealization of the structure of the atom. However, although the exact movement of an electron in an atomic stationary state cannot be pictured, recent developments in the investigation of highly excited atoms nevertheless suggest an enticing possibility: the creation of a bound electron wave packet following (in a probabilistic sense) a classical Keplerian orbit about the nucleus.[20] The properties of such a state differ greatly from those of a completely delocalized stationary state for which only the mean orbital radius may coincide with the radius of the corresponding Keplerian (i.e., Bohr) orbit.

Rydberg levels, it should be recalled, are closely separated with the energy interval between any two adjacent electronic manifolds varying inversely as the third power of the principal quantum number. Under appropriate circumstances, it is possible to excite an electron (e.g., with a broad-band laser) into a linear superposition of Rydberg states comprising a wide distribution of principal quantum numbers. The resulting wave packet is then localized with respect to its radial coordinate.

For a superposition of Rydberg states with *low* angular momentum, for which the corresponding classical elliptical trajectory is highly eccentric, the electron probability distribution is delocalized with respect to the angular coordinates. When close to the inner turning point of its orbit, where acceleration is greatest, the electron can emit light most strongly. Far from the nucleus, the electron behaves more like a free particle and cannot emit radiation. One can, therefore, study the motion of the delocalized electron wave packet by monitoring the time dependence of the emitted light. The periodic return of the electron to the ion core—until the wave packet ultimately decays—would be signaled by bursts of light emission at time intervals given by the classical orbital period.

There is much interest in producing an electron wave packet with a broad linear superposition of Rydberg states of *high* values of angular momentum. Such a wave packet would be localized with respect to both radial and angular coordinates and should orbit the nucleus very much like a classically bound particle. The classical trajectory would be nearly circular, however, and at first glance, it would seem that the rate of light emission should no longer depend on the location of the electron in its orbit. How would the periodic motion be discerned experimentally?

Although the electron in a hydrogen atom experiences a purely $1/r$ Coulomb potential, the electrostatic potential field through which an alkali atom Rydberg electron moves includes a small $1/r^4$ contribution resulting from core polarization and relativistic effects. This deviation from the Coulomb potential causes the wave packet to *precess*; that is, after repeated revolutions of the electron, the orientation of the packet relative to some fixed direction in the laboratory slowly changes. This effect is quite similar to the precession of the orbit of Mercury about the Sun as a result of an analogous deviation from the $1/r$ potential of Newtonian gravity predicted by Einstein's theory of general relativity.

An electron wave packet comprising many states of large principal and orbital quantum numbers, but with only the highest magnetic quantum numbers contributing, will have a well-defined alignment relative to some arbitrary, but fixed, axis. Suppose the Rydberg atoms were subjected to a pulsed electric field that can ionize the highly excited electron (i.e., provide enough energy to separate it entirely from the atom). The rate of ionization would then depend on the orientation of the wave packet relative to the electric field and be greatest when the packet is aligned along the field. In principle, therefore, the precession of the wave packet could be inferred from the time dependence of the ionization signal. Because of the very long timescale of the precession (order of milliseconds), the effect, as far as I am aware, has not yet been observed. Its accomplishment, however, is literally only a matter of time.

Planetary Atom Update

In the period following publication of *And Yet It Moves*, techniques for "sculpting" electron wave packets and producing atoms with Keplerian orbits have become well developed with numerous applications to explore. In contrast to an energy eigenstate in which the electron density is fully delocalized about the nucleus, a wave packet that evolves in time in a manner analogous to the classical motion of an electrically or gravitationally bound particle must be localized in both radial and angular coordinates. One way that was found to achieve this is by a two-step process employing a pulsed electric field of extremely short duration.[21]

In the first step, using laser excitation or other excitation methods from the atomic physicist's toolbox of interactions, one generates a Rydberg atom in a so-called circular state of principal quantum number n. A circular state is a single-electron state of highest angular momentum allowed within a given electronic manifold, i.e., a state with angular momentum quantum numbers $L = |m_L| = n - 1$ and a corresponding electron density distributed uniformly around a circle of radius $n^2 a_0$. The second step is to subject this state to a picosecond (i.e., 10^{-12} s) electric field pulse that drives transitions to circular states of neighboring electronic manifolds. The exact shape of the pulse is not important; the feature of primary interest is its duration—and in this regard the shorter the better. (The shorter the pulse, the more symmetric is the resulting population of circular states about the initial state.)

The effect of the electric field pulse is to create an electron probability distribution that in the course of time evolves in shape from something like a circular caldera to a form strongly peaked about a point on the Kepler orbit. Elliptical wave packets can be produced as well by subjecting the Rydberg atom to a weak static electric field that deforms the circular orbit into an ellipse. Figure 5.2 shows the evolution of a circular-orbit wave packet (for $n \sim 50$) at different times denoted as a multiple of the period T of the corresponding Kepler orbit. Thus, after about three Keplerian periods, the circular electron density evolves into a localized wave packet, which then makes its way around the orbit in another interval T.

To understand the way in which the pulsed electric field creates a localized electron wave packet, one must again recall that the electron probability distribution does not describe the dynamics of a single electron, but an ensemble of numerous similarly prepared electrons. A classical analogy would be a collection of electrons distributed uniformly around a circular orbit (i.e., the original circular Rydberg state) all moving with the same angular velocity. Upon mixing in circular Rydberg states of higher and lower principal quantum numbers, the pulsed electric field creates a spread in electron energy and momentum values. As time goes on, faster electrons will overtake slower ones, and at certain intervals of time, most of the electrons will be concentrated within a small segment of the circle. These are the intervals of time that correspond to the peaked probability distributions in Figure 5.2.

The creation of localized electron wave packets in an atom provides an experimental basis for a number of fascinating investigations of the

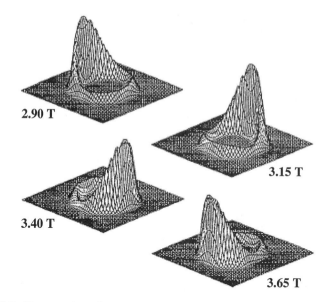

Figure 5.2. Progression of a wave packet comprising a distribution of circular Rydberg states (centered on principal quantum number $n = 50$) around the corresponding Kepler orbit with classical period T. Numbers below each figure signify the time (in units of T) following initial preparation of the circular state $n = 50$. [Adapted from Z. Gaeta, *et al.*, *Physical Reviews Letters* **73** (1994) 636.]

fundamentals of quantum mechanics. One of the most interesting I have encountered involves the creation of a single-electron wave packet comprising *two* initially disjointed subpackets effectively localized *on opposite sides* of the Keplerian orbit.[22] The two subpackets result from sequential excitation by two phase-coherent laser pulses. As they evolve in time, the subpackets progressively overlap, giving rise to a quantum interference effect manifested experimentally—when probed by a third laser that ionizes the atom—as a series of fringes analogous to the "Ramsey interference fringes" produced by irradiating an atom with two phase-coherent radiofrequency fields (as mentioned in Chapter 4).[23] In effect, one has created wave packets that move in quasiclassical fashion around a ring, yet give rise to a uniquely quantum interference effect.

There is, of course, no classical analog to this dual-packet electron configuration, for, after all, there is only *one* electron in the orbit.

Notes

1. Zeno proposed four famous paradoxes purportedly showing that motion cannot occur. In essence, to move from one point to another, an

object must first traverse one-half the total distance. Similarly, every half-interval can be decomposed into two further half-intervals that must be traversed. The object, according to Zeno, cannot cover an infinite number of spatial intervals in a finite amount of time, and hence cannot move.

2. Edna E. Kramer, *The Nature and Growth of Modern Mathematics*, Hawthorne, New York, 1970, p. 578.

3. Werner Heisenberg, *The Physical Principles of the Quantum Theory*, Dover, New York, 1930, p. 1.

4. C. T. R. Wilson, On a Method of Making Visible the Paths of Ionising Particles Through a Gas, *Proceedings of the Royal Society London* **A85** (1911) 285.

5. It is the *corresponding* components of the coordinate and linear momentum operators that do not commute (e.g., $xp_x - p_x x = i\hbar$). Unlike components commute (e.g., $xp_y - p_y x = 0$). Thus, quantum mechanics does not put a limit on the precision to which one can measure simultaneously the location of a particle along the x axis and the motion of the particle along the y axis.

6. P. Carrruthers and M. M. Nieto, Phase and Angle Variables in Quantum Mechanics, *Reviews of Modern Physics* **40** (1968) 411.

7. In quantum mechanics, the time derivative of an operator is calculated from the commutator of that operator with the Hamiltonian, H, which is ordinarily the sum of the operators representing the kinetic and potential energies. To obtain an explicit expression for the velocity operator defined in relation (5.3c) once the Hamiltonian is known, one must evaluate $v = -(i/\hbar)[xH - Hx]$.

8. As discussed previously in the context of the Aharonov–Bohm effect, the fundamental fields of quantum electrodynamics are the vector and scalar potentials and not the electric and magnetic fields. A function known as the Lagrangian is constructed from these basic fields. If the Lagrangian remains invariant when each field is multiplied by an arbitrary phase factor of the form $e^{i\phi}$, where ϕ is a real infinitesimal constant, the conservation of electric charge results.

9. J. G. King, Search for a Small Charge Carried by Molecules, *Physical Review Letters* **5** (1960) 562.

10. Imagine two stationary clocks A and B placed on the ground a distance d apart. A third clock C moving at a speed v parallel to the ground passes over A at which moment the A and C times are recorded. When C passes over B, the B and C times are likewise instantaneously recorded. Although the difference in A and B readings shows the passage of a time interval d/v, the C clock will have advanced by only $d/v\gamma$.

11. The neutrino, of which three varieties are known (electron, muon, and tau), was long thought to be massless. Recent research, motivated in part by an observed dearth in solar neutrinos reaching the Earth, provided convincing evidence that neutrinos could change from one variety to another periodically in time. Quantum physics can account for this only if at least one neutrino variety has mass. At the time that I am writing this, no one has yet measured the mass of a neutrino, but it is expected to be orders of magnitude smaller than the mass of the electron (or positron), which is

the least massive of the elementary particles for which a definite rest mass has been determined.

12. I have also heard a sequel to the story, which seems more plausible to me, that Rabi, having dinner at the time with a group of physicists in a Chinese restaurant, uttered the remark in apparent confusion over the arrival of an unexpected dish.

13. I recall a physics seminar many years ago with the title "Is the Moon a Heavy Electron?" The misprint on the flyer must surely have attracted people who came to learn of the marvelous new astronomical "discovery."

14. Yukawa's particle, termed the pi-meson or pion, was found in 1947. There are three varieties. The neutral pion, π^0, decays naturally with a lifetime of about 10^{-16} s; the charged pions, π^+ and π^-, have a natural lifetime of about 10^{-8} s.

15. E. Fermi and E. Teller, The Capture of Negative Mesotrons in Matter, *Physical Review* **72** (1947) 339. A "mesotron" is the old designation of "meson"—or medium-mass particle. At the time of the article, the distinction between muons and pions was not yet clear.

16. B. Rossi and D. H. Hall, Variation of the Rate of Decay of Mesotrons with Momentum, *Physical Review* **59** (1941) 223. We still have "mesotrons," rather than muons, because at the time of the experiment the pion was not yet discovered.

17. D. H. Frisch and J. H. Smith, Measurement of the Relativistic Time Dilation Using μ-Mesons, *American Journal of Physical* **31** (1963) 342.

18. I. Blair, H. Muirhead, T. Woodhead, and J. Woulds, The Effect of Atomic Binding on the Decay Rate of Negative Muons, *Proceedings of the Physical Society* **80** (1962) 938.

19. R. W. Huff, Decay Rate of Bound Muons, *Annals of Physics* **16** (1961) 288.

20. G. Alber and P. Zoller, Laser Excitation of Electronic Wave Packets in Rydberg Atoms, *Physics Reports* **199** (1991) 231.

21. Z. D. Gaeta, M. W. Noel, and C. R. Stroud, Jr., Excitation of the Classical-Limit State of an Atom, *Physical Review Letters* **73** (1994) 636.

22. M. W. Noel and C. R. Stroud, Jr., Young's Double-Slit Interferometry within an Atom, *Physical Review Letters* **75** (1995) 1252.

23. See M. P. Silverman, *Probing The Atom: Interactions of Coupled States, Fast Beams, and Loose Electrons* (Princeton University Press, Princeton, NJ, 2000) and *More Than One Mystery: Explorations in Quantum Interference* (Springer-Verlag, New York, 1995) for a comprehensive discussion of the interaction of an atom with two phase-coherent fields.

CHAPTER 6

Reflections on Light

6.1. Exorcising a Maxwell Demon

On the outside wall of the University, facing the Rue Vauquelin, is a
small plaque with the words:

> EN 1898 DANS UN LABORATOIRE
> DE CETTE ECOLE
> PIERRE ET MARIE CURIE
> ASSISTES DE GUSTAVE BEMONT
> ONT DECOUVERT LE RADIUM

The institution is the Ecole Supérieure de Physique et Chimie
Industrielles (ESPCI) in Paris. The historical laboratory in which the
Curies discovered radius—a drafty wooden shed dismantled long
ago—stood in the courtyard practically below the window of my office
in what is, today, the Laboratory of Physical Optics, a building of
vintage appearance itself. A small engraved stone marker at the edge
of a parking place is all that indicates the location of the Curies' shed.
Nevertheless, whenever I looked out my window, I could imagine
watching Marie and Pierre tirelessly processing the tons of pitchblende
ore that arrived periodically from the St. Joachimsthal mines in
Bohemia.

Except for singular occasions like the ESPCI centenary anniversary
in 1982, the thought of the Curies, if it exists at all, is probably a
distant one in the minds of the researchers and students scurrying
up steps and through corridors of the many buildings that border
the courtyard. However, history can be fascinating and instructive.
Appointed to the Joliot Chair of Physics at the ESPCI, I was curious
to know more about the distinguished son-in-law Frédéric Joliot, in
whose eponymous suite of rooms furnished by the University I was
living with my wife and young children. Did he and his wife, Irène
Curie, both of whom served as assistants to Marie, return home
covered with radioactive dust? Were Irène's cookbooks radioactive like

those of her mother? Was my family now breathing in these toxic exhalations of a past age of scientific glory? Happily, this was not the case. To my relief, Joliot was an "ancien élève" and not a professor at the ESPCI, and he and Irène never lived in the "Apartement Joliot." There can be solace in history—at least if one keeps the facts straight!

It was not the discovery of radium, nor for that matter anything connected with nuclear physics, that brought me to the ESPCI, but another event much less well known and celebrated. There, in the 1960s, my French colleague, Professor Jacques Badoz, and his students developed the photoelastic modulator (PEM), an ingenious and versatile optical device for examining the property of light known as polarization.[1] The years may have passed, but, in a tradition seemingly typical of the French "grandes écoles," members of the original group were still there, no longer students, of course, but researchers leading their own groups. A center of expertise for constructing and using PEMs, the ESPCI Laboratory of Physical Optics was as likely a place as any to exorcise what Jacques and I had come to regard as our "Maxwell demon."

The Scottish physicist James Clerk Maxwell is one of the giants of 19th-century classical physics.[2] His theory of electromagnetism unified under one set of laws the hitherto separate branches of physics constituting electricity, magnetism, and optics. Maxwell's great synthesis provided the foundation upon which contemporary quantum physicists continue to build in their efforts to unify all known physical interactions, a point to which I shall return in Chapter 9. A principal contributor as well to the science of thermal phenomena (thermodynamics and statistical mechanics), Maxwell, as I have already related in Chapter 1, once described how the Second Law of Thermodynamics might be circumvented by means of a molecular-sized sentient being—the Maxwell demon. In point of fact, such a violation of physical law is not possible—but all that is actually irrelevant to this essay. The issues discussed here have nothing to do with thermodynamics; they concern, instead, a devilishly difficult experiment with light and the implications of the experiment for Maxwell's electromagnetic theory.[3]

That light behaves like a wave had been demonstrated by means of diffraction and interference experiments long before the culmination of Maxwell's work in the 1860s. However, the early developers of wave theories of light did not know what was actually "waving." The pinnacle of Maxwell's achievement in this area was to deduce from his four basic laws of electricity and magnetism the equation of a wave whose calculable speed of propagation, whether in vacuum or in a material medium, was numerically equal to the corresponding speed of light. This speed was expressed in terms of properties of the medium—the electric permittivity (or dielectric constant) ε and the magnetic permeability μ—that can be determined by electric and magnetic experiments not in any way directly involving the properties of light. From Maxwell's theory (applied to an optically isotropic, homogeneous medium), it readily followed that the index of refraction n, which is the ratio of the speed of light in vacuum to the speed of light in the medium, should be equal to $\sqrt{\varepsilon\mu}$.

Light, then, was shown to be an *electromagnetic* wave: the propagation (even through materially empty space) of synchronously oscillating electric and magnetic fields. In the absence of matter or in the presence of an optically isotropic material, the oscillating electric and magnetic fields of a light wave are perpendicular to each other and to the direction in which the wave is propagating.

That there could still have been any doubt that Maxwell's theory provided a complete and correct description of classical electromagnetic phenomena came as a surprise to me when I first became aware of it in the early 1980s. Nevertheless, as my colleague Jacques remarked in referring to another surprising revelation, "Il y a donc encore des taches blanches sur les cartes du continent scientifique . . .". In the case at hand, however, these blank spaces were found not in some far-flung impenetrable valley of the scientific continent as one might have expected (e.g., the realm of quantum gravity), but in the developed urban area of physical optics. At issue were fundamental principles governing the reflection of light—a subject that one might well have thought was laid to rest over a century a half ago.

A good controversy in science is ordinarily a noisy affair, at least within the discipline affected, accompanied by academic teeth-gnashing and *ad hominem* aspersions. (The well-informed reader has but to recall the recent controversies over the extinction of the dinosaurs and the alleged discovery of "cold fusion.") However, that was not the situation here; the controversy passed quietly through the journal pages without creating any furor at all.

Yet, the issues involved were fundamentally of momentous import. Within the context of more-or-less routine investigations, there arose questions with extraordinary implications. Had one found an area of *classical* optics that fell outside the scope of Maxwell's theory? Did Maxwell's theory lead to violation of the law of energy conservation? Were the Maxwellian boundary conditions—the mathematical expressions describing the behavior of electromagnetic waves at an interface between different media—wrong? Quiet and unnoticed, the controversy effectively embraced the restructuring of classical electromagnetism.

What theoretical inadequacy or experimental observation could possibly have led to such far-reaching implications? The problem, in fact, took its origin in theoretical attempts to answer a deceptively simple question: How does light reflect from a left- or right-handed material? The statement of the problem may draw a skeptical raise of the eyebrow from a reader unfamiliar with chemical structure who recalls childhood jests about left-handed spanners. However, such chiral, or handed, materials exist and manifest intriguing phenomena collectively known as optical activity.

My interest in light reflection from chiral media arose as the natural extension of a series of experiments begun several years earlier addressing a completely different controversial issue centered on what outwardly would appear to be an even more provocative question: Can a *greater* amount of light be reflected from a surface than is incident upon it? Although the correctness of Maxwell's theory of electrodynamics was not formally at issue, the experiments nevertheless rigorously tested that theory in a domain far removed from common experience in which prior theoretical and experimental efforts gave rise to confusing and conflicting results. Although I did not articulate the solution as such at the time, in a way it was an exorcism of another Maxwell demon.

My studies of the interaction of light at the surface of different media did not lead to new or modified laws of electrodynamics. However, I had learned once again that what seemed to be well known was not necessarily well understood, and even so venerable a subject as classical optics still had its surprises.

6.2. Enhanced Reflection: How Light Gets Brighter When It Is up Against a Wall

A light beam incident upon a transparent material is partially transmitted through the surface and refracted (i.e., deviated from its original direction) and partially reflected from the surface. The exact division of light energy between these two processes was first worked out in the early 1820s—that is, long before the electromagnetic theory of light—by the French physicist and engineer, Augustin Fresnel, whose name, like that of Maxwell's, is associated with a variety of inventions, discoveries, and principles.[4] Along with the Englishman Thomas Young, Fresnel was a major contributor to the perception of light as a wavelike phenomenon.

Fresnel regarded light waves as a type of elastic wave like that of sound passing through air or of ripples spreading on the surface of water. Since all elastic waves require a medium, Fresnel assumed that the light propagated through an extremely tenuous hypothetical medium, the ether, that permeated all space and penetrated all objects. Not until many years later, after Einstein developed the theory of special relativity in 1905, were most physicists fully prepared to dispense with the concept of the ether. Nevertheless, Fresnel's elastic theory enabled him to predict or account for many aspects of the behavior of light. In the course of his investigations of light polarization, Fresnel deduced the amplitudes (relative to an incident light wave of unit amplitude) of light specularly reflected and refracted at the surface of a transparent medium. (Specular reflection is "mirrorlike" reflection from a smooth surface, in contrast to diffuse reflection from a rough surface.) These amplitudes are ordinarily designated the Fresnel relations or Fresnel coefficients. The relative intensity of the reflected or transmitted light is proportional to the square of the corresponding Fresnel coefficient and is termed the reflectance or transmittance, respectively. Ironically, although Fresnel pioneered the investigation of light polarization, it was precisely this attribute of light that his elastic theory was incapable of treating adequately.

Within the framework of electromagnetic theory, the polarization of light is defined by the orientation of the oscillating electric and magnetic fields. Light waves, for which the electric and magnetic fields oscillate in planes, are said to be linearly polarized; the direction of polarization is, by convention, the direction of oscillation of the electric field. Light waves, for which the electric and magnetic fields rotate about the direction of propagation, are said to be circularly (or, more generally, elliptically) polarized; the sense of the rotation, clockwise or counterclockwise to someone looking toward the light source, defines the type of circular polarization as right or left, respectively.

Light waves propagating through a vacuum (or any optically isotropic medium) are transverse waves: The electric field, magnetic field, and propagation direction are all mutually orthogonal. It was one of Fresnel's signal achievements to recognize the transverse nature of light, but it was this very point that constituted a serious flaw in his derivation of the Fresnel relations. These equations are correct, but they would *not* have been had Fresnel implemented consistently the boundary conditions that pertain to the passage of an elastic wave from one medium to another. Correctly applied, the elastic theory of light gives rise to a refracted wave that is not purely transverse, but has a longitudinal component (i.e., corresponds to an oscillation of the medium parallel to the direction of wave propagation). There is no experimental evidence for the existence of such a longitudinal wave.

It was Maxwell's theory of electromagnetism that provided the first self-consistent derivation of the Fresnel relations (without the extraneous longitudinal wave),[5] and the experimental verification of these relations correspondingly constitutes an important confirmation of the *particular dynamics* of Maxwell's theory. Certain kinematical aspects of reflection and refraction, such as the laws governing the angles at which light is reflected and refracted, were well known, at least since the 17th century, and are characteristic of any wave theory. The above emphasis on dynamics underscores the fact that the mathematical form of the Fresnel amplitudes is not characteristic of all wave theories, but depends sensitively on the specific laws governing the electric and magnetic fields.

Although reflection and refraction amplitudes can be derived for any state of light polarization, what is ordinarily designated *the* Fresnel relations, refers to two basic types of linear polarization, s and p, defined with respect to the plane of incidence, i.e., the plane formed by the incident light ray (the direction along which the incident wave propagates) and a line normal (perpendicular) to the reflecting surface.[6] This plane is always perpendicular to the reflecting surface (Figure 6.1). For s-polarized waves, the electric field is perpendicular to the plane of incidence and, therefore, parallel to the reflecting surface irrespective of the angle of incidence at which the light ray intersects the reflecting medium. For p-polarized waves, the electric field oscillates within the plane of incidence; the angle that it makes with the surface depends on the incident angle of the light, which is ordinarily measured with respect to the surface normal.

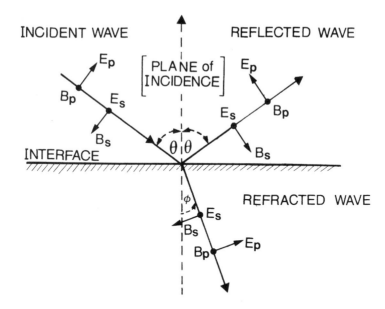

Figure 6.1. Geometry of waves specularly reflected and refracted at a planar interface (perpendicular to the page) between two dielectric media. The plane of incidence (lying in the page) is defined by the incident light ray and the normal to the interface. Waves are designated s-polarized or p-polarized, depending on whether the electric field is perpendicular (E_s) or parallel (E_p) to the plane of incidence. The angles of incidence and reflection (θ) are equal; the angle of refraction (ϕ) is given by Snel's law.

The exact variation of the Fresnel coefficients with angle of incidence depends on the relative index of refraction of the two media forming the interface at which the light is reflected and refracted. Nevertheless, for most familiar dielectric (nonconducting) materials, the following general behavior is observed. The intensity of reflected s-polarized light at normal incidence is low and increases nonuniformly, but continuously, with the angle of incidence until it reaches 100% at exact grazing incidence (i.e., for a light ray skimming the surface at 90° to the normal direction). The intensity of reflected p-polarized light equals that of s-polarized light at normal incidence, but with increasing angle of incidence, it drops to exactly 0% at a special angle called the Brewster angle,[7] after which, as in the case of s-polarization, the reflectance continuously grows to 100% at grazing incidence (Figure 6.2). An incoherent mixture in equal measure of s- and p-polarized beams generates a beam of unpolarized light.

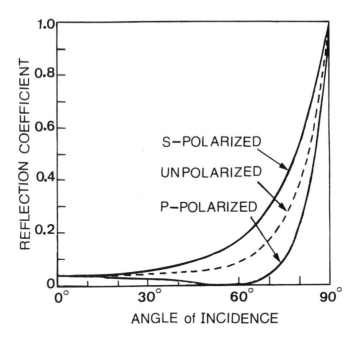

Figure 6.2. Variation in reflectance as a function of the angle of incidence for light reflected from a medium with refractive index higher than that of the medium within which the light originates. The reflected intensity of s-polarized waves increases smoothly between normal and grazing incidence, whereas the intensity of p-polarized waves drops to zero at the Brewster angle. If the refractive index of the reflecting medium is lower than that of the incident medium, then the reflectance reaches 100% at the critical angle θ_c and remains unchanged over the range from θ_c to 90°.

It is very likely that nearly everyone—whether familiar with the Fresnel relations or not—has experienced in one way or another the optical phenomena embraced by these expressions. The dim image one sees while standing before a clear window pane at night readily corroborates that only a small fraction of normally incident light is reflected.[8] By contrast, the bright glare on a glossy magazine held at an oblique angle to a light source irritatingly confirms that a large fraction of light near grazing incidence is reflected.

Polarized light of a well-defined frequency and propagation direction is one of nature's "absolutes"; it cannot be reduced to anything simpler. The light that we ordinarily encounter, such as direct sunlight or light from incandescent and fluorescent bulbs, is largely unpolarized; the net electric field of the constituent waves is distributed randomly in time over all possible orientations perpendicular to the propagation direction. It is of interest to note, then, that specular reflection of *un*polarized light at the Brewster angle results in a 100% s-polarized reflected beam! At first glance, this phenomenon may seem extraordinary—the generation of complete order out of disorder in apparent violation of the Second Law of Thermodynamics. There is no violation, however, for the light and reflecting surface have different temperatures and constitute an open system not in thermodynamic equilibrium. Nevertheless, reflection at the polarizing angle can lead to some remarkable—and easily observable—optical effects.

I can recall one example in my own kitchen where a crumpled cellophane bread wrapper lying on a smooth table and illuminated from behind by an open window gave rise to an impressive array of colors on the table surface. This phenomenon is an example of the "interference colors" produced by a birefringent material, i.e., an optically anisotropic material for which the speed of light—and correspondingly the index of refraction—depends on the direction of light propagation. In the usual classroom demonstration of this effect, a birefringent material is placed between two linear polarizers (sheets of Polaroid plastic, for example). The first polarizer constrains the electric field of the transmitted light wave to a well-defined plane. Upon entering the birefringent material, the light wave is split into two components for which the electric field is either parallel to or perpendicular to a special symmetry axis of the material designated the optic axis. The two components travel through the material at different speeds and thereby incur a relative phase shift by the time they emerge and recombine at the far end of the material. Because the two components are orthogonally polarized—that is, their electric fields are mutually perpendicular—they cannot interfere and manifest any effect of the relative phase shift.[9] The second polariser, however, projects the electric fields of both components onto a common transmission axis whereupon the two waves linearly superpose and interfere. Depending on the phase shift, the two components may reinforce one another and appear brighter, or interfere destructively and be eliminated. Since the phase shift varies with the thickness of the material and the wavelength (color) of the light, the composite effect on an incident white-light beam of the constructive and destructive interference of waves passing through an unevenly thick birefringent material is to produce a bright patchwork quilt of colors when the material is viewed through the second polarizer.

In the "kitchen experiment," the crumpled cellophane constituted a birefringent material of randomly varying thickness, but where were the two polarizers, for none had been explicitly employed? The first polarizer was that of the atmosphere itself. The sunlight streaming through the open window had been polarized by incoherent molecular scattering (Rayleigh scattering) by the oxygen and nitrogen molecules in the air. This effect is greatest for skylight coming from directly overhead when the Sun itself is near the horizon; the light scattered at 90° to the incident light can, in principle (though, because of depolarizing mechanisms, rarely in practice), be polarized 100% with the electric field normal to the plane determined by the incident and scattered light rays. The second polarizer was the smooth table surface, itself, which reflected the light (passing through the cellophane) at the Brewster angle.

There is another angle—known as the critical angle—that plays an important role in light reflection when the reflecting material has an index of refraction *lower* than that of the medium within which the light originates. When light passes, for example, from glass (with an index of refraction $n_1 = 1.5$) to water (with an index of refraction $n_2 = 1.3$) the refracted rays in the water bend *away* from the axis normal to the glass–water interface in accordance with Snel's law of refraction

$$n_1 \sin \theta = n_2 \sin \phi, \tag{6.1}$$

where θ is the incident angle and ϕ is the refracted angle (both angles measured with respect to a line perpendicular to the surface). Starting with a light beam directed normally at the interface ($\theta = 0$), for which the transmitted light beam is undeviated ($\phi = 0$), and gradually increasing the angle of incidence, one eventually reaches the critical angle θ_c at which the refracted rays are parallel to the surface ($\phi = 90°$). It readily follows from relation (6.1) that the critical angle characterizing a particular interface is determined from

$$\sin \theta_c = \frac{n_2}{n_1} \quad \text{(where } n_1 > n_2\text{)}. \tag{6.2}$$

For incident angles greater than θ_c, relation (6.1) leads to no real transmission angle ϕ. Experimentally, no light propagates through the second medium; *all* incident light is reflected irrespective of polarization.

The effect of total reflection is readily demonstrated by a transparent container (preferably one with flat rather than rounded sides) filled with water. Looking up through the bottom of the container, an observer would have no difficulty seeing his finger held just above the surface of the water. When viewed upward through one of the sides, however, the surface is no longer transparent, but appears opaque and reflective like a mirror. The effect can be particularly pronounced to someone submerged in the transparent water of a swimming pool gazing obliquely upward toward the surface and seeing, not the sky or ceiling, but objects at the bottom of the pool.

A fundamental attribute expected of the Fresnel amplitudes is that they must be compatible with the conservation of energy. The formal expression of this for a transparent medium is that the reflectance and transmittance must sum to 100% at any angle of incidence and for any polarization. No light energy can be lost or created by reflection.

It may seem paradoxical, but it is nevertheless the case, that although all incident light energy goes into the reflected wave under conditions of total reflection, the incident wave still penetrates the surface of the reflecting medium. The transmitted wave is not a traveling wave, i.e., it does not propagate undiminished through the medium, but dies off exponentially with depth of penetration. This evanescent wave transports no net energy into the medium. Nevertheless, its effects can be dramatic.

*   *   *

If I were to single out one of the achievements of theoretical physics that has had the greatest impact on 20th-century science and technology, it would be Einstein's prediction in 1917 of *stimulated* light emission.[10] Before Einstein's work, physicists and chemists were familiar—even if they did not understand the mechanism—with the production of light by *spontaneous* emission, the process that gives rise to the spectral lines of excited atoms and the fluorescence and phosphorescence of molecules. Einstein reasoned, however, that in addition to spontaneous emission, there had to exist another light-creating process if a piece of matter were to reach thermodynamic equilibrium with its environment. Whereas spontaneous emission, like the name implies, ordinarily occurs without external provocation at a rate characteristic of the quantum states of the emitter,[11] stimulated light emission is driven by the presence of light.

Some thirty-five years passed before Einstein's prediction was verified in the operation of the ammonia maser, a device that produced microwave radiation by stimulated emission. The term "maser" is an acronym deriving from **m**icrowave **a**mplification by **s**timulated emission of radiation. Shortly afterward, in 1960, the stimulated emission of visible (red) light was first produced in the ruby laser (substitute "light" for "microwave"). Today, there is hardly a branch of science, technology, industry, or medicine that does not employ lasers in one capacity or another. Clearly, the need to amplify light is of great importance.

In the operation of a laser, energy is pumped—for example by optical, electrical, or chemical means—into the atoms or molecules of the lasing material. The material is said to be excited, for its elementary constituents have been driven into their excited quantum states. When the number of atoms (or molecules) in an excited state is greater than that of a lower-energy state to which quantum transitions are possible, the system is said to have a population inversion. Under appropriate conditions—in particular, if the wavelength falls within the emission spectrum of the atoms or molecules—a light wave propagating through the excited material can stimulate the release of this stored energy. Whereas the original mode of excitation (such as an electrical discharge through the lasing medium) may be, in a manner of speaking, disordered, the process of stimulate emission is a highly ordered one. A photon present in the wave stimulates an atom or molecule to emit a second photon with identical properties. In this way, upon multiple passages of the light *through* the material, stimulated emission can, in principle, turn a weak initial wave into an intense, collimated, monochromatic, polarized light beam.

I have emphasized the word "through" above to accentuate a most unusual feature of a mode of light amplification proposed by Russian researchers in 1972 that differs markedly from the laser.[12] The researchers purported to demonstrate theoretically that a light beam can be amplified by specular *reflection* from the surface of an excited substance under conditions of total reflection—that is, where *none* of the incident light propagates through the energetic medium!

The claim stimulated as much contention as light. For one thing, the mathematical analysis of this novel effect suffered a serious ambiguity. For another, experimental tests yielded degrees of amplification far different from those predicted. And third, the physical mechanism for how such an amplification could occur outside the amplifying medium was not entirely clear.

Let us consider first the theory. To someone not familiar with the application of mathematics to physics, it may seem surprising that a properly conducted analysis can lead to ambiguous results. The popular (and not undeserved) image of physics as a mathematically rigorous science would seem to imply that, given the equations of motion for some system, one could, in principle, always (although not necessarily easily) solve them—and if the equations are correct, then the solutions will accurately describe the system. Unfortunately, the situation is rarely that simple. The equations that govern a physical system—and which are ordinarily differential equations relating the temporal and spatial rate of change of dynamical quantities—usually give rise to more than one solution, perhaps to an infinite number, distinguished by the choice of the initial conditions (specifying the state of the system at some fixed time) or the boundary conditions (specifying the state of the system at some fixed place).

The problem with the Russian investigation of light reflection, based on the Fresnel relations for a uniformly excited medium, was that the pertinent equations gave rise to two fundamentally different solutions. To see how this came about, let us examine the standard Fresnel relation for the reflection of s-polarized light at the interface of two ordinarily *unexcited* dielectric media with respective indices of refraction n_1 and n_2 (where the light originates in the medium characterized by n_1):

$$r_s = \frac{n_1 \cos\theta - n_2 \cos\phi}{n_1 \cos\theta + n_2 \cos\phi}.$$ (6.3)

For any incident angle θ chosen by the experimenter, the angle ϕ at which the light is refracted in the second medium is governed by Snel's law [relation (6.1)]. In fact, it is not just the angle ϕ that appears in the amplitude r_s, but rather $n_2 \cos\phi$, and this may be written explicitly in terms of the incident angle as follows[13]:

$$n_2 \cos\phi = \sqrt{(n_2)^2 - (n_1 \sin\theta)^2}.$$ (6.4)

Because the index of refraction of a transparent medium is the ratio of the speed of light in vacuum to the speed in the medium, it must, consequently, be a positive real number. Moreover, if $n_2 > n_1$, as in the case of ordinary reflection (e.g., light originating in air and reflecting from glass), the left-hand side of Eq. (6.4) must also be a positive real number, and Eq. (6.4) leads to a unique angle of refraction ϕ for any θ within the allowed range of 0°–90°.

In the case where the refractive index of the second medium is *lower* than that of the first ($n_2 < n_1$), the right-hand side of Eq. (6.4) is the square root of a negative number, and therefore a pure imaginary number, for incident angles beyond the critical angle θ_c given by relation (6.2). There are two possible square roots differing by a sign, and for neither, can ϕ be interpreted as an angle of refraction. Indeed, as described in the previous section, the light is totally reflected. One of the roots, which is the appropriate one for this physical system, leads to the evanescent wave. The other root, however, gives rise to a wave that grows exponentially with penetration of the medium and, as the presently considered medium has no latent source of energy with which to augment the wave, this root must be discarded as unphysical.

If the second medium is *not* transparent—if, for example, it absorbs light—then the situation becomes somewhat more complicated, for now the refractive index itself is a complex number expressible in the form

$$\tilde{n} = n + i\kappa. \tag{6.5a}$$

The real part of n, the ordinary refractive index, characterizes the phase shift incurred by a wave as a result of its spatial displacement through the medium; the imaginary part κ, the absorption coefficient, characterizes the diminution in amplitude of the wave as a result of absorption. Thus, the amplitude of a plane wave of angular frequency ω that has propagated a distance d through an absorbing medium is proportional to

$$e^{i\tilde{n}\omega d/c} = e^{in\omega d/c}e^{-\kappa\omega d/c}. \tag{6.5b}$$

If the medium is *amplifying* rather than absorbing, then the amplitude of a traveling wave grows as it propagates. As inferred from relation (6.5b), an amplifying medium is one for which the absorption coefficient κ is negative. In any event, when the refractive index of the second medium is complex, then Eq. (6.4) leads to two complex square roots (one the negative of the other) and, therefore, to two different expressions for the Fresnel amplitude r_s. However, in this case, it is not so obvious which root to retain.

Upon substitution of the complex expression for $n_2\cos\phi$, which has both real and imaginary parts,

$$n_2\cos\phi = q' + iq'', \tag{6.6a}$$

the resulting reflectance can be written in the form

$$R_s = |r_s|^2 = \frac{(n_1\cos\theta - q')^2 + (q'')^2}{(n_1\cos\theta + q')^2 + (q'')^2}. \tag{6.6b}$$

One of the two roots of Eq. (6.4) generates a positive q' and a negative q'', in which case it is clear from Eq. (6.6b) that the Fresnel coefficient is less than unity for all angles of incidence, except at grazing incidence ($\theta = 90°$), where the light is totally reflected. There is no amplification. The second root, however, leads to a negative q' and positive q'', from which it correspondingly follows that, exclusive of 90°, the reflectance (6.6b) is greater than unity for all angles of incidence. Thus, if the second root is the correct one, light amplification is predicted at all angles, except for total reflection at grazing incidence. Is the light amplified or not? How is one to know which root to select?

In their analysis of enhanced reflection, the Russian authors made the following arbitrary selection: Choose the first root (no amplification) for angles of incidence below critical angle and the second root (amplification) for angles of incidence above critical angle. At the critical angle itself, there is a discontinuity in the Fresnel coefficient. A similar problem and resolution applies to the reflection of p-polarized light.

An examination of the actual wave forms [similar to that of relation (6.5b)] corresponding to the two roots leads to the following interpretation. The wave for which q' is positive and q'' is negative represents an *amplified* wave propagating *away* from the interface into the second (the excited) medium with a population inversion, as in the example of the laser described previously. The wave for which q' is negative and q'' is positive represents a *decaying* wave that originated infinitely deep within the excited medium and is propagating *toward* the interface. According to the common perception that light reflection occurs at the outside surface of the second medium, the existence of such a wave, especially under conditions of total reflection, is problematical. From where and how did this wave arise? I will return to this question later.

When determined by the foregoing arbitrary selection of roots, the Fresnel coefficients for a uniformly excited medium lead to a maximum intensity enhancement of about e^2, or less than 10. Does such an amplification of reflected light actually occur?

At about the same time the theory was published, a different group of Soviet scientists reported a most interesting experiment.[14] Light from a neodymium laser was directed at normal incidence through a glass prism in contact with a solution of organic dye (known as Rhodamine 6G). The composition of the solution was adjusted so that its index of refraction was a little less than that of the overlying glass in order that total reflection could occur at the glass–dye interface. Upon absorbing the light, the dye molecules became excited, and a population inversion was established. The excited dye molecules, undergoing spontaneous radiative transitions back to the ground level, isotropically emitted light (fluorescence). A portion of the fluorescence emerging from one end of the prism was directed by a mirror back onto the glass–dye interface at a narrow range of angles spanning the critical angle. These rays then reflected from the interface and were received at a distant photographic film. From measurement of the extent of exposure of the film (by a densitometer) as a function of the angle of incidence, the researchers concluded that the reflected fluorescence was enhanced by about a factor of 25. In a subsequent experiment, a maximum enhancement in excess of 1000 was reported!

By the time I first learned of the controversy over enhanced reflection, the waters surrounding it were rather muddied. Theoretical attempts to account for what the Soviet group may have actually observed were not satisfactory. Nor did there seem to be experiments by other researchers. Assuming the phenomenon of enhanced reflection actually existed, theorists did not, in the main, agree about some of its basic attributes. According to some, amplification could be produced at any angle of incidence; according to others, the phenomenon occurred only for incident angles beyond the critical angle.

As the controversy bore on conceptually subtle issues of both theoretical and practical importance, one of my graduate students and I decided to examine the problem systematically. We were not interested in trying to explain an experiment performed under conditions that we only vaguely understood, and that may not have corresponded at all to the theoretical analyses worked out by others. Rather, we set out to devise an experiment in which all conditions and parameters were clearly ascertainable and to compare our observations with a theoretical analysis directly applicable to the experiment.

Recognizing that an infinitely deep uniformly excited medium, such as initially treated, is an idealization that one does not encounter in an actual experiment, we examined the type of "gain profile"—i.e., the spatial distribution of excited molecules—one is likely to engender by pumping a dye solution with a laser. If the intensity of the laser pump is not too great, so that the effective ground-state lifetime (the reciprocal of the pumping rate) is much longer than the lifetime of the excited states, then the gain profile has the same spatial dependence as that of the pump beam in the dye solution. (Recall from the discussion of quantum beats that this is the condition required to avoid multiple cycles of absorption and stimulated emission during the passage of the pump light pulse.) It has long been known (and referred to as Beer's law) that the intensity of a weak light beam passing through an absorbing medium diminishes exponentially with depth of penetration z as follows:

$$I(z) = I(0)e^{-z/\delta}. \tag{6.7a}$$

The characteristic depth of penetration δ depends on the number of absorbing molecules and the effectiveness (or so-called absorption cross section) with which a molecule can absorb a photon from the pump beam. Where molecules have been excited, the dielectric constant ε (and, correspondingly, the refractive index) incurs a negative imaginary part which, in consequence of Eq. (6.7a), also follows an exponential decay law within the medium.

$$\varepsilon(z) = \varepsilon(0)(1 - i\gamma e^{-z/\delta}). \tag{6.7b}$$

Here, $\varepsilon(0)$ is the dielectric constant of the (transparent) medium in absence of excitation and γ, the gain parameter,[15] is a measure of the strength of the pump beam or, equivalently, the extent of population inversion.

The solutions of Maxwell's equations for the waves of s and p polarization that propagate through a nonmagnetic ($\mu = 1$) medium with dielectric constant (6.7b) were (with no pun intended) quite illuminating. For one thing, the results tended to vindicate the Russian theorists' arbitrary selection of roots in the case of a uniformly excited medium. Far from the interface—i.e., for a penetration large in comparison to the characteristic depth δ—the excited medium is effectively transparent rather than amplifying (since the amplitude of the pump beam is too low to effect much of an excitation). In this region, a wave that had entered the surface at an angle below θ_c behaved, as expected, essentially like a plane wave traveling undiminished away from the interface. For angles of incidence above the critical angle, however, the character of the wave changed; it became an evanescent wave attenuated exponentially with distance of penetration. So far, so good.

It was in the excited layer within a few penetration lengths δ of the interface that the physics became interesting. Examination of the exact solutions[16] (which are complicated mathematical expressions involving, or resembling, Bessel functions) showed the waves to be decomposable into a linear superposition of two basic components. One component, analogous to the amplitude derived from the first root of Eq. (6.4), represented an amplified wave that travels away from the interface into the excited medium; the other component—as the reader may have surmised—characterized a decaying wave traveling toward the interface, such as derived from the second root of Eq. (6.4).

In the present case, however, no question arises concerning which component to keep, nor is there any discontinuity in the reflectance at critical angle. For any light polarization, there is only one physically acceptable solution—namely the solution that remains finite at the interface; this solution can contain both components. The two components do not necessarily contribute to the total wave form equally, but depend on the angle of incidence. Nevertheless, their relative magnitudes are automatically provided by the theory. In order for enhanced reflection to occur, the component traveling *toward* the interface has to be present in the excited layer. This occurs for angles beyond a certain threshold angle which ordinarily lies close to the critical angle. In this way, therefore, the prediction of enhanced reflection from a uniformly excited medium can be considered justified.

Calculations of the enhancement expected for various experimental conditions believed achievable in the laboratory led to peak values of less than 10 over a very narrow range of angles—a few hundredths to a few tenths of a degree—in the vicinity of critical angle; outside of this range, the amplification fell off rapidly. Also, other things being equal, the enhancement increased as the critical angle approached grazing incidence. The reason for this will become clearer shortly. Was such amplification observed?

Indeed it was, although the stringent requirements posed by theory on the stability of the experiment made for no mean task. As in the Soviet experiment, we constructed a dye cell with an organic dye (Rhodamine B) as the medium to be excited (Figure 6.3). A thick fused quartz window with flat sides overlay the dye. Varying the composition of the dye solvents permitted a coarse adjustment of the critical angle to a value of about 88°; fine adjustment was subsequently achieved by careful control of the dye cell temperature. To create a population inversion, the dye was excited by light from a pulsed dye laser directed at normal incidence through the transparent quartz window, and the highly collimated, monochromatic beam furnished by a separate helium-neon laser served as a probe with which to measure the amplified reflection. After reflection from the quartz–dye interface, the probe beam was sent through a monochromater (an instrument employing a diffraction grating to remove extraneous fluorescence) and then into a photodetector.

The dye cell was mounted on a rotatable stage so that the reflectance could be measured over a range of incident angles. Because amplifi-

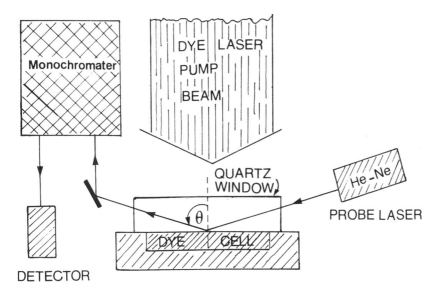

Figure 6.3. Schematic diagram of the enhanced reflection experiment. Light from a pulsed dye laser creates a population inversion in the dye solution. The probe beam from a helium–neon laser reflects at the interface between the dye solution and a quartz window (of slightly higher refractive index), passes through a monochromater which filters out fluorescent light from the dye, and is received at a photodetector. Amplification is expected for incident angles in the vicinity of the critical angle.

cation was predicted for only an extremely narrow range of incident angles close to critical angle, precise control and measurement of the angle of the probe beam were critical to the success of the experiment. This was achieved by employing a second helium–neon laser located at a screen about 5 m from the dye cell. Light from this calibration laser reflected from the quartz surface, leaving a sharp red spot on the screen from which the direction of normal incidence could be inferred. By removing the photodetector and determining the location and size (~1 mm) of the probe beam spot on the screen, one could measure the angle of incidence to within one-hundredth of a degree—a precision adequate for testing the theory of enhanced reflection.

The intensity of the reflected probe beam was then measured, first with the dye unexcited and then while the dye was being pumped, and the reflectance was determined as a function of incident angle for different choices of critical angle θ_c (in the vicinity of about 88°) and penetration length δ (ranging from about 40 to 100 wavelengths of the 633-nm helium–neon red light). With peak amplifications that reached 200–300%, depending on experimental conditions, we were greatly satisfied to find that the results agreed well with the theoretical expectations of our model (Figures 6.4a and 6.4b).

Although the amplification of light by reflection is now well established, the phenomenon may yet raise some puzzling questions in the minds of those who have not thought about the matter of light reflection before. How does amplification actually occur, if, in contrast to the laser, light does not propagate through the excited medium? As I pointed out earlier, total reflection does not imply that the incident light is in no contact whatever with the excited molecules. There is the evanescent wave whose depth of penetration, depending on the angle of incidence, can be substantial. Heuristically speaking, enhanced reflection may be attributable to stimulated emission by this evanescent wave.

Considering, again, the simplest case of reflection from a uniformly excited medium, I note that the component of the plane wave travel-

---

▶

Figure 6.4. (a) Reflectance as a function of incident angle for light reflection from the unexcited dye (of Figure 6.3) at a wavelength for which the dye is largely transparent. The critical angle is at 88.12°. As expected, there is no amplification. The solid line denotes the theoretical reflectance of a transparent medium. The broken line shows the small modification that occurs when the finite divergence of the probe beam is taken into account. (b) Example of the reflection curve obtained from a dye solution with population inversion. With a critical angle of 88.79° and characteristic penetration length of about 58 wavelengths (of 633-nm helium–neon laser light), the intensity of the reflected beam is amplified nearly twofold. The solid line shows the theoretically predicted reflectance.

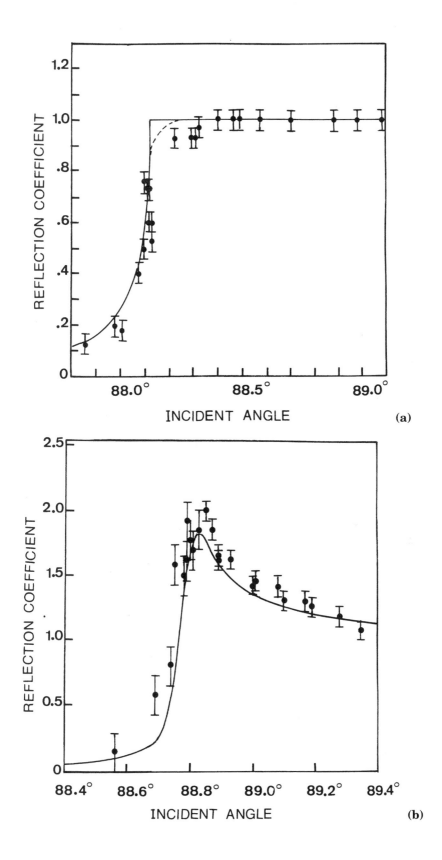

(a)

(b)

ing through the medium in a direction *normal* to the surface has the simple form

$$E(z) \sim e^{i(\omega/c)(n_2 \cos\phi)z}. \tag{6.8a}$$

Making use of Eqs. (6.2) and (6.4), one will see that for angles of incidence beyond the critical angle (where $n_2 \cos\phi$ is a complex number), the expression above reduces to that of an exponentially damped wave:

$$E(z) \sim e^{-z/\delta'}, \tag{6.8b}$$

where the penetration depth δ' for a wave of wavelength $\lambda = 2\pi c/\omega$ is

$$\delta' = \frac{\lambda/2\pi n_1}{\sqrt{\sin^2\theta - \sin^2\theta_c}}. \tag{6.8c}$$

As the angle of incidence θ approaches the critical angle θ_c, the penetration δ' becomes infinitely large. The degree of enhancement depends on the extent to which the incident wave—even though totally reflected—penetrates the medium. Hence, greater amplification is expected for incident angles close to the critical angle. Note, too, that the incident wave also has a component traveling *parallel* to the surface. The closer the critical angle is to grazing incidence, the more of the excited medium the parallel component traverses—just like sunlight passes through more of the Earth's atmosphere when the Sun is near the horizon than near the zenith.

Clearly, the above "explanation" is not complete, for one may well wonder why there is no enhancement for angles below the critical angle where a transmitted traveling wave could conceivably stimulate the medium to radiate. The enhancement occurs only when there is present inside the excited medium that wave component traveling toward the interface. From where does it come?

The equations describing the process of enhanced reflection from a laser-pumped medium are complicated, and it must be admitted frankly that not every feature predictable by the theory is correspondingly amenable to a simple interpretation. Nevertheless, there is a subtle and significant issue here that bears profoundly on the nature of light reflection in general, which, when once appreciated, may help clarify the existence in the reflecting medium of a seemingly puzzling wave form.

Why, for example, does the reflection of light at the outside surface of a material depend at all on the optical properties of the interior region? The answer, by no means trivial to demonstrate, is given by what is termed the Ewald–Oseen extinction theorem. The basic idea is that at an atomic level, the reflected and transmitted waves, although apparently generated at the boundary, actually originate from *within* the reflecting medium by the coherent radiation of molecular dipoles that have absorbed (extinguished) the penetrating incident light. The waves radiated by individual molecules superpose constructively in the directions for which reflected and transmitted waves are predicted to exist by macroscopic physical optics. For other directions, the waves superpose destructively. Thus, the reflected wave, although existing in medium 1, bears the imprint of the optical properties of medium 2. In a similar way, the refracted wave, whether traveling or evanescent, originates ultimately within the interior of the medium and not exclusively at the interface.

To my knowledge, this microscopic picture of light reflection has been implemented rigorously only in a few tractable cases such as reflection at the interface of transparent media. Even then, the analysis is not simple.[17] Were a corresponding microscopic treatment of enhanced reflection to be given, I believe it very likely that a satisfactory molecular explanation of the wave form within the gain region would emerge. In any event, it is worth emphasizing that the decomposition of a given wave form into various components is a mathematical stratagem that can usually be effected in different ways for different purposes. In reality, there is only *one* wave of specified frequency and polarization within the amplifying region and no problematic wave propagating toward the surface from the infinite depth of the material.

*   *   *

Having satisfied myself that the amplification of *linearly* polarized light by reflection from an excited medium was possible and followed in a self-consistent way from the laws of classical electrodynamics, I wondered next whether one could selectively amplify *circularly* polarized light by the same method. There are certain materials that interact asymmetrically with left and right circularly polarized light when unexcited, and I expected that they would do the same when pumped to higher quantum states. I also expected that the theoretical analysis of this process would be easily accomplished—at least for the special case of a uniformly excited medium. All I had to do was start with the appropriate Fresnel relations for a transparent unexcited material and then, following the approach taken by the Russian theorists, replace the real refractive indices with complex ones with negative imaginary parts.

I was wrong. Six years would pass before I could return to the problem of enhanced reflection from a medium that reflects left- and right-handed light differently. I was to discover first that the simplest case I imagined—a problem that ought to have been solved during the 19th century—seemed to have no physically acceptable solution!

6.3. Left- and Right-Handed Reflection

That light waves could exist in right- and left-handed forms was a bold hypothesis initially proposed and experimentally confirmed around 1825 by Fresnel as part of his investigation of a curious phenomenon today referred to as optical rotation. Many naturally occurring substances, corn syrup for example, have the capacity to rotate the plane of vibration of a transmitted linearly polarized light beam. Some fourteen years earlier, the French physicist, Dominique F. J. Arago,[18] first observed this effect—the nature of which he did not understand—in the passage of linearly polarized sunlight along a particular axis (called the optic axis) of a quartz crystal. The sunlight had been polarized by reflection from glass at the Brewster angle. By viewing the light through a plate of Iceland spar (calcite), Arago saw two solar images in complementary colors. He had seen such colors before when the light passed through plates of mica or gypsum instead of quartz, but in those cases, the colors of the images changed when the plate was rotated in its plane. The orientation of the quartz plate about the optic axis, however, had no effect on the images.

It was another French "optician," Jean Baptiste Biot, who recognized shortly afterward that the effect resulted from the rotation of the linear polarization of the light in quartz. Biot carried out extensive investigations of the phenomenon discovered by Arago—his first written memoir in 1812 read before the Institut de France covered some 400 pages!—showing that optical rotation occurred not only in crystals but also in liquids such as turpentine and oils of laurel and lemon, and in their vapors. On occasion, the demonstrations were accompanied by spectacular optical effects of an unanticipated nature, as when Biot set up his gas-phase optical rotation experiment in an ancient church then serving as the orangery for the house of peers. Turpentine vapor, issuing from a boiler, was conducted into a 30-m long iron tube with glass ends. Just when the effect of optical rotation was beginning to be observable, the boiler exploded, setting fire to the church! Unfortunately, Biot was not able to measure the extent of the optical rotation (although I have no doubt that city officials readily quantified the extent of the damage).

In general, the degree to which the plane of polarization is rotated is proportional to the quantity of substance through which the light travels and inversely proportional to the wavelength of the light. For most natural products, the rotation is usually modest, perhaps a few degrees or tens of degrees per millimeter of substance, although there are materials for which the rotation can be much larger.[19] (In a special class of materials known as cholesteric liquid crystals, the rotations can be enormous, on the order of 100,000° per millimeter.) What is the explanation of optical rotation?

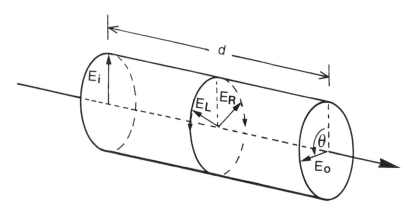

Figure 6.5. Optical rotation of a plane-polarized wave in an optically active medium. The electric field ($\boldsymbol{E}_i$) of the incident linearly polarized wave is a superposition of left ($\boldsymbol{E}_L$) and right ($\boldsymbol{E}_R$) circularly polarized components which advance through the medium at a rate depending on the respective refractive indices n_L and n_R. Upon leaving the medium, the superposition of the phase-shifted fields $\boldsymbol{E}_L$ and $\boldsymbol{E}_R$ results in a linearly polarised field $\boldsymbol{E}_0$ rotated by angle θ with respect to $\boldsymbol{E}_i$. The degree of rotation is proportional to the circular bire-fringence ($n_L - n_R$) and the sample length (d).

The essence of Fresnel's interpretation, still valid today, is that, upon entering the material, linearly polarized light is decomposed into two coherent beams of opposite *circular* polarization. The right- and left-handed waves propagate with different speeds and incur a relative phase shift. Emerging from the far side of the material, the two circular polarizations—no longer rotating in synchrony—superpose again to yield a linearly polarized beam with a rotated plane of polarization (Figure 6.5). For a light beam of wavelength λ passing through a substance of thickness d, the degree of rotation can be expressed as

$$\theta = \frac{\pi(n_L - n_R)d}{\lambda}, \tag{6.9}$$

where n_L and n_R are the different indices of refraction for left and right circular polarizations.

One of the most striking demonstrations of optical activity that I know—and which would no doubt have greatly pleased Fresnel—consists of passing the red beam of a helium–neon laser through a long vertical transparent glass tube of corn syrup, the bottom of which (where the light enters) is covered by a rotatable linear polarizer. Without the linear polarizer, the tube of syrup is more or less uniformly reddish along its length when regarded from the side. With the polarizer, however, the light appears distributed around the axis of the tube,

much like the red spiral of a barber pole. Indeed, by rotating the polarizer in one sense or the other, one can make stripes of light wind
upward or downward around the tube axis.

This remarkable effect is a combination of both optical rotation and
molecular light scattering. At the plane of entry through the linear
polarizer, the electric field of the light is oscillating along a well-defined
axis perpendicular to the light beam. As the beam propagates upward
through the syrup, this axis is progressively rotated. From a classical
perspective, the electric field of the light wave, which is oscillating at
some 10^{14} cycles per second, causes electrons of the sugar molecules in
the syrup to vibrate about their equilibrium positions and, therefore,
to radiate electromagnetic waves at the same frequency as that of the
incident light. By this process of absorption and reradiation, the sugar
molecules redirect or scatter a part of the incident light. The theory of
molecular light scattering—the same theory (Rayleigh scattering) that
accounts for the polarization of skylight—predicts that the light scattering is greatest at right angles to the axis along which the electric
charges are oscillating. Thus, as the syrup continuously turns the
electric field of the advancing light wave, it turns, as well, the orientation of the induced electric dipoles in the medium and the direction in
which the incident light is maximally scattered. Hence, the observed
"barber pole" effect.

Optical rotation is one of a complex of phenomena, more generally
termed optical activity, the mechanisms of which entail an asymmetric
interaction with right- and left-handed light. For a substance to interact asymmetrically with the two forms of circularly polarized light, it
must, itself, be built of units that have a handedness or chirality. (A
chiral object like a glove is one that, in general, cannot be superposed
on its mirror image.) This can occur in two basically different ways.

A material may be optically active because its fundamental chemical unit, or molecule, is chiral. This is the case for the corn syrup,
which is composed of sugar molecules with a characteristic right- or
left-handedness. In most organic compounds, there is at least one
carbon atom whose chemical bonds are directed outward toward the
vertices of a tetrahedron (i.e., with an angle of about 109° between any
two bonds). If a molecule contains one or more "asymmetric" carbon
atoms—a carbon atom bonded to four different substituents—it will
not be superimposable on its mirror image. Optical activity deriving
from intrinsic molecular structure can occur in any state of matter:
solid, liquid, or vapor.

Even if the molecules themselves are not chiral, however, a substance may still manifest optical activity if—as in a crystal, for
example—the molecules are arranged in a well-defined chiral structure such as that of a helix. The optical activity of crystalline quartz
comes from the helical winding of achiral silicon dioxide molecules

about the optic axis. If melted or dissolved in a solvent, such a substance would lose its optical activity.[20] Fused quartz, therefore, is not optically active.

It is worth noting that free atoms—i.e., atoms not bound in molecules nor subjected to static electromagnetic fields—are spherically symmetric and should have no preferential handedness. Correspondingly, the laws of electrodynamics (both classical and quantum) strictly forbid individual atoms from exhibiting optical activity. It turns out that they do, anyway—but about this I will comment in the next chapter.

In addition to optical rotation, another common manifestation of optical activity is that of circular dichroism, in which incident linearly polarized light is converted to elliptically polarized light by propagation through a light-absorbing chiral substance. Elliptical polarization may be thought of as an unequal linear superposition of left and right circular polarizations. The phenomenon arises from the differential absorption—rather than phase shift—of the component left and right circularly polarized waves. An expression for the so-called ellipticity of the light, which, like optical rotation, is also expressible as an angle, would resemble Eq. (6.9) except that chirally asymmetric absorption coefficients (κ_L, κ_R) would replace the indices of refraction.

The two phenomena, optical rotation and circular dichroism, are, in fact, closely related. As illustrated in the previous section [relation (6.5a)], one can regard the index of refraction as a single complex number of which the real part characterizes phase shifts and the imaginary part characterizes light absorption. Which phenomenon may occur for a given substance depends on the frequency (or wavelength) of the light. If the frequency corresponds to a Bohr transition frequency of the chiral system, the light is absorbed (presuming the transition is allowed) and circular dichroism results; if the frequency lies outside the regions of the spectrum where light is absorbed, then the substance is transparent and optical rotation occurs instead.

Besides modifications of the polarization of *transmitted* light, there is yet another process by which a chiral material may manifest optical activity, namely light *reflection and refraction*. Although reflection and refraction may sound like two separate processes, they are basically dual manifestations of a particular example of light scattering at an interface. To derive the amplitudes of either the reflected or refracted waves, one must analyze both processes together. At this point, it is appropriate to return to Fresnel and to inquire just how he tested his hypothesis of the existence of circularly polarized light.

There is a subtlety to the nature of circular polarization not encountered with linear polarization. Polaroid plastic, routinely found in sunglasses and commonly available from scientific supply houses, did not exist in Fresnel's day. Instead, birefringent materials such as Iceland spar were used to create and analyze linearly polarized light. An incident unpolarized light beam passing through a calcite crystal is decomposed into two transmitted beams with orthogonal linear polarizations. These two emerging beams would produce spots of equal brightness on some distant screen. If the incident light is linearly polarized, however, the resulting spots would be of unequal intensity, depending on the angle that the electric field of the incident light wave makes with the optic axis of the crystal. For an incident beam polarized either parallel or perpendicular to the optical axis, only one beam of corresponding polarization would emerge from the crystal; that is, there would be only one spot on the screen.

Circularly polarized light passing through Iceland spar *also* gives rise to two spots of equal intensity, from which one might have erroneously inferred that the incident light was *un*polarized. Yet, there is a world of difference between circularly polarized light, for which the electric field at any point in the wave rotates uniformly in one sense or the other about the direction of propagation, and unpolarized light, in which case the orientation of the electric field fluctuates randomly and rapidly during the time the polarization is observed. How is one to tell the difference?

Fresnel rightly understood that circularly polarized light may be thought of as a superposition of two linearly polarized light waves oscillating out of phase with one another by 90°. Indeed, he was able to create circularly polarized light by directing a linearly polarized light beam into a specially shaped glass prism—today called the Fresnel rhomb—and causing it to undergo total reflection at two opposing glass–air interfaces before emerging. With the electric field of the entering light oriented at 45° to the plane of incidence, the amplitudes of the reflected s- and p-polarized components incurred a relative phase shift of 45° for each total internal reflection. To show that the beam subsequently emerging into the air was circularly polarized, rather than unpolarized, Fresnel passed it again through a similar rhomb. After two additional total internal reflections, the resulting light (now with a phase shift of 180° between s- and p-polarized components) again became linearly polarized. By contrast, unpolarized light entering a Fresnel rhomb leaves as unpolarized light irrespective of the number of internal reflections.

Fresnel verified his interpretation of optical rotation—and correspondingly confirmed the existence of circularly polarized light—by means of adroit application of reflection and refraction within another of his ingenious prisms. If the refractive indices of right- and left-handed light are different in an optically active medium, then the two circular polarizations ought to refract at different angles. Thus, an initially linearly polarized light beam obliquely penetrating the surface of an optically active material should split into *two* beams of opposite circular polarization. The expected effect, however, is extremely small. Yellow light (589 nm) from a sodium lamp, for example, propagating along the optic axis of a quartz crystal experiences a difference in refractive index, $|n_L - n_R|$, of about 7×10^{-5}.

With his customary inventiveness, Fresnel circumvented this difficulty by concatenating segments of left- and right-handed quartz prisms to make a composite prism. Linearly polarized light entered one end of the prism. At each interface between optically active quartz segments of opposite chirality, the deviation between refracting left and right circularly polarized light waves was enhanced. From the opposite end of the composite prism, there emerged, as Fresnel predicted, two circularly polarized light beams sufficiently separated so as to leave no doubt about the existence of this "new" form of light.

There is something both beautiful and ironic about Fresnel's researches on light reflection, polarization, and optical activity. I have often wondered whether Fresnel ever concerned himself with the problem of light reflection from an optically active medium. It would, after all, have been a natural thing for someone to do who derived the laws of reflection for ordinary (achiral) dielectrics and who employed the differential refraction of circularly polarized light to elucidate the nature of optical rotation. Nevertheless, I never found any mention of such an investigation in Fresnel's collected *Oeuvres*. In retrospect, this is perhaps not so surprising. Given the contemporary state of technology within which he had to work, any sought-for chiral effects could well have been impossibly weak to observe—even for Fresnel. (Indeed, with the photometric methods available through the 1820s, it was not possible to measure reflectance with sufficient precision to test the original Fresnel coefficients for achiral media.) What surprised me more, however, when I was just beginning to turn my attention to the study of optical activity, was to find that over a century and a half after Fresnel, this fundamental problem was apparently *still* not solved.

Although Fresnel's interpretation of optical rotation was fine as far as it went, it did not explain the origin of chiral refractive indices. How do left- and right-handed molecules or crystals specifically affect light differently? The answer to this question can be rigorously provided by a quantum mechanical description of the interaction of chiral systems with light.[21] Nevertheless, the following more easily visualizable classical model provides a heuristic explanation that embodies the seminal features of the quantum treatment. Imagine a linearly polarized light wave incident upon an arbitrarily oriented helical molecule in a large sample of identical molecules (i.e., all the helices have the same handedness). Responding to the oscillating electric field, electrons in the molecule suffer periodic displacements about their equilibrium locations and give rise to an oscillating electric dipole moment. In addition, the alternating flow of electric charge along the helix engenders an oscillating magnetic dipole moment.[22] Similarly, the oscillating magnetic field of the light wave produces (by Faraday's law of induction) a time-varying electric field at the molecule that also gives rise to induced electric and magnetic dipole moments.

The key feature to note is that the relative orientation of the induced electric and magnetic moments depends only on the *sense* of the helical winding and not on the orientation of the helical axis. For example, the moments may be parallel for a right-handed helix and antiparallel for a left-handed helix. Oscillating electric and magnetic dipoles are themselves sources of electromagnetic radiation. However, the electric and, correspondingly, the magnetic fields of the waves emitted in a given direction by these two sources are perpendicular to one another. Thus, the waves superpose to yield a resulting scattered wave with electric and magnetic fields that are rotated with respect to the corresponding fields of the incident wave. The *extent* of rotation of the polarization varies with the orientation of the helix, but the *direction* of the rotation depends only on the relative orientation of the induced dipole moments . . . and this is the same for all the helices in the sample. Consequently, the net forward-scattered wave, produced by the superposition of waves scattered from all the helices, is rotated clockwise or counterclockwise depending on the chirality of the helices.

Keeping in mind that the above description is only a model introduced for the purpose of helping make tangible what, in effect, is a lengthy mathematical analysis—and that it certainly does not account for all aspects of optical activity—one may still ask how the model accounts for "circular birefringence" (i.e., for different indices of refraction for the two forms of circularly polarized light). At no point in the discussion has a light wave been assumed to move at anything other than the vacuum speed of light. Why, then, are the refractive indices for right- and left-handed light different?

The *microscopic* description of light propagation through a transparent material refers mainly to relative phase: the relative phase of the induced dipoles, the relative phase of their radiated waves, the relative phase of the scattered and incident waves. It is only in the *macroscopic* or phenomenological description of optical activity, where light is presumed to interact with a continuum of matter, that the index of refraction is introduced as a means of accounting for the net phase retardation produced in a transmitted wave by all the molecules of the sample. Microscopically, there are only "atoms (or molecules) and the void"; light moves at speed c in the interstices of matter only to be absorbed and reradiated—and therefore apparently slowed—by molecular encounters. The net result is an effective lowering of the speed by an amount that depends, in the case of a chiral material, on the handedness of the molecules.

To recapitulate and generalize the salient points, the origin of optical activity derives from two distinct processes. The first, termed "spatial dispersion", is the variation in the phase of an incident light wave over the extent of a chiral molecule or molecular aggregate. This is to be contrasted with the interaction of a light wave with an atom the size of which (about 10^{-8} cm) is some three orders of magnitude smaller than the wavelength of visible light. To an incoming light wave, the atom is a mere point; the variation in phase of the wave over an atomic scatterer is usually negligible. The variation in phase of an incident light wave over a chiral molecule, however, reveals in the scattered wave the molecular handedness.

The second process is the interference that results from the super-position of scattered waves issuing from the electric and magnetic dipole moments induced in the medium by the incident wave. For elementary constituents that are not chiral, the incident light cannot induce electric and magnetic dipole moments simultaneously, and there would be no optical activity.

In the quantum mechanical treatment of optical activity, the role of symmetry is perhaps more direct and fundamental. The states of a quantum system with a center of symmetry are characterized by sharp values of a quantum attribute termed "parity." If, upon inversion of the coordinates $(x,y,z \rightarrow -x,-y,-z)$ of all particles in the system, the wave function is unchanged, the state is said to have "even" parity. If the wave function changes sign under coordinate inversion, the state is said to have "odd" parity. An oscillating electric dipole can induce transitions between two states—and thereby produce light—only if the two states have *opposite* parities. By contrast, an oscillating magnetic dipole can induce transitions only between two states that have the *same* parity. Electric and magnetic dipole transitions, therefore, cannot occur simultaneously between states of sharp parity, and systems characterized by such states do not manifest optical activity. The quantum states of chiral systems, however, are superpositions of states of even and odd parity; an incident wave induces both electric and magnetic dipole moments as depicted in the classical heuristic model previously described.

From the phenomenological perspective of physical optics, the conceptual problem of describing optical activity can be considered resolved when the so-called constitutive, or material, relations are known. Although it suffices to speak simply of the "electric" and "magnetic" fields of a light wave in vacuum, the situation is more complicated for light in a medium. The electrons of the molecules, induced to oscillate by the electric and magnetic fields of the incident wave, serve as sources of additional internal electromagnetic fields. In all, there are four types of electromagnetic fields, generally designated E (the electric field), B (the magnetic induction), D (the electric displacement), and H (the magnetic field), whose properties one must know in order to predict the response of a material to light. The constitutive relations connect the secondary fields D and H (arising from induced charges and currents within the medium) to the fundamental fields E and B.

The constitutive relations for the simplest optically active medium— one that is intrinsically *non*magnetic, isotropic, homogeneous, and transparent—were derived from quantum mechanics some sixty years ago.[23] I call these the symmetric set of relations because their form remains invariant under a special type of symmetry transformation that effectively interchanges E and H, and B and D. However, an alternative and simpler set of relations—which I designate the asymmetric set—is the set one would most likely find in optics books that treat the subject of optical activity.[24] The chief feature of the asymmetric set is that the optical activity of the medium is presumed to derive exclusively from its dielectric properties; there is no magnetic effect of the light wave on the medium and, as a result, the magnetic fields B and H are considered identical. Clearly, then, the above-mentioned symmetry transformation would not leave the theoretical expressions unchanged—hence the term "asymmetric."

Ironically, the two outwardly dissimilar sets of relations have both successfully accounted for optical rotation and circular dichroism in the transmission of light through an optically active medium. That is, when employed in Maxwell's equations, both sets predict the existence of circularly polarized waves with different refractive indices of the form

$$n_L = n_0(1+f), \tag{6.10a}$$

$$n_R = n_0(1-f), \tag{6.10b}$$

where n_0 is the mean refractive index and the parameter f is a measure of the intrinsic strength of the chiral interaction between the medium and light. In the absence of evidence to the contrary, the symmetric and asymmetric sets of constitutive relations have long been considered physically equivalent. Indeed, this equivalence has been asserted as a consequence of another fundamental symmetry of the laws of classical electromagnetism. It has been argued that the fields D and H for a particular medium (like the vector potential field A of the Aharonov–Bohm effect discussed in Chapter 3) are not unique—that it is always possible to transform a given pair to a new pair of fields D' and H' (that also satisfy Maxwell's equations) by redefining in a prescribed way the induced electric and magnetic dipoles. By means of one such family of transformations, the magnetic dipoles can be made to vanish altogether, in which case the optical properties of the medium derive only from the (redefined) electric dipoles. Such transformations convert the symmetric set of constitutive relations into the asymmetric set, apparently demonstrating that the different mathematical forms superficially mask a fundamental physical equivalence.

As already explained, I first became interested in light reflection from optically active media as a possible means of selectively amplifying circularly polarized light. However, other motivations rapidly developed as well. Ever since the discovery of natural optical activity, the principal experimental methods (optical rotation and circular dichroism) generally involve measurement of polarization changes incurred by transmitted light. I was interested in exploring what new things might be learned by an alternative experimental technique. For example, the study of optical activity by light reflection could confer significant advantages in the investigation of chiral thin films that would be too thin to have much of an effect on a transmitted beam or, conversely, in the investigation of opaque chiral samples through which a transmitted beam would be undetectably weak.

A more exotic potential application relates to the study of life itself. Biochemical processes carried out in the laboratory with substances of nonbiological origin ordinarily lead to equal mixtures of left- and right-handed molecular forms (called enantiomers) that display no residual optical activity. By contrast, the capacity to produce and consume optically active substances (sugars, amino acids, etc.) of a particular chirality is perhaps the most outstanding chemical feature of life on Earth as we know it. Whether such chiral asymmetry in living things arose by chance or evolved deterministically from the laws of physics is not known. The question has far-reaching implications, however. A universal origin of biological homochirality would suggest that nonterrestrial life (if there is any) should display the same chiral preferences. It is conceivable that, some day, the manifestation of optical activity in the light reflected from planetary or asteroidal surfaces may signify the existence of living things.

Unaware of potential subtleties in the description of optical activity, I calculated the theoretical Fresnel coefficients and the resulting reflectances for incident light of linear and circular polarizations. To my surprise, the analyses of light reflection based on the two "equivalent" sets of material relations for an optically active medium gave entirely different results! Indeed, for light striking the surface of the material perpendicularly, the symmetric set of relations led to no difference in reflection at all between left and right circularly polarized light—even though the reflecting medium has an intrinsic handedness. I viewed this result with considerable suspicion and found more satisfying at first the prediction of the asymmetric set that at normal incidence occurred the largest difference in reflection of circularly

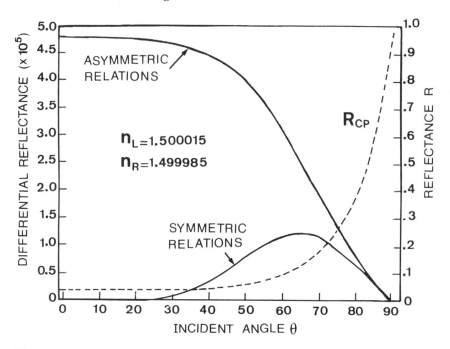

Figure 6.6. Theoretically predicted differential reflectance $D = (R_L - R_R)/(R_L + R_R)$ of left and right circularly polarized light from an isotropic optically active medium. The curve based on the symmetric constitutive relations shows a null D at normal incidence where the curve derived from the asymmetric constitutive relations is maximum. The individual reflection curves for left and right circular polarizations (dashed line designated R_{CP}) are not distinguishable on the scale shown at the right.

polarized light. This satisfaction was short-lived. To my still greater surprise, the reflectance and transmittance deduced from the asymmetric set of relations did not sum to unity, even for a transparent medium. These amplitudes violated the fundamental physical law of energy conservation!

Clearly, the two descriptions of optical activity were not equivalent (Figure 6.6). I wondered which set of Fresnel coefficients, if either, was correct. Was it possible that so basic a problem in the optics of chiral media could have gone unnoticed and untested since the development of classical electrodynamics a century and a half ago?

Untested it seemed to be, but not unnoticed. Several others had also been aware of theoretical inconsistencies in the amplitudes associated with the asymmetric set of material relations. However, as these relations were considered well established by previous studies of optical activity, the origin of the discrepancies was attributed to the structure of classical electrodynamics itself. Proposals were made to change the familiar Maxwellian expressions representing the flow and conservation of energy or the boundary conditions of electromagnetic fields at surfaces.

It is one of the intriguing features of physics, perhaps of other sciences as well, that what is construed to be understood best turns out often enough not to be well understood at all. Then, the importance of experimentation, too often forgotten or ignored as the elements of physics theory become increasingly abstract and remote from experience, must be reasserted. Maxwell, himself, expressed this sentiment eloquently in his 1871 Cambridge lecture celebrating the establishment of the Course of Experimental Physics and the erection of the Devonshire Laboratory at Cambridge University:

This habit of recognising principles amid the endless variety of their action can never degrade our sense of the sublimity of nature, or mar our enjoyment of her beauty. On the contrary, it tends to rescue our scientific ideas from that vague condition in which we too often leave them, buried among the other products of a lazy credulity. . . .[25]

And so I was at the ESPCI in Paris to test with my French colleague whether a widely accepted description of optical activity or whether Maxwell's own electromagnetic theory (in its proper domain of classical optics) was one of the "products of a lazy credulity." The objective of our experiment was to measure the difference in the intensities of left-handed and right-handed light beams reflected by a naturally optically active sample. The difference is ordinarily very small—on the order of the circular birefringence ($|n_L - n_R| \sim 10^{-5}$ or 10^{-6}) itself—and can be masked by a variety of instrumental artifacts. This experiment was to be a difficult one.

One might wonder why a reflection experiment should give rise to so weak an effect, when a transmission experiment can lead to a readily measurable optical rotation. The answer lies in part in Eq. (6.9). In addition to the circular birefringence, the expression also contains the ratio of the sample thickness to the wavelength of light. With an optical path length through the sample many times larger than the wavelength of light, the intrinsically weak circular birefringence can be effectively amplified. The difference in reflection of left and right circularly polarized light from the surface of a bulk (ideally infinitely thick) optically active medium is insensitive to the thickness of the medium and does not depend on the wavelength explicitly (although there is an implicit dependence through the index of refraction). Thus, it cannot be amplified in this way.

Nevertheless, there are ways of enhancing the difference in reflected circularly polarized light. One way is by "index matching" or adjusting the mean refractive index $[n_2 = \frac{1}{2}(n_L + n_R)]$ of the optically active medium to be close to that of the achiral medium (n_1) within which the light originates. As a rough approximation, the difference in reflectances is

$$R_L - R_R \propto \frac{|n_L - n_R|}{(n_2/n_1)^2 - 1} \tag{6.11}$$

for a transparent medium with $n_2 > n_1$. There comes a point of diminishing returns, however, for the closer the indices are matched, the greater is the transmitted light and the weaker is the reflected light—unless the material is absorbing. In the latter case, the refractive index is complex [Eq. (6.5a)], and matching the real part to the index of the achiral medium does not lead to a vanishingly small reflectance. A second way employs multiple reflection of light at the optically active surface. Under appropriate conditions—involving an absorbing medium—the difference in reflectances for circularly polarized light can be made to increase linearly with the number of reflections. Both of these methods were eventually to play a significant role in our experiments.

The instrumental heart of the reflection experiment is the photoelastic modulator (PEM), which makes it possible to determine the difference in reflectance of circularly polarized light nearly instantaneously in a single-step measurement. Because this difference is small compared with intensity fluctuations of the light source, it would be a hopeless endeavor to attempt to measure the reflectance of left and right circularly polarized light separately and then to subtract them. Stripped to its bare essentials, the PEM is a bar of fused (and therefore optically inactive) quartz made to oscillate at a frequency of about 50 kHz along its long axis (Figure 6.7). A light beam, before or after reflection from the surface of an optically active sample, traverses the quartz bar in a direction perpendicular to the axis of oscillation. As a result of the mechanical vibration, the refractive index for light polarized along the axis of the bar also oscillates at 50 kHz. Thus, upon passing through a PEM vibrating at frequency f, the light itself incurs an oscillatory phase shift of the form

$$\phi = \phi_0 \sin(2\pi f t) \tag{6.12a}$$

between the components of the beam polarized parallel to and perpendicular to the axis of the quartz element.

The effect that the modulator has on the light depends on the modulation amplitude ϕ_0 and on the polarization of the incident light. Suppose the incident light is linearly polarised at 45° to the vibration axis of the quartz. The parallel and perpendicular components of the wave (of optical frequency ν) can be expressed as

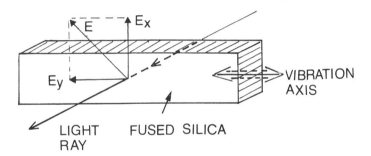

Figure 6.7. Schematic diagram of the PEM, a bar of fused silica made to vibrate at an ultrasonic frequency along its long dimension. Periodic compression and extension of the intrinsically isotropic bar make the bar birefringent with a refractive index oscillating at the same frequency along the axis of mechanical vibration. The electric field E of a light wave passing transversely through the PEM is a superposition of components (E_x, E_y) linearly polarized perpendicular and parallel to the vibration axis. The PEM produces an oscillating relative phase between these two components.

$$E_\parallel = \cos(2\pi v t - \phi), \qquad (6.12b)$$

$$E_\perp = \cos(2\pi v t). \qquad (6.12c)$$

With ϕ_0 set equal to $\pi/2$ radians, the emergent wave oscillates harmonically between left and right circular polarizations as the quartz bar is maximally extended and compressed at the mechanical frequency of 50 kHz. At intermediate positions of the end of the bar, the wave polarization is elliptical (i.e., a linear superposition of left and right circularly polarized components). Other settings of the modulation amplitude produce emergent waves that evolve in time between other states of polarization. For example, if ϕ_0 is set to π radians, then the transmitted wave oscillates harmonically between two orthogonal states of linear polarization. Linearly polarized light, as Fresnel first showed, is also resolvable into a superposition of circularly polarized components. In effect, the PEM makes it possible for the chiral reflector to sample incident left and right circularly polarized light of equal intensity at least once every period of mechanical oscillation [T = 1/(50 kHz) = 0.02 ms], a timescale short compared with that of light-source fluctuations.

Received by a photodetector, the modulated light gives rise to an electric current in the output containing a dc component (0 Hz) as well as components oscillating at the fundamental frequency (50 kHz) and higher harmonics (100 kHz, 150 kHz, etc.). The various components can be measured individually by means of an instrument known as a lock-in amplifier or synchronous detector. A quantitative measure of the difference in reflection of circularly polarized light can then be determined directly from appropriate ratios of these current components.

In principle, at least, that is how the experiment was supposed to work. Were matters actually so simple, the desired data could have been collected within a week. In reality, however, the experiment, first begun in the mid-1980s with my students, was pursued over a period of years as it became necessary to find and eliminate spurious signals that mimicked the effects of optical activity. One of the last and trickiest problems stemmed from the PEM itself. Although reliably employed since the mid-1960s to measure the circular dichroism of optically active materials with a sensitivity of 1 part in a hundred thousand, the PEM now appeared to give rise to a curious false signal two orders of magnitude larger than ever expected when used to test the chiral Fresnel coefficients. Nor was this a "local" phenomenon; the signal persisted for all PEMs tried, whether of commercial origin or home-made. How was it possible for scores of optical physicists using similar devices over a period of more than two decades to have missed so large a systematic error?

It was not possible; this artifact, as Jacques and I were to understand better later, did not show up in experiments employing only one polarizer. In our experiments, however, there were always at least two polarizers: the one that prepared the light beam for passage through the modulator, and the reflecting surface itself (the "hidden" polarizer of my kitchen experiment years earlier).

The solution to the problem, which was to have an intrinsic utility of its own, was traced to the existence of a weak static *linear* birefringence in the quartz bar. This is the type of optical anisotropy found, for example, in calcite, where linearly polarized waves pass through the crystal at different speeds depending on their direction of propagation and orientation. Ironically, it had long been known that stresses induced by the manufacturing process or by pressure from the edges upon which the bar rested generated a weak linear birefringence in the quartz. It was also assumed, however, that the axis of static birefringence lay parallel to the axis along which the quartz oscillated. In fact, we had expressly designed at the outset an experimental configuration for which this type of birefringence should not affect the desired signal.

The chiral reflection experiments showed that the axes of static and oscillating birefringence in the fused quartz were not parallel and that even minuscule angular deviations—which ordinarily would have been inconsequential in measurements of circular dichroism—yielded disturbingly large signals in other experimental configurations. How was one to circumvent a problem that seemed to be an unavoidable consequence of fabricating the most essential part of the experiment! After all, the PEM still remained the most suitable way (short of counting individual photons—a procedure that was not without its own difficulties) of observing weak chiral asymmetries in scattered light.

The reflection experiment was temporarily put aside in order to examine in minute detail, both theoretically and experimentally, the passage of light through an elastic modulator. The comprehensive theory that finally emerged from this detour happily suggested a number of ways to circumvent the stress-induced birefringence of the quartz, if not also to reduce it, at least in the small region through which the light passes. Everything was back on track.

By taking advantage of index matching and multiple reflection, my colleague and I were able to observe for the first time the difference with which a naturally optically active material reflects left- and right-handed light (Figure 6.8). It is a testament to the extraordinary experimental ability of Augustin Fresnel that the complementary phenomenon to the differential circular refraction of light has only relatively recently, after more than one hundred and seventy years, been achievable. As to the theoretical description of optical activity, our results indicated that neither Maxwell's electrodynamics nor quantum theory was likely to be embarrassed. Rather, a number of optics books

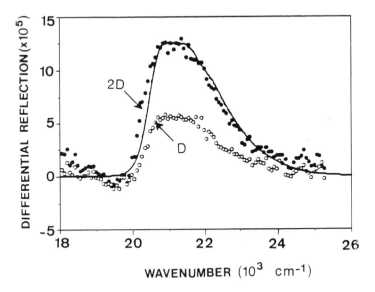

Figure 6.8. Experimentally observed differential reflection of circularly polarized light from an optically active liquid (camphorquinone in methanol) as a function of wave number (reciprocal wavelength, which is proportional to photon energy) at an incident angle of 67°. D denotes single reflection and 2D denotes double reflection from the chiral medium. The solid line shows the theoretically calculated curve.

could well stand some revising. In my own mind, I have no doubt that the symmetric description of optical activity is the correct one.[26]

But what about the argument purportedly demonstrating the equivalence of the two sets of chiral material relations? The argument fails. Although it is indeed true that the prescribed transformation connects the two phenomenological descriptions of optical activity while leaving Maxwell's field equations unchanged, the transformation does not leave the Maxwellian boundary conditions unchanged. If the symmetric set of relations is correct, then use of the asymmetric set together with standard electrodynamics constitutes a theoretically inconsistent calculation. No wonder that the resulting amplitudes violated physical laws.

The episode reminds me somewhat of the theoretical arguments against the Aharonov–Bohm effect (Chapter 3) in which case the vector potential of the confined magnetic field could allegedly, but not actually, be transformed away. In that case, the particular gauge transformation was not legitimate. In the present case, the symmetry transformation, while not disallowed, had been applied to only part of the mathematical framework needed to solve the problem of reflection.

If the amplitudes deriving from the symmetrical material relations are correct, how is one to understand why an intrinsically chiral material reflects left- and right-handed light *equally* at normal incidence? Why is the intrinsic chirality of the medium not manifest? A heuristic explanation of this puzzling behavior may be sought again in the microscopic model of reflection justified by the Ewald–Oseen extinction theorem discussed in the previous section.

Consider first the passage of light through a transparent optically active material, a medium with no distinguishing optical axes. The sense of optical rotation is determined exclusively with respect to the direction of propagation. Suppose a linearly polarized wave propagates 10 cm from right to left, during which the plane of polarization is rotated 45° toward the left-hand side of someone looking at the light source. Let the wave then be reflected and made to propagate 10 cm back again from left to right. The plane of polarization is rotated an additional 45° toward the left-hand side of an observer looking toward the light source—but in this case, the light source is the reflecting mirror and the second observer is facing the direction *opposite* that of the first observer. Thus, the plane of polarization is actually brought back exactly to its original orientation. In other words, the net optical rotation of a light beam that has made an even number of passages back and forth through a naturally optically active medium is zero.

With regard to the problem of chiral reflection, an incident light beam does not interact with the reflecting medium at the surface only, but may be thought of as propagating into the medium, being absorbed, and thereby inducing molecular dipoles to radiate secondary waves that superpose coherently to form the reflected wave. At normal incidence, this interaction is equivalent to a penetration of the wave into the medium followed by reflection as if from a mirror. Upon reflection and propagation back to the surface, the net chiral effect vanishes. At all other angles of incidence, except for grazing incidence (where the difference in reflection of circularly polarized light is again zero), the planes of optical rotation of the incident and reflected waves are no longer parallel and exact cancellation does not occur.

It is worth noting in conclusion that, with a self-consistent theory of chiral reflection at hand, it was possible to return to the question that sparked my study of optical activity in the first place: Can circularly polarized light be selectively amplified by reflection from an optically active medium with a population inversion? The answer is yes, and possibly one day this process may provide a new way to probe the chemical structure and physical interactions of excited molecules or prove useful in devices to amplify light. Hopefully, the first experiment will not turn out to be another Maxwell demon.

Update on Chiral Reflection

Since publication of *And Yet It Moves*, a number of subsequent theoretical studies have confirmed that the symmetric constitutive relations properly describe optical activity and that the asymmetric relations, without further amendment, lead to theoretical inconsistencies. Although the data shown in Figure 6.8 experimentally support the theoretical reflectance and transmission formulas I have derived, it is worth noting that no experiment to my knowledge has yet measured the reflectance from a naturally optically active medium (excluding liquid crystals) over the full range of incident angles.

*   *   *

The curious controversy over light reflection from left- and right-handed media has raised several profound theoretical issues about the way in which light interacts with matter and has produced a number of results of conceptual and practical interest to scientists and engineers concerned with the origin and measurement of small chiral asymmetries. One theoretical study in particular had emerged in the course of this research that addressed a most unusual interaction.

Years ago, when I was first introduced to the quantum mechanics of atoms and molecules, I often wondered whether there may be an unsuspected structure to atomic energy levels finer even that the finest structure described in the textbooks—finer than the structure due to the magnetic coupling of electrons and protons or to the interactions of electrons with the vacuum. My interests ranged widely since that time, but old questions often have a way of returning until they are answered. One day I found an answer, and it quite literally lay right under my feet.

Notes

1. M. Billardon and J. Badoz, Modulateur de Biréfringence, *Comptes Rendus de l'Académie des Sciences* **262** (1966) 1672.
2. I discuss Maxwell's development of electromagnetic theory and the insights it provides to current researches in electromagnetism and quantum theory in my book *Waves and Grains: Reflections on Light and Learning* (Princeton University Press, Princeton, NJ, 1998).
3. At the end of one particularly frustrating afternoon, when the light source was fluctuating, an amplifier failed, and background electrical noise was comparable to shouting at a soccer match, Jacques turned to me and asked rhetorically, "Quel démon nous fait faire cette expérience?" The answer came to both of us simultaneously: Maxwell's. (The vortex tube, an operational device that superficially behaves like Maxwell's demon, was discussed in Chapter 1.)
4. Fresnel's life and wide-ranging contributions to optics are discussed in *Waves and Grains: Reflections on Light and Learning* (Princeton University Press, Princeton, NJ, 1998).
5. This derivation was first given by Dutch physicist, H. A. Lorentz, in 1875.
6. The letters s and p derive from the German words for perpendicular ("senkrecht") and parallel. The two polarizations are also referred to as TE (transverse electric) and TM (transverse magnetic).
7. The Brewster or polarizing angle θ_B for light originating in a medium with refractive index n_1 and reflecting from the surface of a medium with index n_2 is determined from the relation: $\tan \theta_B = n_2/n_1$.
8. At normal incidence, the magnitude of the Fresnel amplitude for both s- and p-polarized light is

$$r = \frac{n-1}{n+1}$$

where n is the ratio of the refractive index of the second medium (e.g., glass) to that of the incidence medium (e.g., air). For glass in air, n is about 1.5, and the reflectance is $r^2 \sim 0.04$.
9. The linear superposition of two light waves, represented by their electric field vectors $\boldsymbol{E}_1$ and $\boldsymbol{E}_2$, produces a net wave of amplitude $\boldsymbol{E} = \boldsymbol{E}_1 + \boldsymbol{E}_2$. The interference term in the resulting intensity

$$I = |\boldsymbol{E}|^2 = |\boldsymbol{E}_1|^2 + |\boldsymbol{E}_2|^2 + 2\boldsymbol{E}_1 \cdot \boldsymbol{E}_2$$

vanishes as a consequence of the scalar product if the two fields are orthogonal.
10. A. Einstein, Zur Quanten Theorie der Strahlung, *Physikalische Zeitschrift* **18** (1917) 121.
11. From the perspective of quantum electrodynamics, spontaneous emission is a type of stimulated emission induced by fluctuating electromagnetic

fields of the vacuum. Confining an excited atom or molecule to a sufficiently small enclosure significantly modifies the fluctuations of the vacuum and, therefore, also the spontaneous decay rate of a quantum state; see D. Kleppner, Inhibited Spontaneous Emission, *Physical Review Letters* **47** (1981) 233.

12. G. N. Romanov and S. S. Shakhidzhanov, Amplification of Electromagnetic Field in Total Internal Reflection from a Region of Inverted Population, *Soviet Journal of Experimental and Theoretical Physics* **16** (1972) 209.

13. Substitute the expression $\sin\phi = (n_1/n_2)\sin\theta$, obtained from Snel's law [Eq. (6.1)] into the trigonometric identity $\cos^2\phi = 1 - \sin^2\phi$ and take the square root.

14. B. Ya. Kogan et al., Superluminescence and Generation of Stimulated Radiation under Internal-Reflection Conditions, *Soviet Journal of Experimental and Theoretical Physics Letters* **16** (1972) 100; S. A. Lebedev et al., Value of the Gain for Light Internally Reflected from a Medium with Inverted Population, *Optics and Spectroscopy* **35** (1973) 565.

15. Do not confuse the gain parameter with the γ of special relativity in Chapter 5 or the ratio of specific heat capacities in Chapter 2.

16. The theory and experimental demonstration of enhanced reflection are treated more comprehensively in *Waves and Grains: Reflections on Light and Learning* (Princeton University Press, Princeton, NJ, 1998).

17. See, for example, M. Born and E. Wolf, *Principles of Optics,* 4th ed., Pergamon, Oxford, 1970, pp. 104–108.

18. The scientific work of Arago, his influence on Fresnel, and the virtually life-long conflict between him and Biot are discussed in *Waves and Grains: Reflections on Light and Learning* (Princeton University Press, Princeton, NJ, 1998).

19. A comprehensive treatment of optical activity, including history, experimental techniques, and tables of data, is given in the classic work by T. M. Lowry [*Optical Rotary Power*, Longman, Green, and Co., London, 1935; reprinted by Dover, New York, 1964].

20. Liquid crystals, mentioned previously, would actually constitute a separate category, for the molecules are neither necessarily chiral nor do they form a rigid chiral structure. Rather, these optically anisotropic molecules are partially oriented within stacked horizontal layers with the direction of order turning progressively about the vertical axis from one layer to the next. The optical activity of liquid crystals falls outside the discussion of this chapter.

21. A quantum description of optical activity is given in my book *More Than One Mystery: Explorations in Quantum Interference* (Springer-Verlag, New York, 1995).

22. The separation of charges $+q$ and $-q$ by a distance d constitutes an electric dipole moment of magnitude qd directed from the negative charge to the positive charge. A current I around the periphery of a loop enclosing area A constitutes a magnetic dipole moment of magnitude IA/c (c is the speed of light) oriented perpendicular to the loop in a right-hand sense. (If the fingers of one's right hand encircle the loop in the direction of the current, then the extended thumb points in the direction of the magnetic moment.)

23. E. U. Condon, Theories of Optical Rotatory Power, *Rev. Mod. Phys.* **9** (1937) 432.

24. See, for example, G. R. Fowles, *Introduction to Modern Optics*, Holt, Rinehart, and Winston, New York, 1975, pp. 185–189.

25. L. Campbell and W. Garnett, *The Life of James Clerk Maxwell*, MacMillan, London, 1984, p. 269.

26. A technically more comprehensive account of the chiral reflection experiment and the theory describing the functioning of the photoelastic modulator is given in *Waves and Grains: Refections on Light and Learning* (Princeton University Press, Princeton, NJ, 1998).

CHAPTER 7

Two Worlds, Large and Small: Earth and Atom

As I stood in an unlighted horizontal passageway of the Cashmere Cavern and looked up at the narrow ventilation shaft receding to a small circular opening some twenty meters above me, I felt a shiver of fascination and amusement as I thought of my New Zealand colleagues being raised and lowered by a rope harness. Located under the Cashmere Hills just outside Christchurch on the South Island, the cavern was originally excavated shortly after the start of World War II to serve as a command post in the event of a Japanese invasion. With the passage of time, it had long since faded from public memory until rediscovered by accident. A timely rediscovery, too. With its solid bedrock floor and sheltered environment, the cavern is expected to provide an ideal workplace thirty meters below ground for the University of Canterbury Ring Laser Laboratory. A wide horizontal adit seventy meters in length now gives easy, if less dramatic, access for the construction crew and physicists.

During the summer (i.e., Southern Hemisphere winter) of 1990, while at the University of Canterbury to deliver a series of lectures, I observed the ongoing construction of the new laboratory and testing of the laser system with more interest than merely that of a curious visitor. When completed, this facility may quite possibly be able to confirm a remarkable optical effect that has long interested me: that unbound atoms unperturbed by static electric or magnetic fields on the rotating Earth interact inequivalently with left and right circularly polarized light.

Actually, it has been known for nearly forty years that atoms can be optically active as a result of the weak nuclear interactions.[1] These interactions are not invariant to reflection in a mirror and, therefore, can be expected to engender a left–right asymmetry in the quantum states of atomic electrons. That is, if they did not destroy the integrity of the atom, for the weak interactions are usually associated with particle disintegration processes as in the familiar example of beta decay, the natural transformation of a neutron into a proton, electron, and antineutrino.

In order for the weak interaction to break the chiral symmetry of bound-electron states, without at the same time altering the identity of the atom through some charge-changing process, there must be a way for electrons and nucleons (the constituents of the atomic nucleus) to interact by exchange of a massive neutral particle. Just such an interaction is provided by the so-called "electroweak" theory, a sweeping theoretical synthesis of electrodynamics and the weak interactions unmatched in scope since Maxwell unified all of electricity, magnetism, and optics. Within the framework of this theory, electrons and nucleons can exchange a Z^0 vector boson, a neutral particle with a mass approximately one hundred times the mass of a proton. The existence of such an exchange or "weak neutral current" was demonstrated in 1973 by high-energy experiments involving the scattering of neutrinos by nucleons.

The corresponding existence of atomic optical activity was confirmed in the early 1980s by low-energy experiments on the vapors of a variety of heavy atoms such as bismuth, lead, and thallium.[2] The effect is small; 1 m of dense vapor can rotate the plane of linear polarization of a transmitted light beam by about 10^{-7} radians.

The atomic optical activity that I predicted, however, has nothing whatever to do with the weak interactions. It arises, instead, from the rotation of the Earth and is many times weaker than any which has heretofore been measured. The very existence of this phenomenon, however, captures the imagination. For one thing, the weak interactions aside, the laws of electrodynamics exhibit perfect mirror symmetry from which it follows that optical activity in free atoms—spherically symmetric systems held together by an electrostatic force—is ordinarily strictly forbidden. Second, apart from the issue of atomic handedness, the predicted effect represents a potentially observable influence of planetary spin on the internal workings of an atom.

* * *

It is the essential dichotomy in the application of the laws of physics to systems large and small that makes the thought of such an atom–planet interaction so unusual. To predict the orbit of a comet about the Sun or of the Moon about the Earth, one relies on the laws of classical mechanics as embodied in Newton's equations of motion. Correspondingly, to determine the motion of an electron about a nucleus—in other words, to understand the structure of the atoms and molecules out of which the objects of the macroscopic world are built— one turns instead to the laws of quantum mechanics as embodied in entirely different equations, for example those of Schrödinger, Heisenberg, and Dirac. This recourse to separate and incommensurate theoretical frameworks for deducing the behavior of large-scale and ultra small-scale objects reflects in a profound way the decoupling of the objects themselves. The motion of a single atom hardly influences the daily affairs of a planet; likewise, the motion of a planet ordinarily has no perceptible effect on the internal dynamics of an atom. It is to the extreme smallness of Planck's constant ($h = 6.67 \times 10^{-34}$ J s) that one may effectively attribute this decoupling of the large and small. A fortunate circumstance, too. Were Planck's constant much larger so that the fall of an apple could be accounted for only in the counterintuitive terms of quantum physics, one might well wonder whether even an Isaac Newton could have made sense of the world.[3]

Nevertheless, the decoupling of the very large and the very small is not total. For one thing, as may be expected, even the elementary particles are influenced by the gravitational force of the Earth, for, after all, bulk matter is made up of protons, neutrons, and electrons. The manner in which they are affected, however, has given rise to some surprises. It has been known for centuries that a bulk object of mass m near the Earth's surface is attracted toward the center of the Earth by a force of magnitude mg, where g is the local gravitational acceleration ($g \sim 9.8$ m/s^2). Although individual neutrons fall freely in this expected way, experiments to probe the free fall of electrons[4] through an evacuated vertical cylindrical tube of copper showed that the force of attraction was less than 10% of mg. Are electrons exempt from the law of gravity? Actually, the result was no violation of Newton's law of gravity; on the contrary, it confirmed it in an unusual context.

A metal can be regarded in some ways as a rigid ionic lattice permeated by a mobile electron gas. At moderate temperatures and in the absence of external perturbations, attractive electrostatic forces prevent the electrons from escaping from the metal surface, but within the metal interior, loosely bound (valence) electrons are free to circulate. Before the electron free-fall experiment was performed, two theorists, Schiff and Barnhill, realized that the mobile electrons within a metal should also fall vertically in response to the pull of gravity.[5] The descent terminates, however, when the downward pull of gravity is balanced by the upward electrostatic attraction of the positive ions. The net downward displacement of the negatively charged electrons relative to the positively charged metal lattice creates an electric field directed downward *outside* the metal surface. This electric field exerts an upward force on electrons falling freely through the copper tube and thereby retards their acceleration. Indeed, if the gravitational acceleration of an electron within the metal is the same as that of a free-falling electron in empty space, then the magnitude of the electrostatic field outside the metal surface should be about mg/e (where e is the electron charge). This field would, in principle, counterbalance the pull of gravity on a free-falling electron, which, if dropped down a vertical metal tube, should then not fall at all! With account taken of nonidealities in the metal surface, this is effectively what was observed.

It is of interest to note that if a positron, the positively charged antiparticle of an electron, were introduced into the tube, the force of gravity and the Schiff–Barnhill force would now both be pulling downward. Thus, a positron would be expected to fall through the tube at *twice* the acceleration of gravity. I am not aware that this experiment has ever been done. However, had such an effect been observed in ignorance of the Schiff–Barnhill effect, the apparently unsymmetrical action of gravity on particles and antiparticles would doubtlessly have created much excitement within the physics community, for it would seem to violate Einstein's equivalence principle, one of the seminal principles upon which the present understanding of gravity is based. In fact, anomalies in the interaction of the Earth with normal matter (rather than with antimatter), reported some fifteen years ago, appeared to manifest just such a violation.

In effect, one version of the equivalence principle (of which there are various inequivalent versions) maintains that mass, alone of all the conceivable attributes of matter, determines the force of gravity that an object experiences. Actually, there are two conceptually different types of mass. One is inertial mass, appearing in the definition of linear momentum (mass × velocity) and, consequently, in Newton's second law of motion:

$$\text{Force} = \text{Inertial Mass} \times \text{Acceleration} = m_I\, a. \qquad (7.1a)$$

The other is the gravitational mass introduced in Newton's law of gravity, a special case of which is the familiar expression

$$\text{Force} = \text{Gravitational Mass} \times \text{Free-fall Acceleration} = m_G\, g \qquad (7.1b)$$

for the force of gravity near the Earth's surface. The so-called weak principle of equivalence affirms as an exact identity the experimentally observed numerical coincidence of the inertial and gravitational masses. It would then follow from relations (7.1a) and (7.1b) that if m_I and m_G are always equal, two lumps of matter should fall freely at the same acceleration in response to the pull of gravity, irrespective of differences in mass, isotopic composition, chemical structure, or physical state (e.g., solid or liquid). Although a particle and its antiparticle may differ with respect to such properties as the sign (but not magnitude) of electrical charge or the relative orientation of spin angular momentum and magnetic moment, they are believed to have exactly the same mass and, therefore, if the equivalence principle is valid, to behave identically in a gravitational field.

Galileo is alleged to have been the first to test the equivalence principle by dropping different objects from the Leaning Tower of Pisa, although it is questionable whether he really performed such an experiment. Instead, credit for the first precision tests is usually accorded to Baron Roland von Eötvös of Hungary, whose series of experiments begun in the late 1880s and completed by 1922 remained the state of the art until the early 1960s.[6] Eötvös constructed a torsion balance—a device in which two masses of different composition were suspended at opposite ends of a horizontal bar supported at the center by a thin fiber about which the bar could turn. Because of the Earth's rotation, each mass is subjected not only to the vertical pull of gravity but also to the centrifugal force, which accelerates the mass outwardly in a direction perpendicular to the axis of the rotation. Although sometimes designated a fictitious or pseudo force, the centrifugal force gives rise to real enough physical consequences as judged by a corotating observer. Anyone who has felt himself thrown outward as he drove a motor car around a turn in the road has experienced centrifugal force. The centrifugal force on an object constrained to rotate with the Earth is proportional to the inertial mass of the object, the perpendicular distance of the object from the rotation axis, and the square of the angular velocity of rotation.

Concerning the Eötvös balance, if the ratio of the inertial to gravitational mass were different for each of the two suspended masses— that is, if the centrifugal force acted on one object proportionately greater than did the force of gravity—there would be a torque (i.e., a twisting effect) on the rod causing it to rotate about, and therefore to twist, the fiber. The small angle through which a torsion fiber is twisted can be measured to high precision, for example by an optical technique whereby an incident light beam is reflected from a small mirror affixed to the fiber. The angle of reflection, which is twice the angle through which the fiber is twisted, may be small, but the linear deviation of the reflected light beam from the incident direction increases in proportion to the distance between the detector and the mirror and can be made measurably large. However, since one cannot turn off the rotation of the Earth, the equilibrium position of the balance arm provides no information about the relative influence of gravity and inertia; one cannot tell what the orientation would have been if gravity, alone, acted on the masses.

The key point to recognize is that in the event that the equivalence principle is violated, the equilibrium orientation would depend on which mass is located on which side of the balance. Suppose the torque on the fiber for a given configuration of the masses orients the balance at equilibrium along an east–west-directed line. Exchanging the two masses or, equivalently, rotating the entire apparatus (including the frame to which the fiber is mounted) by 180° would then cause the fiber to twist in the opposite sense, and the balance would no longer lie along the east–west line. If gravitational and inertial mass were truly identical, there would be *no* differential torque and the balance would maintain the same orientation irrespective of the side on which each mass was located. By searching for such a change in equilibrium orientation, Eötvös was able to establish the equivalence of gravitational and inertial mass for a number of dissimilar materials—readily recognizable ones like copper, water, and platinum, as well as puzzling oddities like "snakewood"—to a few parts in a billion. Subsequent tests by other researchers in which Eötvös's basic procedure was implemented in a torsion balance that responded to the gravitational force of the Sun and the centrifugal force of the Earth's motion around the Sun established the equivalence of inertial and gravitational mass to precisions some three orders of magnitude beyond what Eötvös obtained.

Ironically, a re-examination in 1986 of Eötvös's definitive paper of 1922 sparked a lively controversy when the examiners concluded that, contrary to the long-held interpretation, the data in the paper actually provided evidence for a composition dependence of the gravitational acceleration.[7] The origin of this effect (the establishment of which is far from certain) has been attributed to an attractive "fifth force," a new fundamental interaction complementing gravity, electromagnetism, and the strong and weak nuclear interactions, that depends not just on total mass but on certain properties of the "heavy" elementary particles (the baryons) of which a mass is composed. Protons and neutrons are the principal baryons composing ordinary matter. In the contemporary theory of elementary particles, each baryon is ascribed quantum numbers with whimsical names like baryon number, isospin, hypercharge, or strangeness that play an important role in the various interactions resulting in particle transformations. The origin and significance of these numbers are of no concern here, but it is relevant to note that the baryon number for protons and neutrons is +1, the number for the corresponding antiparticles (antiproton and antineutron) is −1, and the net baryon number of a sample of ordinary matter is just the sum of the protons and neutrons.

The examiners of Eötvös' paper claimed to have found that recorded differences in the accelerations of two masses—which, ideally, ought to have been zero, but which, of course like all experimental data, exhibited uncertainties due to the limitations of measurement—were not statistically random, but correlated with differences in the baryon number per unit mass of the sample. Although the baryon number is the same for all the baryons that make up ordinary matter, the baryon number per unit of mass is not necessarily the same for dissimilar materials because the packing of the baryons can be different. One finds that the baryon density is greater for elements around iron in the middle of the periodic table than for elements at either end. Thus, if the fifth force exists, the net interaction between the Earth and a certain mass may depend on whether that mass comprises more protons than neutrons or more neutrons than protons, or even whether it is made of antiprotons and antineutrons rather than of protons and neutrons.

That the postulated fifth force may have eluded physicists for so long can be explained in part by its intermediate range of action estimated to be a few tens to hundreds of meters—a distance scale enormous in comparison with that over which nuclear forces prevail (on the order of 10^{-13} cm) and negligible in comparison with the supposedly infinite range of gravity. Numerous studies of particle collisions in high-energy accelerators have yielded information about the interactions within and between nuclei. Correspondingly, astronomical observations of objects within and beyond the Solar System have long probed the effects of gravity. However, virtually no experiments were designed in the past to test specifically for the existence of an intermediate-range interaction between matter. Actually, subtle manifestations of the fifth force may have already appeared in high-energy experiments with a peculiar family of particles (the K mesons), as well as in discrepancies between satellite and terrestrial measurements of the local gravitational acceleration g.

As one can well imagine, the prospect of finding a new force in nature was bound to stimulate a flurry of new experiments. Unfortunately, in the aggregate, the results of the efforts to detect the fifth force were contradictory and inconclusive, with some experiments leading to positive results and others to null results. Although it is difficult to know with certainty what lies at the root of the discrepancies, one obvious possibility, given that all experiments were performed terrestrially, is the unaccountable influence of nearby masses. Indeed, a number of the experiments depended on the presence of naturally occurring large concentrations of mass (like cliffs or mountains) to produce a differential effect on suspended test masses.

One approach, different from any that has yet been tried, occurred to me shortly after the controversy first began; it was to search for a deviation from Newton's law of gravity by means of a satellite experiment. The basic principle exploits a well-known, but nonetheless extraordinary, property of any force whose magnitude diminishes as the inverse square of the distance from its source (in this case, a point mass). This attribute is shared by both the Coulomb force and (to the extent that general relativistic effects can be neglected) the force of gravity. In his *Principia*, Newton demonstrated mathematically that a test object outside of a spherical distribution of mass is gravitationally attracted as if all the matter of the sphere were concentrated at the center. Suppose, however, the sphere were hollow—a shell rather than a solid sphere—and the test mass lay *inside*. With what gravitational force would it be attracted to the walls of the shell?

Clearly, on the basis of symmetry alone, one could see that no force at all acts on a test mass at the center of the shell. All directions leading away from the center are equivalent; there can be no preferred direction of acceleration. What is perhaps less obvious is that the net gravitational attraction of the test mass by the shell is null *everywhere* in the shell interior! Newton understood this, too. The result holds for an arbitrarily thick shell and for a test mass of arbitrary shape and size, as long as it is entirely contained within the cavity of the shell.

The vanishing of the gravitational force within a spherical shell may be understood heuristically in the following way. Suppose that the test mass is an ideal mass point (one of the most frequently found items in the physicist's stockroom of imaginary objects) and the shell is very thin. Extend a straight line drawn through the test mass in both directions until it intersects the shell at two locations. Unless the mass is at the center of the shell, in which case we already know that it experiences no net gravitational force, one point of intersection is closer to the test mass than the other. Move the line (keeping the test mass fixed) so that each of the two segments generates the shape of a narrow cone with a circular base traced out on the inside surface of the shell. Consider the gravitational force exerted on the test mass by just those two portions of the surrounding shell contained within the circular bases.

The gravitational force exerted by a minute chunk of mass (let us call it an atom, although the argument does not depend in any way on the discreteness of matter) in the closer section is greater than the corresponding force exerted by an atom in the farther section. On the other hand, the surface area of the farther section is larger in proportion to the square of the distance from the test mass and, therefore, contains more atoms. For the case of an inverse-square force law *only*, the stronger attraction by the atoms of the nearer region is exactly counterbalanced by the greater number of atoms of the more distant region, with the result that there is no net gravitational force on the test mass from those two sections of shell.

Because the orientation of the line originally drawn through the test mass (i.e., the generator of the two cones) is entirely arbitrary, the net force on the test mass from *any* two sections of shell so delineated will cancel. In their entirety, all such mass sections constitute the whole of the thin spherical shell which, therefore, exerts no net force on the point mass inside regardless of its location. And since this conclusion holds for a thin spherical shell of any radius, it must be valid as well for any number of *concentric* thin shells or, equivalently, for a single shell of arbitrary thickness. Furthermore, the net force on a test mass of *finite* size contained within the shell must also vanish if each point mass of which it is composed experiences no force.

The spatial dependence of the suspected fifth force is not purely inverse square, but is thought to diminish exponentially with distance. To express the mathematical form of a fundamental interaction, it is often more convenient (and sometimes absolutely necessary) to consider energy rather than force. The potential energy of two point masses, m_1 and m_2, separated by a distance r and interacting through both gravity and the fifth force, may be represented by the formula

$$U(r) = -\frac{Gm_1m_2}{r}\left(1+\beta e^{-r/b}\right). \qquad (7.2a)$$

Newton's universal constant of gravity $G \sim 6.7 \times 10^{-11}\,\mathrm{N\,m^2/kg^2}$ sets the scale of intrinsic strength of the gravitational interaction, whereas β is a dimensionless coupling constant that sets the corresponding scale of strength (relative to gravity) of the fifth force, of which the characteristic range is b. On the basis of both geophysical data and reanalysis of Eötvös' paper, the coupling and range parameters have been estimated to be $\beta \sim -(7.2 \pm 3.6) \times 10^{-3}$ and $b \sim 200 \pm 50\,\mathrm{m}$. A negative β implies that the fifth force is repulsive. To determine the actual force that one mass exerts on the other, one must calculate the negative derivative of $U(r)$ with respect to r. The resulting expression, which need not be reproduced here, clearly gives $1/r^2$ dependence in the special case when β is zero (no fifth force). One can employ relation (7.2a) to derive the total potential energy of a test mass inside a spherical shell. As before, the force is then calculable from $-dU(r)/dr$. Again, the resulting expression is somewhat cumbersome, but the principal result is easy to state: When β is *not* zero, the net force on a test mass inside the shell does *not* vanish.

To understand why, consider again the special case of a point mass in a spherical shell. As a result of the exponential factor $\exp(-r/b)$, the forces exerted by the two patches of shell formed by conical projections from the test mass no longer cancel. The force of each patch diminishes with distance to a greater extent than the patch area increases; the mass patch closest to the test mass therefore exerts a greater force than does the more distant patch.

An interesting and useful consequence of this, which follows from relation (7.2a) if the fifth force is repulsive, is that a test mass located anywhere within the shell (for a shell size small compared with the range b) will be pushed toward the center with a strength linearly proportional to its displacement from the center. This is the type of restorative force, referred to as Hooke's law, which gives rise to periodic motion about a point of equilibrium. The frequency f with which a test object of inertial mass m_I and gravitational mass m_G would oscillate within a spherical shell of inner and outer radii R_1 and R_2, respectively, and mass density μ can be shown to be

$$f = \frac{1}{2\pi b} \sqrt{\frac{4\pi}{3} G\left(\frac{m_G}{m_I}\right) |\beta| \mu (R_2^2 - R_1^2)}.$$ (7.2b)

Thus, the occurrence of a harmonic oscillation at a frequency proportional to the square root of $|\beta|$ and inversely proportional to b, where no Newtonian gravitational force would be expected at all, would be an experimental signature of the putative new interaction.

The above considerations apply, of course, in the absence of forces from outside the shell. Here is where a satellite could prove useful. Orbiting the Earth—or some other parent body—the spherical shell and all its contents are in a permanent state of free fall. To a first approximation, therefore, the gravitational influence of the entire Earth has been eliminated. Like the astronauts in an orbiting space station, a test mass within such a shell would be weightless, its motion relative to the shell deriving ideally from those interactions that deviate from an inverse-square spatial dependence. Moreover, the contribution of the fifth force of the planet should be negligibly small for a satellite located at an orbital radius many times larger than the range of the force.

If, as in the Eötvös experiment, one employs *two* objects of different composition (i.e., differing in the proportion of baryons to inertial mass), the objects will oscillate about the center at different frequencies. The possible advantage of a satellite experiment may then be seen in the following. Compared with the fractional difference in acceleration $\Delta a/g$ that these two masses would undergo in a terrestrial Eötvös experiment, the fractional difference in oscillation frequency $\Delta f/f$ (where f is the mean oscillation frequency) can be shown to be

$$\frac{\Delta f}{f} \sim \left(\frac{2R}{3|\beta|b}\right)\frac{\Delta a}{g},$$ (7.2c)

where R is the radius of the orbited body. In the case of a satellite orbiting the Earth ($R = 6.4 \times 10^6 \, \text{m}$) and a force characterized by the coupling and range parameters specified earlier, the above fractional difference in oscillation frequency is more than *one million* times greater than the fractional change in acceleration.

Testing for the new force by satellite is not without its own problems, and it is unlikely that such an experiment will ever be undertaken soon. For one thing, the differential effect of the Earth on the test mass and shell would be entirely eliminated only if the gravitational field of the Earth were perfectly uniform. As this is not the case, one must take account of the residual "tidal" force (the same type of force responsible for the occurrence of ocean tides) resulting from the variation in the strength of the Earth's gravity throughout the interior of the satellite. Second, the incentive to find a fifth force has considerably waned, for further examination of the Eötvös paper by other researchers seemed to show that the original discrepancies might well have had a far more mundane explanation: air currents in Eötvös's laboratory!

Nevertheless, one never knows when or where something wholly new may crop up. I, for one, would still like to know whether an anti-neutron falls upward.

* * *

Within the framework of quantum mechanics, the Earth's gravita-tional field can affect the wave function of an elementary particle in ways for which no classical interpretation in terms of forces can be given. One striking example of this is the effect of gravity on neutrons moving *horizontally*.[8] Imagine a neutron beam incident upon a beam-splitting device that either transmits a neutron or reflects it vertically upward with 50% probability (Figure 7.1). The vertically reflected neutron encounters a perfect mirror that reflects it horizontally so that it propagates a distance L exactly parallel to, but at a height H above, the path followed by a transmitted neutron. The transmitted neutron, after propagating a horizontal distance L, also encounters a perfect mirror that reflects it vertically upward. The two neutron paths, which together form a rectangle, meet at another beam-splitting device that transmits an "upper" neutron and reflects a "lower" neutron with equal probability (50%) horizontally into a detector. The detected neutrons are counted—but under the circumstances, the experimenter cannot

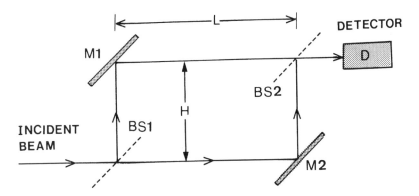

Figure 7.1. Schematic diagram of a neutron Mach–Zehnder interferometer. Beam splitters BS1 and BS2 transmit and reflect a neutron with 50% proba-bility. The plane of the interferometer is vertical so that the path segment of length L between mirror M1 and BS2 is a height H above the corresponding segment between BS1 and mirror M2. The components of a neutron wave transmitted and reflected by BS2 are coherently recombined at the detector D. Only one neutron at a time traverses the apparatus.

know whether a particular neutron has followed the upper or lower horizontal path. The experimental configuration constitutes the neutron counterpart to what in optics is called a Mach–Zehnder interferometer. A classical light wave, however, partitioned at the first beam splitter, traverses both routes to the second beam splitter. It is worth emphasizing, therefore, that the neutron flux is ordinarily low enough that only one neutron at a time passes through the interferometer.

According to standard quantum mechanical procedure, to determine the probability of receipt of a neutron at the detector—or, equivalently, the neutron count rate—one must add the probability amplitude for passage of a neutron along one or the other of the two indistinguishable paths. During the time t that it follows the upper horizontal path, an initially reflected neutron of mass m maintains a gravitational potential energy higher by mgH than that of an initially transmitted neutron that has followed the lower horizontal path. Neutrons following the upper or lower pathways experience no differential effect of the *force* of gravity because both routes include a vertical segment of length H over which gravity does work on the particles and a horizontal segment of length L over which no work is done. Nevertheless, the two spatially separated components of the neutron wave acquire a relative phase difference ϕ of the form

$$\phi = \frac{2\pi U t}{h},\tag{7.3a}$$

where $U = mgH$ is the difference in potential energy and h is Planck's constant.

Moving with mean (nonrelativistic) speed v, a neutron has linear momentum mv and covers the distance L in a time $t = L/v$. The corresponding neutron wave function, representable to a good approximation by a plane wave, is characterized by the de Broglie wavelength λ in terms of which the speed may be expressed by the relation

$$p = mv = \frac{h}{\lambda}.\tag{7.3b}$$

Substitution into Eq. (7.3a) of the expressions for U, t, and v permits one to write the relative phase in terms of experimentally accessible quantities

$$\phi = \frac{2\pi m^2 gHL\lambda}{h^2}\tag{7.3c}$$

[where it is actually the product of the inertial and gravitational masses that enters expression (7.3c) as m^2]. The probability amplitude for arrival of a neutron at the detector by two indistinguishable pathways is then of the form

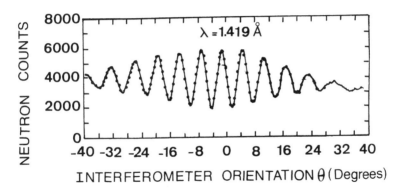

Figure 7.2. Quantum interference of neutrons induced by the gravitational potential of the Earth. The neutron wavelength is approximately 1.42×10^{-8} cm. Rotating the interferometer of Figure 7.1 by an angle θ about the incident beam produces a vertical separation $z = H\cos\theta$ between path segments M1–BS2 and BS1–M2. Each experimental point is the result of a total counting time of about seven minutes. [Conceptually inessential differences between the actual interferometer and the idealized interferometer analyzed in the text lead to an interference pattern of the form $a + b\cos\phi$, where, in contrast to Eq. (7.3e), a and b are unequal.] [J.-L. Staudenmann *et al.*, *Physical Review* **A21** (1980) 1419.]

$$\psi \sim 1 + e^{i\phi}, \tag{7.3d}$$

from which it follows that the neutron count rate, proportional to the probability of arrival $P(\phi)$, must depend on the acceleration of gravity g and vary harmonically with the height difference H, according to

$$P(\phi) = |\psi|^2 = \frac{1}{2}(1 + \cos\phi). \tag{7.3e}$$

(The normalization factor $\frac{1}{2}$ restricts the maximum probability to unity.)

The observation of this neutron interference phenomenon (Figure 7.2) demonstrates convincingly that the Earth's gravity can affect the motion of elementary particles under circumstances where it is not the gravitational force itself but the difference in gravitational potential energy that has direct physical significance. Interestingly, it illustrates as well that the equivalence principle may be of questionable validity in the realm of quantum mechanics. As a consequence of the equality of inertial and gravitational masses, a classical object moves through a gravitational field along a mass-independent trajectory. However, the relative phase shift ϕ depends on mass, and the probability of particle arrival, therefore, is not the same for all particles.

* * *

In addition to the force of gravity, which acts whether the Earth turns or not, and the centrifugal force, which any object on the rotating Earth experiences even if stationary relative to the Earth's surface, there is yet another interaction, the Coriolis force, that affects objects *in motion* on the surface of the rotating Earth. The Coriolis force deflects a moving object from apparent straight-line motion, as judged by an observer at rest on the Earth, and, like gravity and the centrifugal force, is independent of all intrinsic chemical and physical properties of an object except that of mass. The resemblance in this way of the Coriolis and centrifugal forces to gravity is illustrative again of Einstein's equivalence principle, the version which asserts that gravity and accelerated motion are locally indistinguishable.

The Coriolis force is another example of a pseudo force in the sense that an observer in an inertial (nonaccelerating) reference frame does not need to invoke it to explain physical events. Imagine two ball players on diametrically opposite ends of a large rotating platform like that of a carrousel. One throws a ball toward the other. From the bird's-eye view of a stationary observer above the carrousel, the ball moves in a straight line across the surface as the two players rotate with the platform. However, from the perspective of the intended receiver, with respect to whom the thrower has remained motionless, the ball follows a curved path away from the center, as if acted upon by some force— the Coriolis force. Under just the right circumstances, the thrower, himself, can rotate into position to catch the thrown ball. From *his* perspective, the ball has followed a trajectory outward and back again like a yo-yo without a string! In the rotating frame of reference, the Coriolis force has physical consequences.

The Coriolis force on an object of mass m moving with speed v along a surface that is rotating about a perpendicular axis with angular frequency ω is proportional to $mv\omega$. The direction of the force depends on the direction of motion of the object and on the sense of rotation of the frame. On the Earth, which spins at an angular rate of 360° in 24 hours, or about 7.3×10^{-5} radians/s, the Coriolis force can markedly affect the patterns of global airflow, although it is ordinarily too weak to influence the local motion of relatively small objects over a timescale short enough that someone would likely have the patience to watch it. Nevertheless, it does have perceptible effects on small objects over sufficiently long intervals of space or time. In the Northern Hemisphere, for example, a directly aimed cannon shot will fall to the right of the target if deflection by the Coriolis force is not taken into account in the design of the sighting mechanism. British sailors rediscovered this fact during a naval engagement with Germany near the Falkland Islands off the southeastern coast of Argentina early in the First World War. Their sighting mechanisms had been constructed for warfare in the Northern Hemisphere, and, consequently, their projectiles fell to the left of the German ships by some 100 m, *twice* the Coriolis deflection.

In a more mundane example, countless visitors to science museums each year are likely to notice a Foucault pendulum. First devised in the mid-19th century by the French physicist, Jean Léon Foucault, the pendulum shaft—sometimes extending several stories—is suspended vertically above the floor on which is depicted a calibrated ring. As the bob swings back and forth across the ring, the plane of oscillation appears to precess slowly relative to the fixed reference marks. To an inertial observer, it is the floor that rotates under the pendulum at a rate that depends upon the latitude of the site.

That the rotation of the Earth can also affect the motion of an elementary particle was demonstrated in a beautiful experiment, again involving the quantum interference of neutrons.[9] As is clear from the effect of gravity on neutron interference or the effect of a confined magnetic field on electron interference (Chapter 3), the concept of energy retains a physical significance under conditions where it would be meaningless to speak of a force. This is the case with the "neutron Sagnac effect."

The Sagnac effect, which was first demonstrated with light by the French physicist M. G. Sagnac in 1913, is a phase shift in the interference of two coherent waves as a consequence of the rotation of the interferometer. The geometrical configuration of a Sagnac interferometer resembles that of the Mach–Zehnder interferometer described previously except for one critical detail. The second beam splitter is replaced by a mirror so that the waves reflected and transmitted at the first (and only) beam splitter propagate in opposite directions completely around the interferometer and overlap again at their place of entry. If the interferometer were stationary (or moving at a uniform velocity relative to some other inertial reference frame), the time required for a light wave to complete one circuit about the interferometer would be the same for either direction of propagation. When the interferometer rotates, however, the beam splitter rotates toward one of the waves and away from the counterpropagating wave. Suppose that the interferometer is rotating clockwise according to an inertial observer suspended above it. The wave propagating counterclockwise would then complete a circuit in a time interval shorter than that of the clockwise propagating wave. A relative phase difference would therefore develop between the two waves given by $2\pi(\Delta t/T)$, where T is the period (reciprocal of the frequency) of the waves and Δt is the difference in time for the two counterpropagating waves to complete a circuit.

For an interferometer of area A (i.e., the area enclosed by the counterpropagating beams) rotating at angular frequency ω radians/s about an axis inclined at an angle θ to the direction normal to the plane of the interferometer, the time difference Δt is given by the approximate expression

$$\Delta t \sim \left(\frac{4A\omega}{v^2} \right) \cos\theta, \qquad (7.4a)$$

in which v is the speed of the wave relative to the nonrotating laboratory. This approximation is good to the extent that one can neglect the square of the ratio of the speed of rotation to the speed of the wave, or $(\omega R/v)^2$, where R is a characteristic size of the interferometer (e.g., the radius, if the light beam followed a circular path). For the case of counterpropagating electromagnetic waves, the speed of propagation is the universal constant c, and the Sagnac phase shift, ϕ_S, expressed in terms of the wavelength $\lambda = cT$ becomes

$$\phi_S = 2\pi \frac{\Delta t}{T} = \left(\frac{8\pi A\omega}{c\lambda} \right) \cos\theta. \qquad (7.4b)$$

Because neutrons have wavelike properties, the rotation of a neutron interferometer should also lead to a phase shift between counterpropagating components of a split neutron beam. In this case, the quantity corresponding to the period of the neutron wave is λ/v, where the speed v is not a universal constant but is related to the wavelength through relation (7.3b). Substitution of the factors appropriate to a massive particle leads to a phase shift

$$\phi_S = 2\pi \frac{\Delta t}{T} = \left(\frac{8\pi m A\omega}{h} \right) \cos\theta \qquad (7.4c)$$

that depends on the (inertial) mass of the particle but is totally independent of velocity and wavelength.

In the neutron Sagnac experiment, the Mach–Zehnder type of interferometer was employed again, but oriented so that the incoming neutron beam and the plane of the interferometer were vertical. From the symmetry of the configuration, it should be clear that turning the plane of the interferometer about the vertical axis does not alter the height above ground—and hence the gravitational potential—of any point of the neutron pathways through the interferometer. Such a rotation, therefore, would not change the gravity-induced phase shift. It does, however, reorient the plane of the interferometer (specified by its normal direction) with respect to the rotation axis of the Earth. Thus, the entire variation in the neutron count rate for different settings of the angle θ should be attributable to the Sagnac effect. This intensity variation is expressed by a relation analogous to Eq. (7.3e), but with the (now constant) gravitational phase shift augmented by ϕ_S. Actually, because an individual neutron, in going from the first to the second beam splitter (and then to the detector), does not make a com-

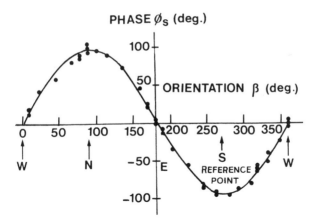

Figure 7.3. Influence of Earth's rotation on the self-interference of neutrons (neutron Sagnac effect). The interferometer, positioned so that the incident beam is along the vertical direction, is turned about the vertical by an angle β (relative to a reference direction) to vary the angle θ between the normal to the interferometer plane and the rotation axis of Earth. The graph shows the predicted variation in rotational phase shift ϕ_S as a function of β (full line) and the corresponding experimental points. The angle θ in Eq. (7.4c) is related to β by $\cos\theta = \cos\theta_L \sin\beta$, where θ_L is the latitude of the experimental site. [Adapted from J.-L. Staudenmann et al., Physical Review **A21** (1980) 1419.]

* * *

plete circuit around the interferometer, but only one-half a circuit, the theoretically predicted Sagnac phase shift should be one-half that of relation (7.4c). The experimental results solidly confirmed this predicted effect of the Earth's rotation (Figure 7.3).

* * *

At the time I began to study the problem of light reflection from an optically active medium (Chapter 6), I also became interested in the effects of the Earth's rotation on quantum mechanical systems. Though outwardly quite different topics, there is an important point of contact that relates them. Both phenomena involve chirally asymmetric interactions.

Consider the rotating Earth. Because the Earth turns toward the east, the Coriolis force on a person running due east along the equator (or counterclockwise to someone looking down upon the North Pole) is directed radially outward away from the center of the Earth. If the runner changes direction and proceeds due west along the equator (clockwise to the observer above the North Pole), the direction of the Coriolis force on him will be radially inward toward the center of the Earth. The Coriolis force distinguishes between clockwise and counterclockwise—between right-handed and left-handed—senses of motion. To an observer confined to a rotating reference frame like the Earth, the Coriolis force is a chirally asymmetric force.

It will be recalled that chiral objects or processes are not superposable on their mirror images. The transformation of an east-bound runner into a west-bound runner is effected by reflection of the Earth and runner in a mirror. In such a reflection, both the direction of the runner *and* the sense of rotation of the Earth are reversed, in which case the Coriolis force continues to point outward. However, an Earth that turns toward the west simply does not exist—although presumably there is no physical law forbidding such a historical possibility. In any event, the original scene and its mirror image are not superposable.

Although the neutron is believed to be composed of three basic particles (quarks), the internal structure of the neutron has, nevertheless, played no significant role in the self-interference experiments sensitive to the gravity and rotational motion of the Earth. These experiments prompted me to wonder, however, about the quantum effects of gravity and rotation on the internal dynamics of composite quantum systems like atoms and molecules. Because the Coriolis force distinguishes left- and right-handed senses of motion, could it by any chance give rise to optical activity in atoms? Would such an effect be observable?

Discussions of the physics of atoms almost always take for granted at the outset that the frame of reference is not accelerating. The laws of quantum mechanics were initially formulated for inertial frames, and actual experiments on atoms are ordinarily executed under such conditions that this assumption would appear adequate. The Earth is not, of course, a true inertial reference frame. However, the Coriolis force of the Earth's rotation on an atomic electron is smaller than the electrostatic force binding the electron to the nucleus by a factor of about 100 billion billion (10^{20}). This is very small, indeed! (The centrifugal force of the Earth on a bound electron is at least four orders of magnitude smaller.) To detect an influence of the Coriolis force of the Earth in the optical properties of an atom would be tantamount to observing one of the weakest interactions by far in which an atom has participated.[10] Still, this prospect may not be entirely hopeless.

It is interesting to speculate that, weak though it may be, the effect of the Earth's rotation on individual atoms could conceivably be connected with one of the outstanding unsolved problems in the life sciences: the origin of biomolecular chirality. Why living things make and use molecules of specified handedness such as right-handed sugar molecules or left-handed amino acids, is not known. Perhaps, over the eons during which the molecules of life evolved, the chirally asymmetric effect of the Earth's rotation may have led to a preferential molecular handedness in one hemisphere that, through the random accidents of history, spread over the entire planet.

Specifically, how can the rotation of the Earth influence the structure of an atom? Although classical mechanics does not, in general, serve as an adequate basis for understanding the dynamics of an atom, there are instances when the imagery of classical physics provides insight, at least when the quantum mechanical degrees of freedom involved have classical counterparts. It is worth stressing at the outset that it is the *internal* dynamics in which we are interested here (both classical and quantum mechanical)—i.e., in the motion of the electrons relative to the nucleus, not the motion of the center of mass of the atom. The center of mass of a system of particles—which need not correspond to the location of any actual particle—moves in response to the net *external* force as if all the mass of the system were concentrated at that hypothetical point. The rest frame of the system is the reference frame in which the center of mass is stationary. For ordinary atoms (in contrast to exotic atoms) in which all bound particles are electrons, the center of mass coincides with the location of the nucleus to good approximation.

Consider, for simplicity, a planetary atom with a single electron in circular orbit about the nucleus at an angular frequency ω_0 radians/s (as determined theoretically for an atom at rest in an inertial frame). In fact, to take the simplest case possible, locate the atom at the North Pole so that the axis about which the electron revolves coincides with the rotation axis of the Earth. To an inertial observer suspended above the North Pole, the angular frequency of the electron is ω_0, irrespective of the sense (clockwise or counterclockwise) of the revolution. However, to an observer fixed on the Earth, which turns, let us say, at ω radians/s, the angular frequency of the electron is $\omega_0 - \omega$ if the electron revolves in the same sense as the Earth rotates and $\omega_0 + \omega$ if the electron revolves in the opposite sense. Even though an observer cannot actually "see" the motion of an electron in an atom, he would, nevertheless, draw the preceding conclusions by correlating the frequency and circular polarisation of the spontaneously emitted radiation. A circulating charged particle is undergoing periodic acceleration and, according to classical electrodynamics, emits along the rotation axis electromagnetic waves with transverse electric fields that rotate in the same sense and at the same frequency as the orbital motion of the electron. A real atom, of course, does not continuously radiate, or it would collapse practically instantaneously. We will see, however, that quantum mechanics sustains the foregoing picture of chirally inequivalent orbital motions.

In addition to the characteristic spontaneous emission of radiation, the optical response of an atom to incident radiation can also be influenced by the rotation of the Earth. The index of refraction of a material, as I discussed in the previous chapter, was shown by Maxwell to be effectively equal to the square root of the dielectric constant in the case (relevant to the present discussion) that the material is not intrinsically magnetic. The dielectric constant is, itself, a measure of the extent to which the bound electrons of the sample are displaced from their equilibrium positions by an external electric field, such as the electric field of an incident light wave. The greater the displacement, the greater will be the electric dipole moment of an individual atom (which is the displacement multiplied by the electric charge), the greater will be the resulting dielectric constant (which grows with the number of electric dipoles in the sample), and, hence, the greater will be the corresponding refractive index of the material.

It is the atomic polarizability α that expresses the proportionality between the displacement of a bound electron from its equilibrium position and the strength of the applied electric field. To determine the polarizability (still within the framework of classical mechanics), one solves Newton's equations of motion for the *forced* motion of the electron at the frequency of the incident light wave. This is a standard and relatively elementary problem for an atom in an inertial reference frame. The electron is subject to the electrostatic binding force, possibly some damping force that takes account of energy loss by spontaneous emission, and the driving force of the electric field of the light. The magnetic field of the light wave can usually be neglected, for it results in a force weaker than that of the electric field by the ratio of the electron speed to the speed of light. Disregarding the effects of damping and magnetism and assuming an incident light wave of angular frequency Ω, one arrives at the following simple expression for the atomic polarizability:

$$\alpha(\Omega) = \frac{(e^2/m)}{\omega_0^2 - \Omega^2},\qquad(7.5a)$$

where e is the electron charge and m is the electron mass. Note that the polarizability increases as the frequency Ω of the light approaches the resonance frequency ω_0 of the atom (or molecule). Thus, the index of refraction of a transparent material like glass, for which the resonance frequencies typically fall in the ultraviolet portion of the spectrum ($\omega_0/2\pi \sim 10^{15}\,\text{Hz}$), is larger for blue light ($\Omega/2\pi \sim 6.3 \times 10^{14}\,\text{Hz}$) than for red light ($\Omega/2\pi \sim 4.4 \times 10^{14}\,\text{Hz}$); blue light will be refracted to a correspondingly greater extent than red light as it enters the glass from air. To a good approximation, the index of refraction (n) of a sufficiently rarefied sample of atoms that behave independently of one another is related to the atomic polarizability in the following way:

$$n \sim 1 + 2\pi N\alpha,\qquad(7.5b)$$

in which N is the number of atoms per unit of volume.

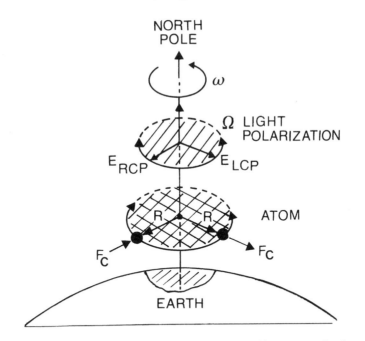

Figure 7.4. Heuristic model of the effect of the Earth's spin on the dynamics of a classical atom in the simple case where the Earth and bound electron rotate about a common axis. When driven by the electric field (E) of an incident left circularly polarised (LCP) light wave of angular frequency Ω, the electron orbits the nucleus in the same sense as the Earth spins; the Coriolis force (F_C) on the electron acts radially outward. When driven by an incident right circularly polarized (RCP) wave of the same frequency, the electron experiences a Coriolis force radially inward. To an inertial observer above the Earth, there is no Coriolis force, but the frequencies of the two light beams are respectively $\Omega + \omega$ and $\Omega - \omega$, where ω is the spin angular frequency of the Earth.

Return now to the classical atom at the North Pole of the rotating Earth (Figure 7.4). As analyzed by an Earth-bound observer, the orbiting electron is subject not only to the forces described above but also to the Coriolis and centrifugal forces. Also, to repeat, it is not the characteristic motion of the electron that is of concern now, but only the motion engendered by the electric field of the incident light wave. Imagine a light wave of frequency Ω, as measured by the *Earth-bound* observer, propagating upward along the common rotation axis of the Earth and electron. If the wave is left circularly polarized, it drives the electron about its center of attraction, the atomic nucleus, in the same sense as the Earth rotates; a right circularly polarized wave drives the electron in the opposite sense. If the electron revolves in

response to a left circularly polarized wave, the Coriolis force accelerates it radially outward (i.e., outward along the radial line from the
nucleus to the electron, not from the center of the Earth to the electron), thereby increasing the displacement of negative and positive
charges within the atom. This leads to a larger electric dipole moment.
Conversely, the Coriolis force accelerates an electron moving in
response to a right circularly polarized wave radially inward, leading
to a smaller electric dipole moment. The centrifugal force on the
electron is directed radially outward, irrespective of the sense of
circulation.

In sum, from the perspective of an Earth-bound observer, the sample
of atoms exhibits a larger index of refraction for left circularly polarized light than for right circularly polarized light. This is exactly what
is required for the atoms to be optically active. The existence of atomic
circular birefringence (difference in chiral refractive indices, $n_L - n_R$)
has been inferred for the special orientation of an atom at the North
Pole, but the conclusion holds generally for any location on the Earth
although the strength of the predicted effect varies with the relative
orientation of the light beam and the Earth's axis.

I have noted previously that the Coriolis force is termed a fictitious
force originating in the acceleration of the reference frame. Does this
mean that the predicted optical activity is, itself, fictitious? Would the
inertial observer suspended above the North Pole agree that the atoms
exhibit a chiral asymmetry? To answer the question let us examine the
expressions derived by the Earth-bound observer for the polarizability of a rotating atom. The two expressions for left and right circular
polarizations are similar in form to that of relation (7.5a) derived for
an atom in an inertial reference frame:

$$\alpha_L = \frac{e^2/m}{\omega_0^2 - (\Omega + \omega)^2}, \tag{7.6a}$$

$$\alpha_R = \frac{e^2/m}{\omega_0^2 - (\Omega - \omega)^2}. \tag{7.6b}$$

To the inertial observer, however, the above relations are, in fact, the
same relation as Eq. (7.5a)—only evaluated for different frequencies.
If the frequency of the left circularly polarized light wave propagating
upward along the rotation axis is Ω relative to the Earth-bound
observer (who is himself rotating in the same sense at the frequency
ω of the Earth), then the inertial observer would declare the light frequency to be $\Omega + \omega$. Similarly, the right circularly polarized wave, also
of frequency Ω to the Earth-bound observer, would present a frequency
$\Omega - \omega$ to the inertial observer.

The inertial observer might therefore say: "Of course the index of
refraction is different for the left and right circularly polarized waves.

Their frequencies are different, and it is well known that a higher frequency leads to a larger refractive index. The atoms, however, are chirally symmetric." To this, the Earth-bound observer could in truth reply: "The frequency of both waves is the same. The Coriolis force produces chirally asymmetric atomic polarizabilities." Both interpretations are correct. Still, it is useful to keep in mind that, as denizens of a rotating reference frame, physicists ordinarily interpret the results of their measurements in terms of the apparatus and interactions in their own stationary Earth-bound laboratories and do not feel constrained to consult inertial colleagues suspended above the planet.

Since the internal dynamics of actual atoms are not accurately described in terms of electron trajectories influenced by Newtonian forces, one might wonder whether the foregoing classical analysis is in any wave reliable. In brief, the answer is basically affirmative—with one important *caveat*. It is understood, of course, that where an objectively real physical quantity like an orbital radius might appear in the mathematical expressions of classical mechanics, the analogous quantum mechanical expressions would contain matrix elements (i.e., integrals) of a corresponding operator, providing a measure of the likelihood that the atom can undergo certain transitions between its states. If the matrix elements connecting particular quantum states of interest vanish, then quantum mechanics does not permit the designated process to occur even though classical mechanics may have yielded a seemingly respectable non-null result. I shall give an important example of this shortly.

Within the framework of quantum mechanics, the interactions that affect the internal state of an atom are incorporated in the appropriate equation of motion (e.g., the Schrödinger equation) not as forces but as contributions to potential energy. For an atom rotating with the Earth, the effects on its constituents of both the centrifugal and Coriolis forces may be shown *classically* to derive from an extra energy term (U_R) that involves a coupling of the internal angular momentum (L) of the atom to the angular frequency (ω) of the Earth as follows:

$$U_R = -\omega L \cos \theta, \qquad (7.7a)$$

where θ is the angle that the atomic angular momentum makes with the rotation axis of the Earth. Again, by "internal" angular momentum I mean the orbital motion of the electron about the atomic nucleus, not the movement of the whole atom about the axis of the Earth. Classically, the angular momentum of an object of mass m moving about a center of attraction with speed ω in an orbit of radius R has the magnitude mvR and is oriented perpendicular to the plane of the orbit in a right-hand sense. That is, if one wraps the fingers of his right hand about the orbit so that they point in the direction in which the object circulates, then the extended thumb gives the direction of the angular momentum.

In stark contrast to the classical picture of an atom, however, the details of electron motion within a quantum mechanical atom cannot be pictured. How, then, is the electron angular momentum to be oriented? In fact, quantum mechanics does not permit one to know this orientation. One can know only the magnitude of the angular momentum and the projection of the angular momentum along an arbitrary axis. In the present case, it is convenient to choose this axis to be the axis of the Earth. For an electron angular momentum of magnitude $L = \sqrt{\ell(\ell+1)}\hbar$, where ℓ is an integer-valued angular momentum quantum number, the projection $L\cos\theta$ must then take values $M\hbar$ in which the azimuthal or magnetic quantum number M spans the range of $2\ell + 1$ integers from $-\ell$ to ℓ in steps of 1. Projections that differ only in sign refer to electron states that differ only in the sense of electron circulation about the quantization axis.

If an atom in an inertial reference frame is not subject to external perturbations, then all directions of the quantization axis are equivalent, as are also the two senses of rotation about the axis. One would, therefore, not expect the energy of a quantum state to depend on the orientation of the quantization axis or on the azimuthal quantum number. On the spinning Earth, however, matters are different, for now there is a distinctive sense of motion about a particular direction.

Substitution of the potential energy (7.7a) into the Schrödinger equation to determine the energy eigenstates of the electron from the perspective of an Earth-bound observer yields the following interesting result. The state energy E, expressible in the form

$$E = E_0 - M\hbar\omega \tag{7.8}$$

(where E_0 is the corresponding energy on a nonrotating Earth), now depends on the component of the electron's orbital angular momentum along the rotation axis of the Earth. If the electron circulates in the same sense as the Earth rotates (i.e., M is positive), the energy of the quantum state is lowered by $M\hbar\omega$. Conversely, the energy is raised by $|M|\hbar\omega$ when the electron and the Earth rotate in opposite senses and M is negative. This result is in complete analogy to the previous classical treatment leading to orbital frequencies $\omega_0 \pm \omega$ and, again, may be tested experimentally by examination of spontaneous emission. A quantum mechanical atom radiates when the bound electron undergoes a "quantum jump" to a lower-energy state. If an electron in a quantum state of nonzero angular momentum undergoes an electric dipole transition to a lower state of zero angular momentum, the emitted photons will have two possible frequencies: $\omega_0 \pm \omega$, where ω_0 is the corresponding frequency in an inertial reference frame.

The quantum mechanical analysis of a rotating atom interacting with incident left and right circularly polarized light involves the use of a mathematical procedure termed perturbation theory which will not be described here. I note only that the calculation justifies the classical picture of the inequivalent action of the Coriolis force on countercirculating electron orbits to create chirally asymmetric polarizabilities. The end result is the prediction of a very weak circular birefringence $n_L - n_R$ that can be on the order of about 10^{-18} for light falling in the visible and ultraviolet regions of the spectrum. A circular birefringence of this magnitude would lead to the minute optical rotatory power of about 10^{-11} degrees per meter of material! As stated before, this optical activity is far smaller than that already observed in atoms as a consequence of the weak nuclear interactions.

Is so weak an optical effect detectable? The answer, a guarded affirmative, brings us back to the Cashmere Cavern and the Canterbury ring laser. In a ring laser, as in the Sagnac interferometer, two coherently produced light waves traverse the same closed path in opposite directions (Figure 7.5). What distinguishes the laser from a passive

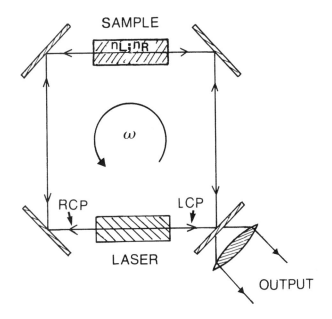

Figure 7.5. Schematic diagram of a rotating ring laser interferometer with a sample that displays circular birefringence (different refractive indices n_L and n_R for the two forms of circularly polarized light). RCP and LCP light waves propagate in opposite senses about the ring and give rise to a beat frequency produced in part by the interferometer rotation and in part by the circular birefringence.

(i.e., no gain) interferometer, however, is the presence within the ring of a medium with a population inversion such as discussed in Chapter 6. Although the laser may emit light over a range of frequencies, only those oscillations are sustained which satisfy a resonance condition whereby an integral number of wavelengths span the perimeter of the ring. Those waves that do not satisfy this requirement are effectively suppressed by destructive interference. In fact, all waves (light, sound, water, etc.) in a closed container or cavity are subject to the imposition of boundary conditions.

When the ring laser is rotating, the effective length of the trip around the cavity—and hence the resonant wavelength and frequency—is different for the clockwise and counterclockwise modes. Upon recombination at a detector, the two modes, no longer synchronized, produce a beat frequency similar in principle to the beat heard when one strikes two neighboring keys on a piano. There is an advantage to measuring a beat frequency in a ring laser compared with a phase shift in a passive interferometer. For one thing, frequency is an experimental quantity that can be measured with relative facility and to very high precision. (As one example, the frequency corresponding to the hyperfine splitting in ground-state hydrogen can be measured to better than one part in 10^{12}.) For another, the factor relating the beat frequency to the optical path-length difference is larger (by the ratio of the speed of light to the circumference of the ring) than the corresponding factor relating the Sagnac phase shift to the optical path-length difference.

In addition to providing a highly sensitive monitor of rotation, a ring laser permits one to measure small optical anisotropies, such as the birefringence of a material. Suppose that the natural modes of the ring laser are circularly polarized and that the laser could be excited bidirectionally with left and right circularly polarized waves traversing the ring in opposite directions. The presence of a sample of optically active matter in the ring would likewise give rise to a frequency difference, or beat frequency, because the left and right circularly polarized waves are retarded by the sample to different extents. Clearly, the case of two counterpropagating waves retarded unequally with respect to a stationary laser is equivalent, in principle, to that of a moving laser "gaining" on one wave and "receding" from the other. This frequency shift is linearly proportional to the circular birefringence and the mean operating frequency of the laser.

For the problem at hand, it is the rotation of the Earth—and hence of the ring laser fixed to the Earth—that *generates* the optical activity in the sample. Since (as in the case of the Eötvös experiment) one cannot stop the Earth from rotating, how will the experimenter know that a shift in beat frequency has occurred? The solution to the problem is conceptually the same as that employed by Eötvös: Make a comparative measurement. Although the Earth's rotation should induce optical activity in all terrestrial matter, the magnitude of this circular birefringence depends on both the quantum level structure of the material and the frequency of the light.

Circular birefringence is a nonresonant phenomenon; that is, the material is essentially transparent to the light employed. Nevertheless, the circular birefringence of a material, like the atomic polarizability, ordinarily increases as the frequency of the light approaches, but never falls within, the region of the spectrum where absorption occurs. Introduction into the ring, therefore, of a sample material for which the laser frequency is close to an atomic transition should shift the beat frequency attributable to rotation alone by a small but significant amount. Subsequent reversal of the direction of the counterpropagating left and right circularly polarized waves—in analogy to the exchange of masses in Eötvös's torsion balance—can be shown to shift the beat frequency in the opposite direction. Unlike the Eötvos experiment, however, a null result is not expected.

Theoretical analysis of the ideal performance of a ring laser[11] suggests that a laser of the type and size (area of $1 \, m^2$) being developed at the University of Canterbury should be able to detect a shift in frequency smaller than the operating frequency itself by a factor of about 10^{19}. If this ideal performance is actually realizable, the predicted atomic circular birefringence would just marginally fall within the resolution capacity of the laser.

If the optical activity induced by the Earth's rotation is far weaker than the atomic optical activity resulting from nuclear interactions, how is it to be distinguished from the latter? Actually, the two types of optical activity differ significantly in their symmetry properties. Because weak neutral current interactions between the nucleus and orbiting electrons actually mix close-lying atomic states of opposite parity (e.g., the S and P sates within a given electronic manifold), the resulting optical activity is similar in nature, although different in magnitude, to the molecular optical activity (discussed in Chapter 6) associated with chirally asymmetric three-dimensional chemical structures. In this type of optical activity, the sense of chiral asymmetry is defined with respect to the direction of propagation of the light beam, there being no other preferred direction in an optically isotropic material. One consequence of this, pointed out earlier, is that a plane-polarized light beam reflected back upon itself to its point of entry in the optically active medium shows no net optical rotation.

The optical activity generated in atoms by the Earth's rotation is different—and illustrates by that difference an illuminating connection between rotation and magnetism. There is a well-known theorem in classical mechanics known as Larmor's theorem (derived in 1897 by J. J. Larmor, who in that same year also derived the Larmor formula employed in Chapter 4) which states that the effect of a constant magnetic field $\boldsymbol{B}$ on a system of particles of mass m and charge q is to superimpose on its normal motion (i.e., in the absence of the magnetic field) a uniform precession. The angular velocity of precession, the magnitude of which is designated the Larmor frequency, may be shown to be

$$\boldsymbol{\omega}_L = -\left(\frac{q}{2mc}\right)\boldsymbol{B}. \tag{7.9a}$$

The theorem is not exact, but rather an approximation valid to the extent that one can neglect terms of order B^2 and higher in the equation of motion. Because of the negative sign in Eq. (7.9a), the orientation of the angular velocity is *opposite* the orientation of the magnetic field if the particles are positively charged. Another way of expressing this is to say that the precession occurs in the same sense as the current of negative electrons that generates the magnetic field in the first place. In any event, the basic idea is that one can sometimes simplify the analysis of a system of particles in a magnetic field by eliminating the field and placing oneself in a frame of reference rotating at the Larmor frequency.

The influence of the rotation of the Earth on atoms can be regarded in some ways as the *converse* of Larmor's theorem. In other words, the effect of a uniform global rotation at angular velocity $\boldsymbol{\omega}$ on a system of particles in an environment free of electric and magnetic fields is equivalent to that of a constant magnetic field (which, for consistency, I shall call the Larmor field $\boldsymbol{B}_L$) obtained by rearranging relation (7.9a):

$$\boldsymbol{B}_L = -\left(\frac{2mc}{q}\right)\boldsymbol{\omega}. \tag{7.9b}$$

Thus, the rotational separation of magnetic substates degenerate in an inertial frame [relation (7.8)] is the analog of the Zeeman effect, the splitting of degenerate states by a static magnetic field. In fact, the potential energy term (7.7a), with ω replaced by the equivalent expression from Eq. (7.9b), has exactly the form of an interaction of a magnetic field with a magnetic dipole moment of magnitude $qL/2mc$. (This is the magnetic dipole moment one would expect for a current loop consisting of a single charged particle in circular orbit with angular momentum L.) Likewise, the phase shifts produced in the Sagnac effect by rotation and in the Aharonov–Bohm effect by a magnetic flux are analogous—the connection being particularly direct when expressed in terms of the vector potentials from which the corresponding magnetic fields derive. In a similar way, rotational optical activity has a magnetic analog, the Faraday effect, discovered by Michael Faraday in 1845. The plane of polarization of a linearly polarized light beam transmitted through an isotropic dielectric in a static magnetic field is rotated by an amount linearly proportional to the magnetic field strength and the path length through the medium.

From a classical perspective, the Faraday effect is produced by the inequivalent action of the Lorentz force on oppositely circulating electron orbits. Replacing the Lorentz force by the Coriolis force or, equivalently, substituting the Larmor field $\boldsymbol{B}_L$ for the actual magnetic field appearing in the calculation of the Faraday rotation, yields the classical mechanical expression for rotational optical activity. The optical rotation, however, of the Faraday effect occurs with respect to the fixed magnetic field, not with respect to the light propagation direction. The practical consequence of this is that the net optical rotation of a light beam reflected back upon itself to its point of entry into the sample is *twice* that of a single passage—not zero as in the case of structural optical activity. This same symmetry holds for optical activity engendered by the Earth's rotation, for which the rotation axis of the Earth replaces the magnetic field direction and helps serve to distinguish the predicted effect from optical manifestations of the weak nuclear interaction.

The weak nuclear interactions are not only weak, but of extremely short range. I have mentioned previously that, according to the electroweak theory, the nucleons and bound electrons of an atom are coupled by the exchange of a Z^0 boson, a particle whose mass is about 100 times the proton mass. In quantum mechanics, the range of a force can often be estimated quickly by means of the uncertainty principle. The linear momentum of an exchanged (or virtual) particle of mass m can span a range of values from 0 to about mc. Hence, the uncertainty in its spatial location is approximately

$$\lambda_C = \frac{h}{mc}, \tag{7.10}$$

the so-called Compton wavelength. The Compton wavelength of the Z^0 boson is about 10^{-15} cm, some seven orders of magnitude smaller than the Bohr radius which sets the scale of atomic size. Consequently, weak neutral currents can directly influence only those atomic states—the S states—for which there is significant overlap of electron and nuclear wave functions. For all states of nonzero orbital angular momentum, the electron wave function has a node or zero amplitude at the nucleus (treated in first approximation as a point mass).

In the electronic manifolds of principal quantum number $n = 2$ and higher, the weak nuclear interactions mix close-lying S and P states, thereby giving rise to an atomic optical activity that increases in strength approximately with the cube of the atomic number. Therefore, by employing light atoms, using light of such frequency as to avoid contributions from electrons in S states, and by taking advantage of the cumulative enhancement of multiple passes through the sample, one might hope to observe the chiral effects of the Earth's rotation in the domain of atomic physics.

* * *

Although a heuristic explanation of the interaction between an atom and the spinning Earth in terms of the classical Coriolis force seems to have led so far to results in basic accord with those of quantum mechanics, this need not always be the case. Quantum mechanics embraces degrees of freedom for which there are no classical counterparts. Consider, for example, an unexcited hydrogen atom. Based on the classically derived potential energy expression (7.7a), the internal dynamics of an atom in a 1S ground state would be entirely unaffected by the rotation of the Earth, as a 1S electron has zero orbital angular momentum. (The center of mass of the hydrogen atom would still, of course, be subject to the Coriolis force of the Earth's rotation.) This expectation is not correct, however.

An elementary particle can have a nonclassical degree of freedom, spin, which also contributes to its angular momentum. The electron and the proton, for example, are both spin-$\frac{1}{2}$ particles. Although one can try to picture the spin of an electron as analogous to the diurnal rotation of the Earth about its axis, this is not really satisfactory. High-energy scattering experiments probing the internal structure of the electron indicate (in contrast to the proton and neutron) that the electron is a "point" particle composed of no more fundamental subunits to within an experimental limit of about 10^{-16} cm. If one models the electron as a spinning charged sphere of radius equal to the so-called classical electron radius, $r_0 = e^2/mc^2 \sim 10^{-13}$ cm—deduced by equating the electron rest mass energy mc^2 to its electrostatic potential energy e^2/r_0—the resulting linear velocity of a point on the "equator" of the electron surface would be[12]

$$v = \frac{1.25c}{\alpha_{\mathrm{fs}}} \tag{7.11}$$

(with α_{fs} the fine-structure constant), which exceeds the velocity of light by a factor of over 170. If a smaller radius is adopted, then the violation of relativity is even greater. No classical model of electron structure, in fact, has proved adequate. It seems, therefore, that spin must simply be accepted and not structurally interpreted.

With account taken of the nonclassical attribute of spin, a completely quantum mechanical analysis replaces (7.7a) with the potential energy expression

$$U_R = -\boldsymbol{\omega} \cdot \boldsymbol{F}, \tag{7.7b}$$

involving the projection onto the quantization axis of the *total* internal angular momentum (F) of the atom comprising a vector sum of the electron orbital angular momentum (L) and angular momenta contributed by electron spin (S) and nuclear spin (I).

In the hydrogen ground level, the total angular momentum derives exclusively from the spins of the electron and proton, which may be oriented either parallel or antiparallel to one another. As a result of the magnetic coupling of electron and nuclear spins, the $n = 1$ level of the hydrogen atom is composed of four hyperfine states. There is a single ground state of zero total angular momentum which, in the notation of Chapter 4, can be designated $1S_{1/2}(F = 0; M = 0)$. In this state, the electron and proton spins are oppositely directed. Above the ground state lie the three states $1S_{1/2}(F = 1; M = +1)$, $1S_{1/2}(F = 1; M = 0)$, and $1S_{1/2}(F = 1; M = -1)$ of total angular momentum $1\hbar$ that result when the electron and proton spins are parallel. In an inertial reference frame, these three states of total quantum number $F = 1$ are degenerate. If the quantum mechanical treatment leading to relation (7.7b) is correct, the rotation of the Earth should split the energy of the two states with azimuthal quantum numbers $M = \pm 1$ by an amount $\hbar\omega \sim 5 \times 10^{-20}$ eV, in accordance with relation (7.8). The hyperfine level splitting E_0 between the two rotationally unaffected 1S states with $M = 0$ corresponds to a microwave photon of frequency 1420 MHz and wavelength 21 cm. (This radiation is of much interest to radio astronomers and astrophysicists who search the skies for, among other things, interstellar clouds of atomic hydrogen gas.)

Despite the fact that, according to relation (7.7b), an atomic 1S state ought to be affected by the spin of the Earth, it does not follow from what has been said so far that atoms in the 1S state must necessarily exhibit optical activity when illuminated with microwave radiation. In fact, at first thought, it might appear that such optical activity cannot occur. All the states of the $n = 1$ ground level have the same (even) parity, i.e., the same behavior under reflection; symmetry rules strictly forbid electric dipole transitions between states of the same parity. The resulting atomic polarizability—and therefore the dielectric constant of a bulk sample of atoms—would not be differentially affected by left and right circularly polarized microwaves.

Nevertheless, quantum theory does predict a rotational optical activity near 1420 MHz. This is one of the occasions where the magnetic field of the incident radiation cannot be neglected. The electron, although a "point" particle, has not only an electric charge, but also an intrinsic magnetic moment. According to classical electromagnetism, an orbiting charged particle constitutes a simple current loop with a magnetic dipole moment. Although no such loop can be envisioned for a spinning electron with zero orbital angular momentum, there is still an intrinsic magnetic moment deriving from (and proportional to) the electron spin angular momentum. To an observer on the rotating Earth, the electron magnetic moment would appear to precess about the Earth's rotation axis—a result in keeping with the previously expressed analogy between rotation and magnetism.[13]

The interaction of this precessing magnetic moment with the magnetic field component of an electromagnetic wave ultimately gives rise to a magnetic permeability μ that is different for left and right circular polarizations. Since the index of refraction depends not on the dielectric constant (ε) alone but on the product $\varepsilon\mu$, a linearly polarized microwave beam passing through the sample of unexcited Earth-bound hydrogen atoms should undergo optical rotation. Surprisingly, under appropriate circumstances, this optical rotation can be several orders of magnitude larger than that attributable to the orbital motion of the electron.

All things being equal—resonance frequency, radiation frequency, numbers of atoms per unit volume, and so forth—magnetic interactions are ordinarily weaker than comparable electric interactions by the square of the fine-structure constant, $\alpha_{fs}^2 \sim 5 \times 10^{-5}$. How, then, can the rotational optical activity associated with electron spin exceed that associated with electron orbital motion? The answer is that all things here are not equal; in particular, the frequencies of (virtual) electric dipole transitions that contribute to rotational optical activity in the visible and ultraviolet are some five or six orders of magnitude larger than the 1S hyperfine transition frequency. The significance of this is as follows.

Recall that as long as the atoms do not absorb the incident light, the birefringence of a sample increases as the light frequency approaches a resonance frequency. The frequency at which absorption can occur, however, is not infinitely sharp. First, the energy levels themselves have a natural width resulting from their finite lifetime (from spontaneous emission). Second, if the atoms are moving about randomly as a result of thermal motion, the absorbed light will extend over a range of Doppler-shifted frequencies. And third, if the sample is sufficiently dense, the atoms will collide with one another, thereby increasing the energy uncertainty of the states.

At low density, it is the Doppler effect that principally determines the range of frequencies over which absorption occurs; the extent of Doppler broadening, it is important to note, is proportional to the resonance frequency. It is the potentially very narrow Doppler width of the hydrogen hyperfine transition that allows one, in principle, to probe the atom with microwave frequencies lying much closer to a resonance than would be possible with visible and ultraviolet radiation. Unfortunately, the resulting optical activity is still extremely weak, and as the technology for state-of-the-art polarimetry is far more advanced for high-frequency electromagnetic waves than for microwaves, the calculated enhancement is unlikely to be of much experimental help at the present.

Any attempt to measure the optical activity of atoms induced by the Earth's rotation will have to overcome some formidable experimental hurdles. One of the most daunting is that of the Earth's own magnetic field, which, with a strength of approximately one-half gauss, gives rise to a true Faraday rotation larger than the sought-for effect by some eleven orders of magnitude.[14] However, the field of quantum magnetometry—the measurement of ultrasmall magnetic fields by means of superconducting quantum devices (SQUIDS)—has already achieved wonders; fields as weak as 10^{-11} gauss can be measured. The question is whether an extant field can be shielded to that low value.

The pursuit of an interaction between the rotating Earth and an Earth-bound atom raises another, perhaps more basic and thought-provoking, question as well. What, in fact, does it *mean* to say that an atom rotates? Mathematically, quantum mechanics provides a formal procedure for expressing any wave function (or operator) in terms of the coordinates of a rotated reference system; this is termed the passive view of rotation. The active view, whereby the wave function itself is rotated with respect to a fixed frame of reference, is considered—again mathematically—to be entirely equivalent. It was by application of such transformations that the equations of motion of a rotating quantum system have been derived and the attendant phenomenon of rotational optical activity inferred.

But does the rotational displacement of an atom—which must necessitate in some way the physical coupling of the atom to its environment through forces of constraint—actually imply as well that the bound electrons are so coupled? This is not an idle inquiry arising from the paradox-laden terminology of classical physics. The issue can be settled in the laboratory: If the frame of reference rotates, but the atom does not, there will be no rotational optical activity. Experiments to search for chiral asymmetries in the interaction of atomic gases or vapors rapidly spun on a laboratory turntable are now under development. At the rates of mechanical rotation achievable (about 100 revolutions per second), the expected optical rotary power (or companion phenomenon of circular dichroism), if it exists, should be measurable.

For nearly two centuries, optical activity has been a catalyst in the progress of science both through the research undertaken to understand it and as an experimental tool to investigate other phenomena. It does not appear even now to be an exhausted subject.

Notes

1. F. C. Michel, Neutral Weak Interaction Currents, *Physical Review* **B138** (1963) 408.
2. M. A. Bouchiat and L. Pottier, Optical Experiments and Weak Interactions, *Science* **234** (1986) 1203.
3. An exposition of the exaggerated relativistic and quantum phenomena to be expected if the speed of light were much smaller, and Planck's constant much larger, than their present values may be found in George Gamow's delightful Mr. Tompkins books combined in *Mr. Tompkins in Paperback* (Cambridge University Press, London, 1971).
4. F. C. Witteborn and W. M. Fairbank, Experimental Comparison of the Gravitational Force on Freely Falling Electrons and Metallic Electrons, *Physical Review Letters* **19** (1967) 1049.
5. L. I. Schiff and M. V. Barnhill, Gravitational-Induced Electric Field Near a Metal, *Physical Review* **151** (1966) 1067.
6. R. von Eötvös, D. Pekar, and E. Fekete, Beiträge zum Gesetze der Proportionalität von Trägheit und Gravität, *Annalen der Physik (Leipzig)* **68** (1922) 11.
7. E. Fischbach *et al.*, Reanalysis of the Eötvös Experiment, *Physical Review Letters* **56** (1986) 3.
8. R. Colella, A. W. Overhauser, and S. A. Werner, Observation of Gravitionally Induced Quantum Interference, *Physical Review Letters* **34** (1975) 1472.
9. S. A. Werner *et al.*, Effect of the Earth's Rotation on the Quantum Mechanical Phases of the Neutron, *Physical Review Letters* **42** (1979) 1103; J.-L. Staudenmann *et al.*, Gravity and Inertia in Quantum Mechanics, *Physical Review* **A21** (1980) 1419.
10. There are, of course, still weaker interactions, as, for example, that of an atom with gravitational waves, the rippling of the space–time continuum itself, produced (among other means) by catastrophic collapse of some massive astrophysical object.
11. G. E. Stedman and H. R. Bilger, Could a Ring Laser Reveal the QED Anomaly via Vacuum Chirality, *Physics Letters* **A122** (1987) 289.
12. The angular momentum of a sphere of radius r and mass M rotating at angular frequency ω about an axis through the center is $\frac{2}{5}Mr^2\omega$. Replacing r with r_0 and setting the angular momentum equal to $\frac{1}{2}\hbar$ leads to an equatorial linear velocity $v = \omega r = 5\hbar/4Mr_0$, which reduces to relation (7.11).
13. The magnetic moment of a current loop in a uniform magnetic field experiences a torque produced by the Lorentz force as a result of which it precesses like a gyroscope about the field direction. Although not describable as a current loop, the magnetic moment arising from particle spin undergoes a similar precession when acted upon by a magnetic field.
14. From Eq. (7.9b), the "Larmor field" corresponding to the Earth's angular frequency of rotation (7.3×10^{-5} radians/s) is 8.3×10^{-12} gauss.

Computers, Coins, and Quanta: Unexpected Outcomes of Random Events

8.1. The Suggestive Power of Fun

Many years ago, I participated in an international conference devoted to improving the teaching of science at all levels of instruction. Although I now recall little of the numerous talks and heated discussions that the conference engendered, there was one event that I have not forgotten. In his introductory remarks, an invited speaker, noted for his compendious study of the life of Isaac Newton, starkly announced that not once, in all the years that Newton engaged in his physical researches, had he (Newton) ever had any "fun." According to the speaker, the pursuit of scientific knowledge for Newton was a solemn and sacred undertaking which the word fun grotesquely trivialized. Moreover, the speaker continued somewhat scornfully, this is precisely how it *should* be; science is too serious a matter to be pursued—*or taught*—with the idea of fun in mind . . . and the sooner teachers grasped this point, the sooner they would be able to teach science more effectively.

I was stunned. I am not a historian, although I have read enough books about Newton to agree that "fun-loving" is not exactly the adjective to apply to a reclusive genius with tendencies toward paranoia. On the other hand, as a scientist—one of very few at the conference in question—I have also read Newton's own writings. It is impossible to read Newton's *Opticks*, for example, and not sense the enormous personal satisfaction and pleasure that its author must have experienced in reflecting upon the deep philosophical problems posed by the behavior of light and in designing and executing simple, yet incisive, experiments to help unravel these mysteries of natural philosophy. Perhaps fun may not be the appropriate word, but any conception of science that ignores the intellectual delight of satisfying one's curiosity, overcoming challenges, and making discoveries has missed a seminal attraction of science both in Newton's time and our own. Indeed, it is precisely this sense of exhilaration and fulfillment in the pursuit of understanding how the world—or a tiny part of it—works that a teacher must communicate to students if they are to appreciate science as something more than a collection of facts and formulas.

Newton was fascinated by the physical behavior of much of what he encountered around him: how objects moved when pushed, how objects fell when released, how objects cooled when heated, how fluids flowed, and what happened to light when it passed through or around various things, to cite but a few of Newton's preoccupations. In the motions and transformations of familiar physical objects, Newton found far-reaching principles waiting to be revealed.

Science has evolved over the past three centuries in ways that Newton could never have imagined, and the objects familiar to many a physicist today now comprise those that can be seen only by powerful microscopes or with satellite-based telescopes or by means of some other kind of expensive apparatus usually requiring the financial support of one government agency or another. In some ways, that is rather unfortunate, although seemingly necessary if the boundaries of scientific knowledge are to expand, for it tends to breed an attitude among at least some scientists and science editors not unlike the attitude of the historian above. The remark of one anonymous wag in the audience of a quantum mechanics conference I spoke at long ago captured this frame of mind precisely. Paraphrasing physicist John A. Wheeler's cryptic assertion that "a phenomenon is not a phenomenon until it is a *measured* phenomenon,"[1] the wag blurted out, "a phenomenon is not a phenomenon until it is a *funded* phenomenon!" Scientists who have ever tried to publish in a premier research journal without having a funding agency to acknowledge as evidence that the submitted work was "serious" science (and *not* fun) will understand the import of the wag's observation.

I have been doing scientific research for over forty years. Much of this research, as recounted in this book and other volumes noted in the Preface, is "serious" science, i.e., part of a carefully planned research agenda. However, a significant fraction of my work was not part of any research plan at all, but undertaken on a whim, for amusement, or out of surprise at some unexpected turn of events. These adventitious projects were often the ones that I enjoyed most and from which I always learned something new and interesting. I cannot believe that a true scientist, including even Newton, does not have fun.

This two-sided nature of scientific motivation—serious and playful—is aptly expressed in Harvey Lemon's vignette of the Nobel Prize-winning American physicist, A. A. Michelson,[2] who, like Newton, was a pioneer in the investigation of light:

When asked by practical men of affairs for reasons which would justify the investment of large sums of money in researches in pure science, he was quite able to grasp their point of view and cite cogent reasons and examples whereby industry and humanity could be seen to have direct benefits from such work. But his own motive he expressed time and again to his associates in five short words, "It is such good fun."

In this chapter, I discuss a project that started as a computer game, but evolved—unexpectedly—into tests of what is perhaps the most fundamental characteristic of the quantum world: the intrinsically unpredictable occurrence of individual quantum events.

8.2. To Switch or Not to Switch—*That* Is the Question

I never heard of the so-called "Monty Hall" problem until a few years ago when I first saw mention of it in a review[3] of a newly published book of mathematical oddities. Even then, having (by choice) no television in my house, the association of the name with the host of a TV game show ("Let's Make A Deal") meant nothing to me. The problem is easy enough to state, but its solution is counterintuitive in the extreme. Indeed, I have read that, when first brought to the American public's attention by a columnist for a popular magazine,[4] it had driven even professional mathematicians to distraction.[5]

There are three closed boxes. Inside one of them is a valuable cash prize and inside each of the others is a banana. The player picks a box, but before its content is revealed, the game master (who is aware of what is inside each box) opens one that he knows contains a banana. Now, the game master offers the player the following option: The player may keep his or her original choice or (for a small fee in one version of the game) choose the other unopened box. What should the player do?

The nearly universal reply—and indeed the reply given by everyone to whom I personally posed this problem—was that it cannot matter which of the two options is selected. With but two choices remaining, there is a 50% chance of winning in either case. (It would, therefore, be ridiculous to *pay* to switch, respondents said.) This, however, is not the case. Players *double* their chance of winning if they switch. Think about that a while, before continuing.

How can one possibly double his chance of winning by choosing the other of only two boxes? The argument is actually quite simple. Assuming that there is an equal likelihood for any one of the three boxes to contain the prize, a player will have a chance of 1/3 of winning if he selects a box and keeps it. This means that there is a probability of 2/3 of not getting the prize on the first selection. However, if the player switches, then 2/3 becomes the probability of winning, for, under the prevailing circumstances, the unopened box to which the player switches *must* contain the prize if the originally chosen box does not. Thus, the odds of winning are twice as great if the player switches.

The preceding reasoning (as well as other more formal arguments) generally elicited a storm of protest from the ordinarily placid students, colleagues, neighbors, and friends on whom I tried the problem. Probability is a measure of present knowledge they all said; once the game master opens a box, the odds of winning jump from 1/3 to 1/2 whether or not the player switches. The fallacy of thinking this way, however, lies in ignoring the *order* in which events transpire, for this order defines the conditions which determine the probability of winning. The probability P_{switch} of winning by switching is a product of two probabilities: (a) the probability $P(A)$ that the player first picks a box with a banana (event A) and (b) the probability $P(B|A)$ that the player next picks the box with a prize (event B) given that event A has occurred[6]:

$$P_{\text{switch}} = P(A)P(B|A). \tag{8.1}$$

Under the rules of the game, the probability of initially selecting a banana is $P(A) = 2/3$ and the conditional probability of selecting the prize *after* the game master has revealed one of the banana-containing boxes is $P(B|A) = 1/1$ (i.e., 100%). Hence, $P_{\text{switch}} = 2/3$. If the game master were to have revealed the content of one of the boxes *before* the player made a first choice, then the probability of winning would have been the same whether the player kept that choice or switched. Order matters.

However, suppose—as one dissatisfied colleague argued—that the player simply flipped an unbiased coin to determine the strategy: heads (H) he keeps, tails (T) he switches. Clearly, in this case there must be a 50% chance of winning the prize either way. That observation, in fact, is true, but it does not conflict with the previous conclusion that the player is better off *choosing* to switch. The "coin-toss" strategy, which underlies the intuitive but misguided reasoning of most players, is again compounded of two distinct sets of probabilities. If P_{toss} is the probability of winning when a coin toss determines strategy, P_{keep} and P_{switch} are the original probabilities of winning by keeping or switching one's initial choice, and P_H and P_T are the probabilities (both 50%) of a fair coin landing H or T, then

$$P_{\text{toss}} = \frac{1}{2} = P_H P_{\text{keep}} + P_T P_{\text{switch}}$$

$$= \left(\frac{1}{2}\right)\left(\frac{1}{3}\right) + \left(\frac{1}{2}\right)P_{\text{switch}}. \tag{8.2}$$

From Eq. (8.2), it again follows that P_{switch} must be 2/3 and, therefore, twice P_{keep} if the overall probability of winning by the outcome of a random process is to be 1/2.

To convince both myself and others that, however unexpected, switching really doubles the odds of winning, I asked my son Chris, a high-school junior at the time, to program the game on a computer, using a random-number algorithm to distribute the prize among the boxes. In the first version of our program, created with the HyperTalk language for the Macintosh, a player picks a box, and the computer, again using the random-number generator, opens one of the two remaining boxes. If the opened box contained the prize, then obviously the player lost—but this event was not included in the dataset from which statistics were compiled, for there had been no option of switching. In a second version of the program, the computer played the entire game itself, executing many rounds of prize distribution and box selection with the opening of a prize-containing box automatically excluded.

The results of 20,000 games—10,000 each for the strategies of keeping or switching—are summarized in Figure 8.1. The fraction of times each box was assigned a prize was very close to 1/3, as was also the fraction of times each box was selected by the "player." The strategy of keeping the original choice resulted in winning the prize in 3359/10,000 = 33.59% of the games. By switching, however, the fraction of wins jumped to a smashing 6639/10,000 = 66.39%.

What more can I say?

8.3. On the Run: How Random Is Random?

Actually, there *is* more to say. It was while programming and playing the game that we noticed that the computer seemed to behave rather oddly at times. Although, on average, each box was assigned the prize in one-third of the total trials, in detail the computer occasionally assigned the prize to the *same* box three or four or more times in succession. Was there a defect in the program? Could it be that the internal random-number generator was not generating random numbers? Or were these outcomes to be expected even in the case of a perfectly random selection process? Thus, began my interest in the matter of "runs."

Random events occur without any assignable cause. Emphasis here is on "assignable," for random occurrences do not represent a suspension of the laws of physics; rather, in the absence of sufficient knowledge of initial conditions, one cannot predict their outcome individually. Consider one of the classic examples of a random process: coin tossing. Certainly, the coin is subject to Newton's laws; however,

Experimental Outcome of 20,000 Games

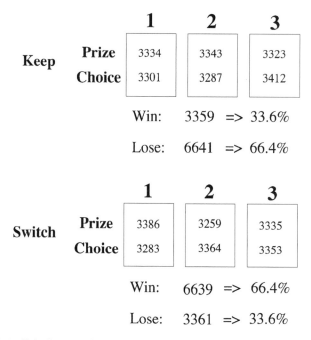

Figure 8.1. Tabulation of "Keep" and "Switch" outcomes. Figures in each box show the number of times the prize was randomly placed in the box and the number of times that the box was randomly chosen by the player (the computer).

too little, if anything, is known about the precise conditions—for example, the magnitude, direction, and point of application of the ejecting force, or the mass and size of the coin, or the pressure and viscosity of the air—under which a coin is launched. Of the two possible outcomes, we would generally expect an unbiased coin to land with head-side up (H) or tail-side up (T) with approximately equal frequency. If, therefore, in a series of tosses this is found to be the case, we are usually satisfied that the flipping was random. However, establishment of randomness is a much more subtle matter than simply the occurrence of all possible outcomes with equal frequency.

If you are a teacher (and therefore have a captive audience), try the following experiment with your class. Divide the class into two groups. For homework, tell one group to toss a coin 256 times and write down in sequence the outcome of each toss; tell the other group to write down what they would *imagine* a typical sequence of 256 random tosses to be (but not actually to do the tossing). The students are not to indicate on their papers whether the recorded data were obtained experimentally or "imaginatively." With a knowledge of the nature of randomness—which presumably the students do not possess—the teacher can determine with approximately 98% accuracy which sets of data were obtained experimentally. How?

I will tell you shortly. The key to the solution, however, involves the idea of a "run," a sequence of binary outcomes of the same kind. Figure 8.2a gives an example of what the results of 256 sequential coin tosses might look like. Each 1 signifies a head and each 0 signifies a tail. Tosses commence with the uppermost left bit (i.e., binary digit) and continue from left to right to the lowermost right bit. Scanning the rows of numbers, one sees, as expected, apparently random fluctuations between short strings of 0s and 1s. But every so often there occur unexpectedly long strings, such as the runs of seven 0s and seven 1s boxed in the figure, which seem to represent islands of order amid disorder. The table in Figure 8.2b displays the total number of runs of 0s and 1s of all lengths that occurred in the sequence of 256 coin tosses. It is essential to recognize that the occurrence of long runs is a natural outcome of randomness and does not necessarily signify any underlying regularity or assignable cause. Indeed, were there to be a deficiency of long runs, the process in question assuredly would *not* be random.

Consider, for example, the question of how likely it is to obtain a run of at least 8H or 8T in 10 tosses. Figure 8.3 illustrates the ways in which such runs can occur. There are five sequences leading precisely to a run of 8H, two sequences leading to a run of 9H, and one sequence leading to a run of 10H. Since these statistics are the same for runs of 8T, 9T, and 10T, there are in all 16 sequences leading to runs of at least 8H or 8T. However, with two possible outcomes for each toss, there are in all $2^{10} = 1024$ distinct sequences of 10 tosses. Thus, the probability of obtaining a run of at least 8H or 8T is 16/1024 = 1.6%. Is this a low probability? Well, let me put it this way: If you knew that the planes of a particular airline went down in flames once in every 64 flights (1024/16 = 64), would you fly on this airline? Small is relative; an unlikely outcome will eventually occur if the number of trials is sufficiently high.

As the total number of tosses in a sequence increases, the enumeration of all individual configurations leading to runs of a particular length or greater becomes very cumbersome. Mathematicians have developed general formulas, but we can deduce the statistics of random runs to good approximation by a simple argument. I designate $\bar{r}^n_{kH}$ to be the mean number of times a run of length k heads occurs in n tosses, and $\bar{R}^n_{kH}$ the mean number of times of a run of *at least* k heads occurs

256 Sequential Coin Tosses

[1] Heads: 138
[0] Tails: 118

(a)

```
1011111010110110101101001000110 0
1101111010000100100111001010010 0
1100110011111000100000101111100 0
1011001000111110011011100111001 0
1111100001101110000000010111110 00
1111011011000000101000001011111 0
1111110011101100101110001011111 0
0111011011110000111111100000110 0
```

(b)

Run Length	Runs of H	Runs of T	Total Runs
1	21	26	47
2	15	17	32
3	7	6	13
4	3	3	6
5	8	3	11
6	1	1	2
7	1	1	2

Figure 8.2. (a) The experimental outcome of 256 sequential coin tosses (starting from the upper leftmost bit) in which 0 represents T and 1 represents H. (b) The table displays the number of runs of each kind for all observed lengths.

in n tosses. (The overbar signifies the average result of a series of experiments, each experiment comprising n tosses.) The corresponding symbols $\bar{r}_{k\mathrm{T}}^n$ and $\bar{R}_{k\mathrm{T}}^n$ apply to tails, and

$$\bar{r}_k^n = \bar{r}_{k\mathrm{H}}^n + \bar{r}_{k\mathrm{T}}^n, \tag{8.3a}$$

$$\bar{R}_k^n = \bar{R}_{k\mathrm{H}}^n + \bar{R}_{k\mathrm{T}}^n \tag{8.3b}$$

give the mean numbers for run lengths of H and T taken together.

Toss Number

1	2	3	4	5	6	7	8	9	10

H	H	H	H	H	H	H	H	T	T/H
T	H	H	H	H	H	H	H	H	T
T/H	T	H	H	H	H	H	H	H	H

H	H	H	H	H	H	H	H	H	T
T	H	H	H	H	H	H	H	H	H

H	H	H	H	H	H	H	H	H	H

Outcomes

Figure 8.3. Tabulation of all possible outcomes for a run of at least 8H or 8T in a sample of 10 coin tosses.

Since $\overline{R}_{k\mathrm{H}}^{n}$ represents the mean number of runs of length k or greater (i.e., the sum of the numbers of runs of length k, $k + 1$, $k + 2$, etc., up to the largest occurring length), it is clear that the difference of $\overline{R}_{k\mathrm{H}}^{n}$ and $\overline{R}_{(k+1)\mathrm{H}}^{n}$ gives the mean number of runs of precisely length k, or

$$\overline{r}_{k\mathrm{H}}^{n} = \overline{R}_{k\mathrm{H}}^{n} - \overline{R}_{(k+1)\mathrm{H}}^{n}. \tag{8.4}$$

Equation (8.4) provides the simplest way to calculate $\overline{r}_{k\mathrm{H}}^{n}$ once one has determined the formula for $\overline{R}_{k\mathrm{H}}^{n}$. The same formula (with T replacing H) applies to tails.

Suppose that we want the mean number of times a run of at least 5H appears in $n = 256$ sequential tosses of a fair coin. In effect, we are asking for the number of times the chain of events (T H H H H H ...) appears in the full sequence. The jump from T to H starts the run, and the dots following the fifth H signify that either T or H can follow. Because each H or T in a sequence of random tosses occurs independently[7] with probability $\frac{1}{2}$, the overall probability of the foregoing configuration of six tosses is $(\frac{1}{2})^{6} = 1/64$. On average, therefore, we should expect to find that a run of at least 5H occurs $(1/64) \times 256 = 4$ times in 256 tosses;[8] likewise for a run of 5T. (Thus, there should occur a total of about 8 runs of length 5 or longer in 256 tosses.)

More generally, for a run of at least $k\mathrm{H}$, where $k \geq 1$, we have

$$\overline{R}_{kH}^{n} \sim \frac{n}{2^{k+1}}.$$ (8.5)

Then, from Eq. (8.4),

$$\overline{r}_{kH}^{n} = \overline{R}_{kH}^{n} - \overline{R}_{(k+1)H}^{n} \sim \frac{n}{2^{k+1}} - \frac{n}{2^{k+2}} = \frac{n}{2^{k+2}}.$$ (8.6)

As an alternative route to Eq. (8.6), one can simply note that the probability of realizing a run of precisely kH is $(\frac{1}{2})^{k+2}$ because the run is terminated on *both* sides by a T, as in the sequence (T H H H H H T) for a run of precisely 5H. Multiplying this probability by n leads directly to Eq. (8.6).

It is also useful to determine the mean count $\overline{R}_n$ of all runs of H and T, irrespective of length (i.e., the sum of runs of length 1 or greater). This single statistic,

$$\overline{R}_n = \overline{R}_{1H}^{n} + \overline{R}_{1T}^{n} \sim \frac{n}{2},$$ (8.7)

provides a quick assessment of whether or not a process analogous to a coin toss is random. Too small a value of $\overline{R}_n$ (i.e., too few runs) is an indication of "clumping"; there is too little change for the process likely to be random. On the other hand, too large a value of $\overline{R}_n$ (i.e., too many runs) signifies too much regularity in the reversal of outcomes. A single statistic, however, is not as reliable an indicator of randomness as having a detailed breakdown of the numbers of runs of each length.

Equations (8.5)–(8.7) are approximate relations, for I have ignored certain configurations such as the possibility of runs occurring at the start or closure of a sequence of tosses, in which case the first or last jump in the chain is absent. For a long sequence, however, the contribution of "end runs" becomes negligible compared to the number of inside configurations ("home runs"). The exact general relations are somewhat complicated,[9] but they reduce to Eqs. (8.5)–(8.7) for the special, though widely applicable, case of large n, small ratio k/n, and equal numbers of heads and tails.

I return now to the homework assignment that I suggested earlier. On the basis of Eqs. (8.5) and (8.3b) and the assumption of equal numbers of H and T, the teacher should expect to find approximately $2(256/2^7) = 4$ runs of 6 or more heads or tails in an *experimentally* produced sequence of 256 tosses of a fair coin. It is rather unlikely that a person unfamiliar with the characteristics of random processes would imagine, in attempting to predict the outcome of 256 tosses, that the coin should land the same way 6, 7, or more times in succession—and not just once but about *four* times. Thus, a quick scan of a student's figures should readily reveal whether the data have been fabricated or not.

The teacher could also count the total number of runs on a student's paper. According to Eq. (8.7), this should be approximately $256/2 = 128$. (This would be considerably more cumbersome and time-consuming, however, unless the teacher could scan each paper into a computer and let the computer do the counting.)

In assessing whether a given process is random or not, it is not sufficient to know only mean numbers of runs; one must be able to estimate as well the dispersion about the mean. This is generally true for the application of statistical reasoning to any problem. For example, upon tossing a coin 100 times, we would expect to get approximately—but not necessarily exactly—50 heads and 50 tails. If a particular experiment led to 54H and 46T, should we suspect that the process was biased toward generating heads?

The usual measure of the dispersion about the mean is the root-mean-square or standard deviation, σ. If the outcome of a random process is some random variable x, which occurs with a mean value $\bar{x}$, then the so-called variance of x is the average (also called the expectation) of the square of the deviations of x from $\bar{x}$, or $\overline{(x-\bar{x})^2}$.[10] The standard deviation is the square root of the variance, $\sigma = \sqrt{\overline{(x-\bar{x})^2}}$. In many cases, it is possible to predict on the basis of a theoretical model of the random process what the mean and standard deviation should be. Then, if a particular experimental outcome falls within the theoretically predicted range $\pm\sigma$ of the expected value $\bar{x}$, there is ordinarily no reason (in the absence of other information) to assume that the results are biased.

It is not difficult to estimate theoretically both the mean and standard deviation for coin-tossing. To return to the above example, if N is the number of tosses, P_H is the probability that a toss yields H, and $P_T = 1 - P_H$ is the probability that a toss yields T, then the mean number of heads expected is (as already illustrated in Note 7) $n_H = NP_H$, and the standard deviation about the mean can be shown (although not here) to be $\sigma_H = \sqrt{NP_H P_T}$.[11] Thus, for 100 tosses of a fair coin ($P_H = P_T = \frac{1}{2}$), one can expect the number of heads to fall within a range of $\pm\sqrt{100/(2 \times 2)} = \pm 5$ of $n_H = 50$. The occurrence of 54H, therefore, does not in itself hint of any bias toward heads (although examination of the corresponding run data may indicate otherwise).

With regard to runs of coin tosses, deriving the standard deviations about $\bar{R}^n_{kH}$ and $\bar{r}^n_{kH}$ is an arduous mathematical task, far more difficult than calculating σ_H about n_H. However, if the sequence of bits is sufficiently large (in principle, $n \to \infty$, but in practice, $n > \sim 20$), the distribution of $\bar{R}^n_{kH}$ converges toward a bell-shaped curve (Gaussian distribution) with a standard deviation approximately equal to the square root of the mean:

$$\sigma(R^n_{kH}) \sim \sqrt{\overline{R^n_{kH}}} \sim \sqrt{\frac{n}{2^{k+1}}}. \tag{8.8}$$

As a consequence of Eq. (8.8) and the definition of variance, it follows that the standard deviation of $\overline{R}_n$ is (again in the limit of large n)

$$\sigma(R_n) \sim \frac{\sqrt{n}}{2}. \tag{8.9}$$

To someone familiar with statistical theory, Eq. (8.8) suggests that the probability of obtaining $\overline{R}^n_{kH}$ or $\overline{R}^n_{kT}$ for a sequence of n tosses can be estimated by a Poisson distribution. The Poisson distribution occurs widely in physics, characterizing phenomena as different as the fall of raindrops and the disintegration of nuclei. It is the distribution that ordinarily describes random processes in which the probability of a desired outcome (e.g., the decay of a nucleus) is very low, but the size of the sample (e.g., number of nuclei) is enormous. A defining characteristic of the Poisson distribution is that the variance (σ^2) of the distributed quantity is equal to the mean (μ). The mathematical form of the Poisson distribution is uniquely determined by the mean alone:

$$P^n_m(\mu) = \frac{\mu^m}{m!} e^{-\mu}. \tag{8.10}$$

In Eq. (8.10), $P^n_m(\mu)$ is the probability of the occurrence of m events out of n tries in which the mean number of occurrences is μ.

Figure 8.4 shows the form of the Poisson distribution for a mean value $\mu = 10$. Note that this is a *discrete* distribution; that is, the diamond plotting symbols show the mathematically meaningful points, whereas the continuous curve through the diamonds merely helps the eye trace out the general shape. The shape is centered about the mean value 10, but is not perfectly symmetric; it rises more sharply on the left and trails more slowly on the right. As μ becomes larger, the Poisson distribution becomes more symmetric, approaching a bell-shaped (or Gaussian) curve.

We now return one last time to the proposed homework assignment. Substituting the mean $\mu = \overline{R}^n_k = 2\overline{R}^n_{kH} = n/2^k$ into Eq. (8.10) enables us to estimate the probability of getting m runs of length k or more heads or tails in n tosses. Note that the probability of getting *no* runs ($m = 0$) is $e^{-\mu}$ and, therefore, the probability of getting *at least one run* of k bits or longer must be $1 - e^{-\mu}$, or explicitly

$$P(R^n_k \geq 1) = 1 - e^{-(n/2^k)}. \tag{8.11}$$

In a random sequence of 256 coin tosses, the teacher should expect to find at least 1 run of 6 heads or 6 tails with a probability $P(R^{256}_6 \geq 1)$ $= 1 - e^{-(256/2^6)} = 1 - e^{-4} = 98.2\%$. If the sequences imagined by students

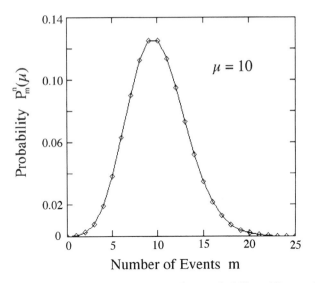

Figure 8.4. Poisson distribution giving the probability (diamond plotting symbols) of obtaining m events out of n tries for a mean $\mu = 10$. The continuous curve through the symbols merely traces out the basic shape of the distribution.

unfamiliar with the characteristics of randomness do not contain long runs, the teacher should be able to distinguish them reliably from the experimentally generated, and presumably truly random, sequences.

If the reader is already surprised at the degrees to which an unbiased coin toss can lead to long sequences of identical outcomes and therefore to the semblance of order, there is an alternative way to view these coin-toss fluctuations that, in the words of one statistician,[12] "not only are unexpected but actually come as a shock to intuition and common sense."

Suppose two players (A and B) are gambling by means of the toss of a fair coin. If the outcome is H, A *receives* $1 from B, and if the outcome is T, A *pays* $1 to B. I have asked people, those with training in physics as well as those with no particular science or mathematics background, to predict what the record of accumulated gain of either player would look like as the number of tosses increased. By accumulated gain, I mean the sum of a player's wins and losses. For example, if the first five tosses yielded H T T H T, then the gain record (in dollars) for player A would be 1, 0, −1, 0, −1, and therefore at the end of the five tosses A would have lost a net $1 to B. Most people (if I assume the replies to me are typical) would guess that the lead in the game fluctuates back and forth frequently between the two players, so that the accumulated gain of either player never diverges too far from 0 and, therefore, each player is in the lead roughly 50% of the duration of the set. After all, the chance of tossing H is exactly the same as the chance of tossing T.

An actual record of accumulation, however, is likely to be *completely different* from this imagined scenario. Indeed, the probability is very low that each player dominates the game for 50% of the time. Much greater by far is the probability that one player will either lead (positive accumulation) or lag (negative accumulation) throughout most of the game.[13] Examples of this seemingly strange behavior are exhibited in Figure 8.5. The three frames represent the records of three sets of 1000 computer-simulated tosses whereby the outcome of each toss was determined by a random-number generator. The generator produced random numbers uniformly distributed over the interval from 0 to 1. A number falling within the interval from 0 to less than 0.5 constituted a head; a number within the interval from 0.5 to 1 constituted a tail. Figure 8.6 shows a sample of the record of binary outputs (H = +1, T = −1) from the 200th through the 600th coin toss. The full record for all three sets of games looks very similar.

As one can see, a player's accumulation rarely returned to zero, the breakeven point, even though the random-number generator produced heads and tails that fluctuated randomly (in accordance with the theory of runs). In the second set, one player has led for close to the entire 1000 games. In the third set, one player has lagged for over 800 games. How can this possibly be? Does this not violate the condition that a player's expected accumulation[14] should be 0 if the probabilities of a coin landing H or T are equal?

In fact, no. The concept of probability is a somewhat slippery one usually based on the idea of frequency. If in a large number n of trials a particular event occurs k times, then the probability of this event should be close to the ratio k/n. The rigorous justification of this intuitive reasoning is known as the law of large numbers, but it is common to read into this law more than it implies. For example, the law of large numbers implies nothing about the fluctuations between H and T *within* a fixed set of games; it asserts only that over many sets of games, the frequency with which the player winning on H dominates is ideally the same as that with which the player winning on T dominates. The expected accumulation of either player over many sets is zero even though within a particular set a player is likely to accumulate or lose a significant sum.

Expressed in terms of gambling, the tendency of one player (either one) to lead in nearly all the games of a set appears extraordinarily peculiar. However, one can regard the entire process from a different perspective that, at least to a physicist, may seem more reasonable. The perspective is that of a random walk. Imagine a drunk by a lamppost who with each step (of equal length) can move at random either

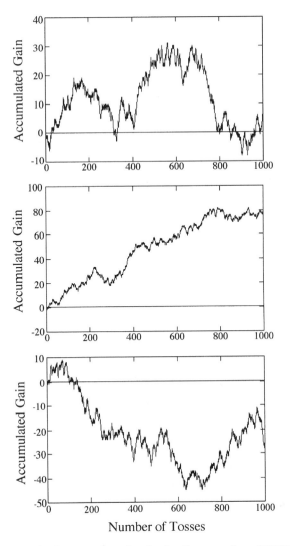

Figure 8.5. Record of accumulated gain for three series of 1000 tosses of a fair coin.

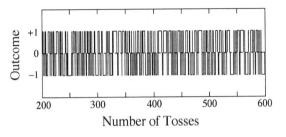

Figure 8.6. Sample of the sequence of binary outputs (H = +1; T = −1) covering tosses 200 through 600 in the first record of Figure 8.5.

to the left or right. Let a coin toss decide the direction: H to the right, T to the left. The analog to accumulated gain is the net displacement of the drunk from the lamppost. After n steps (i.e., tosses), how far from the lamppost would the drunk be expected to be? An elementary analysis of the random walk problem leads to a root-mean-square displacement of $\sqrt{n}$ steps. In other words, if the drunk took 100 random steps of some fixed length, he would likely be found at a distance of approximately 10 steps from the lamppost (either to the right or left with equal probability), rather than at the point of origin (the breakeven point in the gambling game).

If the steps were taken uniformly in time, then Figure 8.5 could very well represent not only the staggering of a drunk but also the (one-dimensional) diffusion of a molecule. Looked at as a diffusion process, the once bizarre outcome of the gambling game may now seem plausible even to someone without a background in physics. After opening a bottle of perfume, most people, I believe, would not expect the preponderance of released molecules to diffuse back to the mouth of the bottle.

It is amazing how a simple change of perspective can transform what was initially a shock to intuition and common sense into a commonplace phenomenon.

8.4. Random Acts of Measurement

Having mastered, or at least acquainted myself, with the theory of runs, I returned to the question, prompted initially by computer simulation of the Monty Hall game, of just how random are the numbers generated by a random number generator. The numbers cannot be truly random, of course, since they are the deterministic output of a mathematical algorithm.

An example of such an algorithm is the iterative relation $x_i = Ax_{i-1}(\mod m)$, in which A and m are integer constants. The formula embodies a procedure by which one begins with a seed number x_0, multiplies it by A, and then divides the result by m, discarding the integer part and keeping only the remainder x_1, which is then treated in the same way to generate the next pseudo-random-number x_2, and so on.[15] Depending on the form of m, the series of numbers generated will eventually repeat. For example, for $m = 2^a$ with integer a, the maximum period before the numbers start repeating is $m/4$. The fact that a period may be large does not necessarily imply that a generated sequence possesses acceptable random properties.

Table 8.1 summarizes the results of the first experiment, the simulation of $n = 16,907,972$ coin tosses using the random number

Table 8.1. Distribution of Runs of Ts in 16,907,972 Coin Tosses Simulated by a Random-Number Generator

Run Length	Observed Frequency	Theoretical Frequency	Mean Number (Experimental)	Mean Number (Theoretical)
1	2,098,949	2,113,714	1.24E–01	1.25E–01
2	1,053,814	1,056,748	6.23E–02	6.25E–02
3	528,146	528,320	3.12E–02	3.13E–02
4	264,486	264,133	1.56E–02	1.56E–02
5	133,645	132,053	7.90E–03	7.81E–03
6	66,374	66,020	3.93E–03	3.91E–03
7	33,492	33,006	1.98E–03	1.95E–03
8	16,887	16,502	9.99E–04	9.77E–04
9	8,309	8,250	4.91E–04	4.88E–04
10	4,178	4,125	2.47E–04	2.44E–04
11	2,091	2,062	1.24E–04	1.22E–04
12	1,100	1,031	6.51E–05	6.10E–05
13	540	515	3.19E–05	3.05E–05
14	278	258	1.64E–05	1.53E–05
15	154	129	9.11E–06	7.63E–06
16	62	64	3.67E–06	3.81E–06
17	27	32	1.60E–06	1.91E–06
18	17	16	1.01E–06	9.54E–07
19	15	8	8.87E–07	4.77E–07
20	4	4	2.37E–07	2.38E–07
21	0	2	0.00E+00	1.19E–07
22	1	1	5.91E–08	5.96E–08
23	1	1	5.91E–08	2.98E–08
24	0	0	0.00E+00	1.49E–08

generator of the oldest computer in my laboratory. I have no idea what algorithm was involved. As in the case of the previous gambling game, random numbers were generated over the range from 0 to 1; if the number was less than 0.5, it was designated a tail (T), otherwise it was designated a head (H).

The data in Table 8.1 pertain to runs of Ts, but the results for Hs were very similar. In the leftmost column is tabulated the length of observed runs, which ranged from $k = 1$ to 23. The frequency with which runs occurred are listed in the second column from the left. It may stretch one's credulity to believe that, in random tossing, 23 tails can land in succession, but remember that the experiment was performed nearly 17 million times! (In fact, the column ends at $k = 24$, because there occurred 1 run of 24H.) The third column from the left shows the corresponding frequencies predicted by the theory of runs [i.e., by Eq. (8.6)]. The numbers in columns 2 and 3 seem to agree rather well. Columns 4 and 5 show mean numbers of runs (i.e., the experimental and theoretical frequencies, respectively, divided by the total number of tosses n); these numbers are characteristic of the

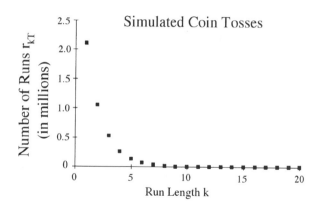

Figure 8.7. Plot of runs of tails of a specified length produced in 16,907,972 coin tosses simulated by a random-number generator. At the scale shown, corresponding experimental and theoretical points overlap one other within the width of a plotting symbol.

particular process and, in principle, independent of the number of experiments performed (in the limit of large n). The numbers in columns 4 and 5 match closely for run lengths shorter than 15. It is, of course, not unreasonable to expect greater fluctuations where the numbers of occurrences are fewer.

A visual condensation of the data in Table 8.1 is shown in Figure 8.7. Here, both the observed and predicted frequencies are plotted on the same graph, but, at the scale of the plot, each pair of corresponding points overlaps within the width of a plotting symbol. Surely, one cannot ask for better agreement than that. To conclude this, however, would be a grievous error. The agreement between experiment and theory is actually bad—*very* bad. In fact, the hypothesis that the simulated coin tosses are random can be rejected as false with virtually 100% certainty! How can this be?

The answer is evident in Figure 8.8, which plots, as a function of run length, the *difference* of the observed and expected frequencies of runs, together with error bars marking ± 1 standard deviation $[\sigma(r_{kT}^{n}) \sim \sqrt{3\bar{r}_{kT}^{n}}]$[16] about the difference. If the observed numbers of runs represented a truly random distribution of Hs and Ts, then the data points should be distributed about the horizontal axis (the line Experiment – Theory = 0) within the length of their associated error bars. The graph shows, however, a significant dearth in runs of length 1 and 2 and an excess of runs of length 5. The distribution of points about the horizontal axis is not random, but shows a clear pattern strongly suggesting an underlying assignable cause (in statistical parlance). The

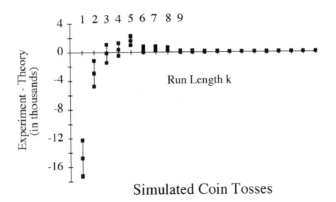

Figure 8.8. Simulated coin tosses. Plot of the difference between the observed and predicted [Eq. (8.6)] number of runs of T, $r_{kT}^{(\text{experiment})} - r_{kT}^{(\text{theory})}$, as a function of run length k. The error bars are of length 2σ in which the standard deviation can be estimated to be $\sqrt{3r_{kT}^{(\text{theory})}}$.

H runs exhibit the same pattern. Further scrutiny of the data would show that the total observed number of runs (R_n) of both T and H differs from the theoretically expected value by nearly 14 standard deviations [Eq. (8.9)]. Standard statistical tests, such as the chi-square (χ^2) test,[17] which enable one to estimate how well a set of data is described by a presumed theory, indicate that the likelihood that the simulated numbers leading to Table 8.1 or Figure 8.8 are random is 0.00%.

Why the coin-toss simulations were not random, I cannot say. Perhaps the random-number-generating algorithm had too short a period. Perhaps it did not generate random numbers uniformly over the interval 0 to 1. In any event, there emerged the important lesson that appearances can be deceiving. One must examine statistical data *carefully*. In subsequent tests with other random-number generators available in powerful mathematical software designed for modern computers, the observed distributions of runs tested reasonably well against the predictions of the theory of runs. This was also the case, I should mention, for the computer used to simulate the Monty Hall game.

At this point, I became curious to know whether physically real—as opposed to algorithmically simulated—coin tosses led to random outcomes. Since the repetitive tossing of a coin is tedious work, an alternative experiment was designed to yield more data per trial than a standard coin toss. I refer to this experiment as a "lottery experiment."

A collection of 256 pennies, each labeled with a small circular sticker numbered from 0 to 255, was placed in a bag and thoroughly mixed. A coin was then selected at random, the sticker number was recorded, the coin was replaced, and the contents of the bag were again mixed. In this way, 32 coins were randomly selected leading to a sequence of 32 decimal numbers. The numbers were converted to binary, thereby producing a string of $8 \times 32 = 256$ 1s and 0s, like the outcomes H and T of a sequence of 256 coin tosses. Because all possible arrangements of eight 0s and 1s—beginning with 00000000 and ending with 11111111—appear equally in the selection of a decimal number between 0 and 255 and because the probability (1/256) of selecting a coin from the bag is the same as the probability of tossing eight coins $(1/2^8)$, the lottery experiment is theoretically equivalent to a coin-tossing experiment but with a felicitous reduction in sampling by a factor of 8.

The experiments were repeated 25 times, leading to $256 \times 25 = 6400$ bits of data, the equivalent of 6400 coin tosses. A full tabulation of the outcomes, which need not be given here, seemed to be in good accord with the expectations of run theory. However, appearances, as we have seen, can be deceiving. For example, one run of 17 1s occurred. Was this a worrisome sign that the coin selection was not random? No, not really. From Eq. (8.11), the probability $P(R_{17}^{6400} \geq 1)$ of obtaining at least one run of length 17 is estimated to be close to 5%—i.e., 1 out of 20 experiments, whereas 25 experiments were performed. An examination of the differences between experiment and theory over the full spectrum of runs, shown in Figure 8.9, strongly supports the hypothesis that the "coin tosses" were random. Further confirmation of the consistency between experiment and theory was provided by a χ^2 test, which indicated that the observed agreement between experiment and theory would be obtained in about 60% of subsequent experiments. In other words, the string of bits resulting from the experiment was in accordance with the theory of runs, and, correspondingly, there was no reason for rejecting the hypothesis that the coin selection process was random.

Intrigued by the simplicity and sensitivity of my newly acquired statistical tool, I looked for other purportedly random processes to check, such as the roll of dice, the draw of cards, or the digits of transcendental numbers like π and e. It was while I was engrossed in the randomness of computers, games, and numbers that I realized that I now had a means to examine the most random process of all—indeed, the "Mother of All Randomness"—quantum mechanics.

Lottery Experiment

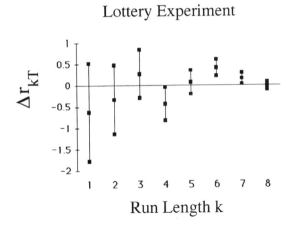

Figure 8.9. Lottery experiment. Plot of $\bar{r}_{kT}^{(\text{experiment})} - \bar{r}_{kT}^{(\text{theory})}$ as a function of run length k. (Runs of length greater than 8 are not shown, since they are indistinguishable from 0 at the scale of the graph.)

8.5. Do Radioactive Nuclei Decay Randomly?

There are numerous ways in which the systems and interactions of the quantum world differ from those of the classical world with which we are more familiar and which are adequately accounted for by Newton's laws of motion and Maxwell's theory of electromagnetism. Some of these distinctions have already been addressed in previous chapters.[18]

Classical particles, for example, move in accordance with a deterministic equation of motion. Given knowledge of their initial locations and velocities, one can, in principle, predict where they will be subsequently and how fast they will be moving. Moreover, the precision with which these properties can be determined is limited only by instrumentation. Quantum particles, by contrast, are described by an equation of motion (e.g., the Schrödinger equation or Dirac equation) that yields a probability distribution, not unique locations and velocities. Passing through apertures or around obstacles, quantum particles are distributed in wavelike fashion, giving rise to interference and diffraction patterns like the electron interferograms recorded in Figure 3.3. Measurements of conjugate quantities like the location and momentum of a quantum particle are subject to the uncertainty principle, a natural limitation prescribed by physical law and not by technology. Physicists now believe, of course, that the real building blocks of the world are quantum particles, not classical ones.

The fundamental unpredictability of individual quantum events greatly disturbed Einstein throughout his lifetime, provoking his desperate remark that

I can, if the worst comes to the worst, still realize that God may have created a world in which there are no natural laws. In short, a chaos. But that there should be statistical laws with definite solutions, i.e. laws which compel God to throw the dice in each individual case, I find highly disagreeable.[19]

Although Einstein never (to my knowledge) reconciled himself to a universe governed by statistical laws, test after test has decisively confirmed the validity of quantum mechanics in accounting for the stochastic nature of phenomena at the atomic and subatomic scales. However now, at the start of a new millennium and some three quarters of a century following the formal creation of quantum theory, how well have physicists actually ascertained that "God does not throw dice"?

Immersed in my explorations of random processes, I thought about that question, and the answer surprised me. Despite countless experiments probing the peculiarities of quantum physics, I knew of very few that specifically examined quantum outcomes for randomness. The occurrence of a particle interference or diffraction pattern, for example, although inexplicable within the framework of Newtonian mechanics, does not by itself demonstrate that particles arrive randomly at the viewing screen. Consider, for example, the single electron interference experiment (Chapter 3) I proposed at Hitachi, whereby electrons coursed one at a time through the barrel of an electron microscope to build up an interference pattern. Captured by a microchannel plate detector, each detected electron gave rise to a sharp white spot on the viewing screen. Examining the video recording of the experiment[20] at a stage when no more than a few to a few thousand electrons had been detected, one could readily believe that the dispersion of white spots looked random enough. However, appearances can be deceiving; recall the "islands" of identical outcomes in a long sequence of coin tosses. Once a sufficiently large number of electrons had been detected, the spatial distribution of spots on the screen was not at all random; electrons were clearly arriving preferentially where bright fringes were forming. Who could say that the electrons were arriving randomly in time? What evidence was there to prove that the emission of one electron did not in some way influence the time of emission of a later one? No one examined, then or afterward, the sequential arrival times of the individual electrons.

The question of whether single quantum events occur unpredictably is applicable to every kind of quantum transition or transformation. There are, for example, in addition to the field emission of electrons from metals exploited in the electron microscope, countless types of quantum jumps of bound electrons in atoms and molecules and numerous kinds of decays of unstable particles. Of the latter, the disintegration of radioactive atomic nuclei provides a particularly clean and accessible testing ground.

Despite the designation "atom," deriving from Greek roots meaning "indivisible," the transmutability of atoms has been known since the researches of Henri Becquerel and Pierre and Marie Curie in the late 1890s. It was the Curies who introduced the term "radioactivity" to characterize the emanation of rays from uranium, which they recognized to be an atomic phenomenon characteristic of the element and not a feature of its chemical environment or physical state. Indeed, one of the principal attributes of nuclear decay is how small an influence, if any, the outside world exerts on the decay rate. There are some exceptions, but, overall, the atomic nucleus, contained by an energy barrier of the order of millions of electron volts (MeV), follows its quantum destiny unperturbed by its thermal, optical, or electronic surroundings.

Nuclei can decay in various ways depending on their mass and distribution of protons and neutrons. One way is by the emission of an alpha (α) particle, or helium nucleus $^4_2$He, to produce a new element with atomic number (Z) diminished by 2 and atomic mass number (A) diminished by 4.[21] In the early days of nuclear physics, before the creation of quantum mechanics, the alpha decay of nuclei posed a seemingly insurmountable puzzle. The alphas should never have been emitted. Scattering experiments with alpha particles on uranium ($^{238}_{92}$U), for example, showed that there was no deviation from the electrostatic (Coulomb) potential around a uranium nucleus for incident alpha-particle energies up to at least about 8.8 MeV. Yet, $^{238}_{92}$U emitted alpha particles with energies of only 4.2 MeV. How could these alpha particles have gotten over a nuclear energy barrier requiring at least twice as much energy?

Although it is risky to ascribe classical imagery to quantum events, there is a certain validity to picturing the alpha particle as having a prior existence within the nucleus where it moves about, repeatedly striking the spherical "wall" of the potential well within which it is confined by the strong nuclear force, until it eventually tunnels through the barrier and escapes as a free particle. In what was essentially the first successful application of quantum mechanics to nuclear physics (in 1928), George Gamow, and independently R. W. Gurney and E. U. Condon, applied the model of quantum mechanical tunneling to derive the transmission probability of an alpha particle.[22] Besides answering the question of how an alpha particle can escape from the nucleus, the theory also accounted, at least approximately, for the extremely sensitive dependence of the lifetimes of alpha-radioactive nuclei on alpha-particle energy. A factor of 2 increase in energy, for example, could lead to a mean lifetime shorter by 20 orders of magnitude![23]

There is no analog to tunneling in classical mechanics. However, the process finds a close counterpart in the optical phenomenon of "frustrated total reflection." A beam of light directed onto the boundary between two media at an angle of incidence greater than the critical angle will be totally reflected at the surface, and no light energy will be transmitted through the second medium. If, however, this second medium, comprising a layer of a few wavelengths, were sandwiched between thick layers of the first medium, some of the incident light will tunnel through the thin layer and emerge in the third. The basis for tunneling is the presence of an exponentially damped evanescent wave in the "forbidden" medium. (The phenomenon of enhanced reflection of light, discussed in Chapter 6, is a consequence of an evanescent wave in an amplifying medium.) Frustrated total reflection can be demonstrated with visible light tunneling through a thin film of air between two layers of glass. An even more dramatic demonstration can be devised using microwaves, which permits a gap between incident and exit layers of the order of centimeters.

Another mechanism for nuclear disintegration is the emission of a beta (β) particle, or energetic electron, to create a new element of the same A but with Z increased by 1. Like alpha decay, the phenomenon of beta decay also posed initially a number of exceedingly difficult problems. For one thing, the disintegration seemed to violate most of the major conservation laws. For example, the total energy, linear momentum, and angular momentum of the detectable particles after decay were less than the corresponding values for the nucleus before decay. (Electric charge, however, was conserved.) Moreover, in marked contrast to the case of alpha decay, the assumption that the beta particle had a prior existence within the nucleus led to serious inconsistencies. According to the uncertainty principle, confining an electron to a region of nuclear size (radius $\sim 10^{-12}$ cm) would result in an energy uncertainty much larger than the depth of the nuclear potential well. An electron, therefore, could not be confined to a nucleus.[24] Furthermore, assuming that a nucleus consisted of protons and electrons (the only known subatomic particles at the time) could not account correctly for the nuclear spin angular momentum. With discovery of the neutron by Chadwick and the hypothesis of the neutrino by Pauli, nuclear beta decay could be explained as the transmutation of a bound neutron

$$n \rightarrow p + e + \bar{\nu} \qquad (8.12)$$

into a proton with emission of an electron (the beta particle) and a ghostly third particle, the (anti)neutrino, which carried off the missing energy and momenta.

Although critical details concerning mechanisms of nuclear decay could not be known until after the creation of quantum mechanics, the fundamentally stochastic nature of radioactivity was recognized in 1905 by the little-known physicist, E. von Schweidler, well before quantum mechanics and, indeed, even before the concept of the atomic nucleus had been formulated by Rutherford (in 1911). Von Schweidler assumed that the probability p for a particular nucleus to disintegrate was independent of the past history and present circumstances of the atom and depended only on the length of time Δt (for short intervals). In other words, one can express the probability as

$$p = \lambda \Delta t, \tag{8.13}$$

in which each radioactive nucleus has a characteristic decay constant λ (not to be mistaken for the symbol for wavelength used elsewhere in this book). If the assumption leading to relation (8.13) is correct, then it is not merely a metaphor, but a mathematical truth, to liken nuclear decay to a coin toss.

If p is the probability that a nucleus decays in time interval Δt, then $q = 1 - p$ is the probability that the nucleus does not decay during this interval. Thus, the probability $P(t)$ that the nucleus survives to the end of n time intervals $t = n\Delta t$ is $q^n = (1 - p)^n = (1 - \lambda t/n)^n$. If there is one "toss" per interval Δt and a head represents decay, then we are, in effect, asking for the probability of tossing n tails in succession. Nuclear disintegrations, however, occur continuously in time. In the limit that the duration of an interval becomes vanishingly small and the number of intervals unboundedly large, the mathematical expression for the probability of survival,

$$P(t) = \lim_{n \to \infty} \left(1 - \frac{\lambda t}{n} \right)^n = e^{-\lambda t}, \tag{8.14a}$$

approaches an exponential function.[25] However, according to the law of large numbers, the probability $P(t)$ is also representable as the ratio of the number N of nuclei surviving at time t and the total number N_0 of nuclei present at the outset. Thus, one arrives by a combinatorial argument (rather than solution of a differential equation) at the well-known exponential expression

$$N(t) = N_0 e^{-\lambda t} \tag{8.14b}$$

for the size of a population that decays at a constant rate.

The half-life of a radioactive nucleus is the time within which one-half of a sample decays. It follows from Eq. (8.14b) that the half-life is related to the decay rate by the expression $T_{1/2} = (\ln 2)/\lambda$. Note, too, that half-life is a statistical concept; it pertains to an aggregate of nuclei, not to a single nucleus (whose lifetime, according to quantum mechanics, is totally unpredictable).

The model of a coin toss in fact yields the full statistics of nuclear decay. From Eq. (8.14a), it follows that the probability that a nucleus does not survive to time t is $Q(t) = 1 - e^{-\lambda t}$. Suppose that we require the probability P_k that precisely k out of N_0 nuclei disintegrate within a certain time interval t. Assuming, again, the independence of decays, we are, in effect, asking for the probability of producing k heads (and therefore $N_0 - k$ tails) in N_0 tosses with a coin for which the probability of tossing H (= decay) is now $Q(t) = 1 - e^{-\lambda t}$ and of tossing T (= no decay) is $P(t) = e^{-\lambda t}$. The probability of any such sequence of tosses is necessarily $P^k Q^{N_0-k}$, but there are

$$\binom{N_0}{k} = \underbrace{N_0 \times (N_0 - 1) \times \cdots \times (N_0 - k + 1)}_{k \text{ factors}} = \frac{N_0!}{k!(N_0 - k)!}$$

ways of selecting the k decaying nuclei out of the total sample of N_0. The desired probability is therefore given by the binomial distribution[26]

$$P_k(t) = \binom{N_0}{k} Q^k P^{N_0-k} = \binom{N_0}{k}\left(1 - e^{-\lambda t}\right)^k \left(e^{-\lambda t}\right)^{N_0-k}. \tag{8.15}$$

At the time I began to consider the problem of the randomness of nuclear decay, I was aware of only one kind of experimental test that had been applied from time to time over the years. This test was to interrogate nature for the answer to the question: "How much time passes between two successive nuclear decays?" No one, of course, can predict precisely when a nucleus will decay; the answer to the question is not an exact number, but a distribution of time intervals. Equivalently, one is asking for the probability that *no* decay occurs within a time t—that is, for Eq. (8.15) in the case $k = 0$. The resulting substitution yields the simple expression

$$P_0(t) = e^{-\mathcal{R}t}, \tag{8.16}$$

in which $\mathcal{R} = N_0 \lambda$ is the rate of nuclear decay, i.e., the number of counts per second that the detector should register if it could detect every decay in the sample. In accordance with one's intuitive sense, Eq. (8.16) confirms that a long time interval between decays has a lower probability of occurring than shorter intervals.

It is significant to note that one does not need to know the exact number of radioactive nuclei in a sample to apply Eq. (8.16). All that really matters is the decay rate $\mathcal{R}$, which can be measured. This is, in fact, the case for any value of k in Eq. (8.15) under the conditions that ordinarily characterize nuclear counting experiments, namely that the number N_0 of nuclei initially present is many orders of magnitude larger than the number k that decay in a specified time interval. In that case, the binomial distribution (8.15) is virtually indistinguishable from the Poisson distribution[27] discussed earlier [Eq. (8.10)], which here takes the form

$$P_k(t) = \frac{e^{-\mathcal{R}t}(\mathcal{R}t)^k}{k!}.$$ (8.17)

The product $\mathcal{R}t$ is the mean number μ of decays in the interval t.

Although measurements of the distribution of zero-decay time intervals led to results in agreement with the predictions of Eq. (8.16), this did not in itself prove that the disintegration of nuclei occurred randomly. Indeed, it is important to stress the fact that no statistical tests can actually prove a process to be random, for no matter how many such statistical tests for randomness the data pass, there may yet be one more that they fail; hence, the importance of applying diverse tests sensitive to different properties of the data. For this reason, I was particularly eager to test nuclear decay against the theory of runs.

Operationally, to determine the distribution of time intervals predicted by Eq. (8.16), the experimenter arranges for a single decay to start a clock, which is then stopped by the next decay. This is repeated numerous times until a statistically significant number of time intervals is collected. The experiments I had in mind worked quite differently with the advantage that all values of k, not just $k = 0$, contributed to the outcome. In the new procedure, one would count the total number of decays occurring in a fixed time interval and repeat this over and over again to obtain a long string of digital numbers. To test a stochastic process for randomness by run theory requires that the process have a binary outcome, like a coin toss. It is not difficult, however, to convert a string of digital counts into a sequence of binary numbers.

One possibility, as illustrated in Figure 8.10, is to exploit the fact that the counts are necessarily integers (there are no fractional counts), and all integers are either even or odd. Thus, replace each digit in the string by 0 if odd and by 1 if even. I refer to this as the generation of runs with respect to parity. In a long string of random counts, one would expect to have approximately equal numbers of even and odd counts, and therefore of 1s and 0s. The ensuing distribution of runs should agree, to good approximation, with the simple relations (8.5) and (8.6) for coin tosses with an unbiased coin.

A second possibility, which I refer to as the generation of runs with respect to a target value, is to compare the digital count in each time interval, or bin, with a predetermined number. The number, or target value, need not be an integer; in fact, it is better that the number not be an integer. Replace the count in each bin with 0 if it is smaller than the target and with 1 if it is greater. One possibility is to chose the mean count as the target value. Since the mean is, in general, not an

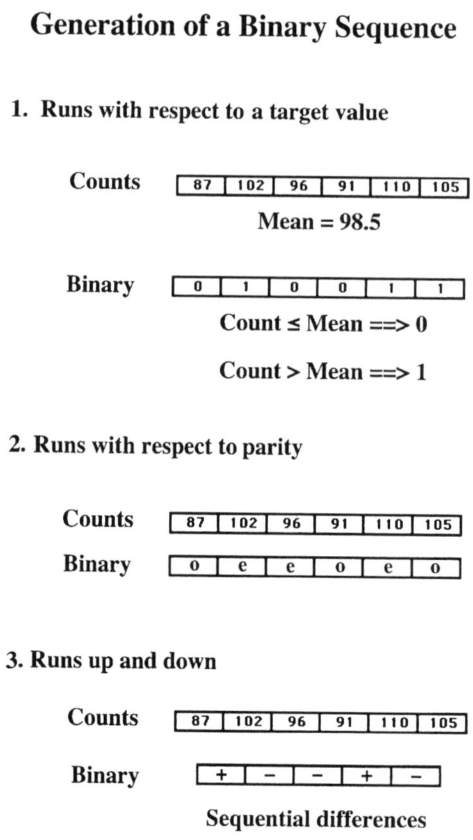

Generation of a Binary Sequence

1. Runs with respect to a target value

Counts

| 87 | 102 | 96 | 91 | 110 | 105 |

Mean = 98.5

Binary

| 0 | 1 | 0 | 0 | 1 | 1 |

Count ≤ Mean ==> 0

Count > Mean ==> 1

2. Runs with respect to parity

Counts

| 87 | 102 | 96 | 91 | 110 | 105 |

Binary

| o | e | e | o | e | o |

3. Runs up and down

Counts

| 87 | 102 | 96 | 91 | 110 | 105 |

Binary

| + | − | − | + | − |

Sequential differences

Figure 8.10. Conversion of a string of digital numbers into binary numbers to generate (a) runs with respect to a target value (e.g., mean), (b) runs with respect to parity, (c) runs up and down.

integer, all counts will be either below or above it, and the outcome is binary. Recall, however, that the Poisson distribution is not symmetric about the mean (although it becomes increasingly symmetric the larger the mean). The choice of mean as the target value leads to an expected frequency of 0s and 1s that is as close as possible, given the discrete nature of the Poisson distribution, but not exactly equal. The results are those of a coin toss with a biased coin. With a target value different from the mean, the bias (i.e., tendency to favor a head or tail) is even greater. Nevertheless, so long as the process is random, the theory of runs applies. One simply needs to know the exact number of 0s and 1s and then apply the formulas in Note 9.

A third possibility, denoted as runs up and down, or difference runs, is to subtract the count in each bin from the count in the following bin. Assign "–" if the difference is negative and "+" if the difference is positive. This leads to a string of binary symbols (+, –) shorter by one element than the original digital string. A succession of +'s constitutes a run up, and a succession of –'s a run down. Actually, if two adjacent bins have the same count, then the difference is zero, and the question arises of how one should treat the null case. The answer, according to statistics articles and books available to me at the time, was to ignore it, because the likelihood of such events was considered comparatively unimportant.

As shown in Figure 8.10, the same series of digital counts gives rise to three different series of binary outcomes. If the original series of nuclear disintegrations is random, then all of the derived series of binary outcomes should likewise be random. Thus, the application of run theory in these various forms poses a stringent, multifaceted test of the data.

Indeed, the test of the data is perhaps even more comprehensive and stringent than what the preceding description may at first indicate, for there is a significant difference in the statistical nature of difference runs compared with that of parity or target runs. The latter two, as I have pointed out, characterize random processes that are equivalent to a coin toss. In statistical parlance, repeated independent trials yielding only two possible outcomes for each trial with fixed probabilities are known as Bernoulli trials. In a coin toss, for example, the probability of tossing H or T is assumed to remain the same for any toss. The probabilities need not be equal (i.e., the coin could be biased), but they must not change from toss to toss if the theory of runs, as embodied in formulas (8.5) and (8.6) [or, more generally, formulas (8.5′) and (8.6′) of Note 9] is to be applicable. The string of binary symbols obtained by taking sequential differences do *not* represent Bernoulli trials.

Consider, for example, a hypothetical string of nuclear decays [87, 78, 89, 92, 96, 103, 110, 75], obtained by counting the number of disintegrations in time intervals of equal length, which form part of a much longer series of counts with a mean of 77 counts per bin. Taking sequential differences yields a sequence of binary numbers [– + + + + + – –] containing a run up of length 5. Each +, starting from the left in this sequence, signifies that a count is larger than the antecedent count and further from the mean. The elements of the string are therefore neither independent nor of constant probability, for the more a count departs from the mean, the less likely it will be that the succeeding count will depart from the mean even further. Thus, the run formulas for Bernoulli trials do not apply to difference runs.

The derivation of the exact expressions for runs up and down will not be given here, but, as in the case of a coin toss, I can give a simple heuristic argument to estimate the mean number of runs up or runs down of length k or greater in a set of n random numbers;

$$\overline{R}_k^n \sim \frac{2n(k+1)}{(k+2)!}, \tag{8.18}$$

and for the mean number of runs up or runs down of precisely length k,

$$\overline{r}_k^n = \overline{R}_k^n - \overline{R}_{k+1}^n \sim \frac{2n(k^2+3k+1)}{(k+3)!}. \tag{8.19}$$

Let us inquire first into the probability that in a long series of random numbers we will find a run up of length k or greater. This is equivalent to asking for the probability of a sequence of $k+2$ numbers of which the first number is not the smallest (otherwise the run would be of length $k+1$), as shown below for the case of a run up of length $k = 3$ or greater:

Numbers	a	b	c	d	e	$k+2=5$
Differences		−	+	+	+	$k+1=4$

There are $(k+2)!$ ways to order a set of $k+2$ numbers. Of these, the number of orderings that lead to a run up of length k is $k+1$. One can perhaps see this more clearly by means of the previous concrete example. Since only comparative magnitudes, and not the exact values of the numbers, matter, let us take the set of five numbers [a b c d e] to be simply [1 2 3 4 5]. Then, there are $k+1=4$ ways to arrange these numbers to produce a run up of length $k = 3$:

	2	1	3	4	5
	3	1	2	4	5
	4	1	2	3	5
	5	1	2	3	4
Differences		−	+	+	+

Note that the number 1 cannot start a sequence, for then the first difference would necessarily be +. Thus, there are $k+1$ ways to start an increasing sequence of $k+2$ numbers.

From the foregoing, it follows that the probability of at least k runs up (and, by symmetry, of runs down) is

$$P(\geq k) = \frac{\text{Number of orderings leading to } k \text{ runs up}}{\text{Number of orderings of } k+2 \text{ numbers}} = \frac{k+1}{(k+2)!}, \tag{8.20}$$

and therefore the mean number of runs (up and down) of length k or greater is $\overline{R}_k^n = 2nP(\geq k) = 2n(k+1)/(k+2)!$, as given in Eq. (8.18). The mean number of runs $\overline{r}_k^n$ of precisely length k is then obtained by directly evaluating $\overline{R}_k^n - \overline{R}_{k+1}^n$ and leads to the result in Eq. (8.19).

Like the runs in Bernoulli trials, number $\overline{r}_k^n$ decreases with run length k, but the falloff is not exponential. In fact, the probability of obtaining long difference runs is considerably lower than that of obtaining equally long runs of Bernoulli trials. For example, we have already calculated that the probability of obtaining a run of at least 5H in a set of random coin tosses is $(\frac{1}{2})^6 = 1/64$. Using Eq. (8.20), however, one finds that the probability of obtaining a run of length 5 or greater in a set of random numbers is $6/7! = 1/840$.

In order that the tests of quantum mechanics for randomness—whatever their outcome—be convincing and general, different kinds of nuclear disintegration processes were selected for measurement. One of these was the beta decay of cesium-137:

$$^{137}_{55}\text{Cs} \rightarrow ^{137}_{56}\text{Ba} + \beta^-$$

in which the decay of a neutron inside the nucleus transmuted cesium into barium with the emission of an electron (whose negative charge is indicated explicitly above). The accompanying antineutrino is not usually included in the reaction equation since it contributes to neither the nuclear charge nor the atomic mass number.

In about 94% of the decays, the resulting barium nucleus is in an excited state, which subsequently returns to the ground state by emission of a 662-keV gamma ray. Because one gamma ray is emitted for each transmutation of cesium into barium, it was experimentally convenient to determine the number of decaying cesium nuclei in a fixed time interval (set to be 0.01 s in all of the experiments described here) by counting the gammas and, correspondingly, shielding the detector from the beta particles.

Gamma rays are quanta of electromagnetic energy of high frequency. By Einstein's relation $E = h\nu$, the frequency of a 662 keV gamma is about 1.6×10^{20} Hz—an enormous frequency compared with about 10^{14} Hz for visible light (with energies in the range 1.5–3 eV). Surprisingly, one can count gamma photons by means of an ordinary photomultiplier tube for visible light. The trick, which nuclear physicists have employed for decades, is to place in front of the tube a scintillating material (e.g., a crystal of sodium iodide) that gives off a burst of visible photons whenever a gamma passes through and is scattered. By suitable arrangement of the radioactive source and detector so as not to overload the electronic counting apparatus, a sufficiently large number of counts—of the order of a million—was obtained in a counting time under 20 hours. Since the half-life of ^{137}Cs is about 30.4 years, no perceptible change in the activity of the source occurred over the duration of the experiment.

Another type of nuclear process that was examined was the electron-capture decay of manganese-54,

$$\ce{^{54}_{25}Mn} + e^- \rightarrow \ce{^{54}_{24}Cr},$$

which is a sort of inverse of beta decay. Here, the nucleus captures one of the bound electrons from the innermost electronic shell (referred to as the K shell) with the resulting transformation of a proton into a neutron.[28] Upon decaying, a ^{54}Mn nucleus is transmuted into an excited chromium nucleus, which returns to the ground state by emission of a 835-keV gamma ray. As in the case of radioactive cesium, the manganese decay events were recorded by counting gamma rays. The half-life of the above process is about 312 days, which is, again, long compared with the duration of an experiment. Interestingly, electron-capture decay is one of very few nuclear processes to be affected by events outside the nucleus. In this case, the chemical environment of the Mn atom affects the electron wave function at the nucleus and, therefore, the probability of electron capture.

Both beta decay and electron-capture decay are examples of the weak nuclear interactions. Alpha decay represents an entirely different kind of interaction, and as an example of this process, the transmutation of the transuranic element americium-241 to neptunium-237,

$$\ce{^{241}_{95}Am} \rightarrow \ce{^{237}_{93}Np} + \ce{^{4}_{2}He},$$

was examined. Neither element occurs naturally on Earth. ^{241}Am, which is the radioactive element ordinarily found in commercially available ionization-detector smoke alarms, has a half-life of about 432 years, which made it particularly suitable for testing quantum mechanics.

Counting alpha particles requires a different kind of detector than counting gamma rays. In place of a photomultiplier tube, the experiment employed a semiconductor device known as a surface-barrier detector. In such a device, the radioactive sample is deposited on the active surface through which alpha particles penetrate, creating a momentary current of electrons and holes.

The fourth and final process examined was a sequential combination of beta decay and alpha decay. All naturally occurring elements with atomic number greater than that of bismuth ($Z = 83$) are radioactive, belonging to chains of successive decays that originate principally with uranium or thorium. Such processes, in fact, are responsible for the molten interior of the Earth, for without the continuous regeneration of heat through the release of energetic decay particles, the Earth would have long ago radiated away its internal energy and solidified all the way to its center. Among the links in the chain of transmutations beginning with uranium-238, the most common isotope of uranium with a half-life of nearly 4.5 billion years, is the formation of the element radium-226 with a half-life of about 1620 years. (This is the element first identified by the Curies in 1902; they began with two tons of pitchblende ore in order to isolate 0.001 g of radium chloride!) Radium-226 eventually gives rise to bismuth-214, which undergoes beta decay to form polonium-214, and the latter undergoes alpha decay to produce lead-210 according to the following reactions:

$$^{214}_{83}\text{Bi} \rightarrow {}^{214}_{84}\text{Po} + \beta^-,$$

$$^{214}_{84}\text{Po} \rightarrow {}^{210}_{82}\text{Pb} + {}^4_2\text{He}.$$

The transmutation of bismuth to polonium actually results in positive polonium ions, rather than neutral atoms, and these ions were electrostatically precipitated onto a surface-barrier detector for counting the alpha particles. Polonium-214, however, has a very brief half-life of 1.64×10^{-4} s, much shorter than the 10^{-2} s dwell time of one bin. The experiment, therefore, was not sensitive to the statistics of polonium decay. However, detection of polonium alphas served as a way of determining the number of beta disintegrations of the parent bismuth-214 whose half-life is about 20 minutes. A desirable feature of this experiment, distinguishing it from the previous three, is that the participating bismuth and polonium nuclei were generated freshly throughout the period of data collection. Thus, these were nuclei created expressly for this experiment with no unknown past history.

What, then, were the results of this expansive effort to see whether God throws dice or not? When I looked over the results of the first experiments, target and parity runs of cesium, I could scarcely believe them. The plots looked very much like Figure 8.8 for runs of coin tosses simulated by a not very good random-number generator. This time, however, the data were real, not simulated, and they seemed to signify that at least one nuclear decay process was not random after all. The analysis showed far too few runs of length 1 and possibly too many runs of length 3. To confirm that the experimental procedure was not at fault, the distribution of time intervals between two successive decays was measured, and the resulting exponential variation, reproduced in Figure 8.11, was in thorough agreement with Eq. (8.16). The runs analyses tested different aspects of the data than did the time analysis. Perhaps Einstein was right; God is not a gambler!

Before calling my travel agent to book reservations for Stockholm, it occurred to me that, although the data were real, the treatment may have been flawed. In the execution of each experiment, nuclear decays were counted in 4096 sequences of 256 bins per sequence, and the mean numbers of 1s, 0s, and their runs of all lengths from 1 to about 20 were then tallied. On average, one would expect—and this was approximately the case—an equal distribution of 1s and 0s. The experi-

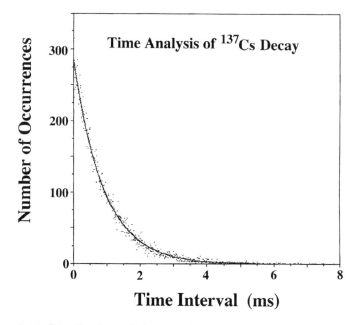

Figure 8.11. Distribution of the number of time intervals between two sequential decays of ^{137}Cs. The exponential shape is in accord with Eq. (8.16).

mental mean numbers of runs of 1s and 0s were then compared at each length to the theoretical values predicted by run theory for 256 Bernoulli trials with a fair coin. However, the fact that, on average, the numbers of 1s and 0s were equal did not mean that they were equal in each of the 4096 separate sequences—and, where the numbers of 1s and 0s differed, the true theoretical values of $\bar{r}_{kH}$ (for H = 1) and $\bar{r}_{kT}$ (for T = 0) could differ significantly from those of a 50–50 distribution. Moreover, given the asymmetry of the Poisson distribution about its mean, precisely equal numbers of 1s and 0s are theoretically not possible.

The 4096 time sequences of data were then carefully rejoined, the exact numbers of 1s and 0s for 1,048,576 bins were counted, and the precise numbers of runs of all lengths were summed. This time a comparison of the experimental and theoretical numbers of runs looked like Figure 8.12. There was no disagreement with run theory. Moreover, the partition of 1s and 0s was found to be almost precisely that predicted by a Poisson distribution with a mean value of 59.91 counts per bin, corresponding to our experimentally observed mean. Identical treatment of data from the other decay processes produced similarly satisfying results. Or perhaps not so satisfying, because

^{137}Cs Target Runs

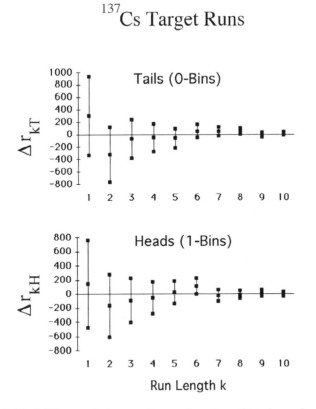

Figure 8.12. Difference between observed and predicted numbers of target runs of ^{137}Cs. The target value of 59 counts per bin led to precisely 511,380 0-bins and 537,196 1-bins or a ratio of 0-bins to 1-bins of 48.77% to 51.23%, in excellent agreement with a Poisson distribution with mean 59.91 counts per bin. (Runs of length greater than 10 are not shown since they are indistinguishable from 0 at the scale of the graph.)

it would have been more exciting to find that quantum mechanics failed.

Having passed tests for randomness by two types of run analyses, the data were examined next for runs up and down. Once again I was subjected to a rude shock. The observed and expected distributions of runs—shown in the upper frame of Figure 8.13 for ^{137}Cs—disagreed markedly, only in this case there seemed to be too many runs of length 1 and 2 and too few of length 3. Already sensitized to the subtleties of statistics, I did not even *think* of Stockholm, but conferred with my colleagues as to what may have gone awry with either the experiment or the analysis.

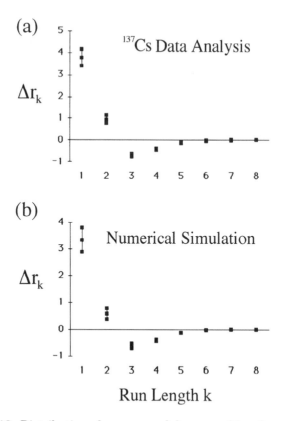

Figure 8.13. Distribution of runs up and down resulting from neglect of null differences. (a) Plot of $\bar{r}_k^{(\text{experiment})} - \bar{r}_k^{(\text{theory})}$ obtained from analysis of ^{137}Cs data. (b) Plot of $\bar{r}_k^{(\text{numerical})} - \bar{r}_k^{(\text{theory})}$ obtained from numerical simulation by computer of Poisson-distributed random numbers.

Having begun as a two-man team, my son and I, our small group by this time had expanded to four; together, we were three professional physicists and a high-school junior. Since Chris was busy with school work and his own special projects, I did not want to disturb him. The rest of us pondered the problem, trying to imagine every conceivable instrumental artifact or theoretical inconsistency, but could come up with no viable solution. Other physicists and mathematicians offered their advice—or regrets—to no avail; the problem would not go away. Was it possible, after all, that the data were telling us something profound about the universe?

One evening, I caught Chris at a free moment and narrated in dreary detail the nature of our problem, more out of a desperate hope that by repeating the story aloud I would somehow gain new insights than that a high-school student (whose passions were art and computers, rather than physics and mathematics) should have any idea of how to resolve the discrepancy.

"So what do you think?," I asked.

"I think the problem is that you are leaving out the zeros," Chris replied.

He was referring to the subtraction of the digital count in each bin from the count in the following bin to arrive at a binary string of +'s and −'s. "I don't believe so," I demurred; "it shows here"—and I tapped a pile of mathematical articles, including one from a book that many physicists considered the "Bible" of computer programming techniques—"that the occurrence of adjacent identical numbers in a random sample of all possible numbers is negligible." Chris reiterated his belief that dropping the zeros changed the statistics, and then left.

"Here, try this," he said, returning later the same evening and handing me a disk with a computer program he had written in HyperTalk. The program generated a string of random numbers and counted runs up and down, keeping track of the number of null differences. The user entered the range N, from which numbers 1 to N were to be chosen, and the number n of random numbers to be generated. I typed in $N = 10$ and had the program generate a string of $n = 1000$ numbers. My eyes widened when the computer counted 97 null differences. I keyed in $N = 100$, and the computer counted 9 null differences. Instantly, like a bolt of lightning, the solution that had hitherto eluded me, my research associates, and the many colleagues whose help we sought finally struck me, a solution that, at least qualitatively, had been intuitively obvious to our high-school student!

What is the probability that, having drawn some number x randomly from a sample of N numbers, another x would be drawn on the next selection? Without thinking about the question too carefully, one might be inclined to answer that this probability is $(1/N) \times (1/N) = 1/N^2$, because the chance of drawing any of the N numbers is $1/N$. However, this is not correct. Since there are N possible values of x, the correct probability is $N \times (1/N^2) = 1/N$, which is much higher. Thus, in a sample of n random drawings, one should expect n/N null differences. This is exactly what Chris's program confirmed.

If the range N were infinitely great, then the expected number of null differences would be infinitesimally small. This was the case for the applications for which the theory of runs had been originally conceived, namely for quality control in the manufacture of objects whose parameters of interest spanned the set of all real numbers. Thus, if a ball bearing picked randomly from a box of bearings had a diameter of 1.2345 cm, it would be exceedingly unlikely that the diameter of the next selected ball bearing would have the same diameter. There would almost always be a positive or negative difference. However, this was decidedly *not* the case for nuclear decay, where the random samples—the digital counts per bin—spanned a practically limited set of positive integers. Although, in principle, any number of disintegrations below the total number of nuclei could occur, in reality the Poisson distribution effectively limited the count per bin to a range of about ±3 standard deviations ($\sigma = \sqrt{\mu}$) about the mean μ. For $\mu = 60$, the case for ^{137}Cs, for example, counts varied from approximately 40 to 80 per bin, a range of only $N = 40$ integers. In a string of 1000 bins, there should occur 25 null differences. (I typed $N = 40$, $n = 1000$ into Chris's program, and the computer counted precisely 25 nulls.)

To verify that neglect of null differences was indeed the origin of the problem, a more sophisticated program was written soon thereafter (by another group member) that counted the numbers of runs up and down in a numerical simulation of Poisson-distributed random numbers. The result, shown in the lower frame of Figure 8.13, is virtually identical to the run analysis of the actual cesium data. In short, the disturbing disagreement with run theory resulted from applying the formulas of difference runs to datasets leading to trinary, not binary, outcomes. The problem was readily remediable, however. One could consider all run distributions that result from replacing each null (0) with each binary value (+, −); if all of these distributions were incompatible with the formulas of run theory, then the hypothesis of randomness had to be rejected. Alternatively, one could assign binary values (+, −) to each zero randomly (e.g., by using an alternative random process or reliable pseudo-random-number generator) and analyze the runs of the resulting sequence. Figure 8.14 shows the result of applying the second method to the cesium data. The anomaly in this case, as well as for the other decay processes, vanished entirely.

Alas, there will be no trip to Stockholm. God is a gambler, after all.

8.6. Mark off Time with Markov

In the long series of tests of the randomness of different nuclear decay processes, I have found no reproducible instance for rejecting the hypothesis that the disintegrations occurred randomly and independent of past history. Yet, it must be emphasized—as I have stated previously in this chapter—that neither these tests nor any others that may be performed subsequently can definitively prove that the underlying cause of nuclear decay is a random process. At best, one can

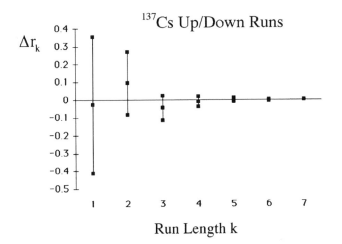

Run Length k

Figure 8.14. Difference of experimental and theoretical values of $\bar{r}_k$ for ^{137}Cs runs up and down with random assignment of a binary value (+, –) in the occurrence of null differences.

demonstrate that the disintegration of nuclei occurs *non*randomly by finding a specific statistical test that the data fail.

Actually, the situation is not quite so simple. If a finite string of data fails to satisfy a statistical test for randomness, does it necessarily follow that the process that generated the data is not random? Consider, for example, two series of binary symbols. One is a string of 100 Hs; the other is obtained by tossing a coin 100 times. The first is obtained from a simple rule: Write the symbol H 100 times. The second is obtained from what is ostensibly a random process. However, the first series, which would clearly fail a test based on the theory of runs, is, like the second series, one of 2^{100} possible outcomes of tossing a coin 100 times and, therefore, has exactly the same probability of occurrence as the second series, namely 2^{-100}. In principle, therefore, one should be no more surprised to obtain the highly ordered first series than the stochastically generated second one. We *are* surprised, of course, because there are relatively few ways to produce ordered sequences containing close to 100 Hs, whereas there are vastly more ways to produce sequences that resemble, to a greater or lesser extent, the disordered sequence. In any event, a random process can give rise to nonrandom-looking results. That, in fact, was one of the principal revelations of run theory. What, then, does it mean to say that something is random?

According to algorithmic complexity theory,[29] a relatively recent approach (compared with classical probability theory) deriving from information theory and computer science, the concept of randomness is related to the idea of "compressibility." A sequence of numbers is deemed truly random if the shortest algorithm needed to generate the sequence is the sequence itself. Such a sequence is not compressible; there is no simpler way to communicate it than by transmitting every symbol. By contrast, a sequence that can be generated by a shorter algorithm, such as the iterative relation illustrating a random-number generator in Section 8.4, is not regarded as particularly random. The randomness of a sequence can be quantified by the minimal number of bits of information required to generate it.

Examined from this point of view, most finite strings of numbers turn out to be random. For example, of all strings of binary numbers of length n, only about one string in a thousand can be compressed into a computer program more than 10 bits shorter than itself; only about one string in a million can be compressed into a program more than 20 bits shorter than itself.[30] However, to demonstrate by the definition of complexity theory that a particular string *is* random, it is necessary to prove that no smaller algorithm exists for calculating it. Unfortunately, one of the far-reaching consequences of Gödel's incompleteness theorem is that such a proof cannot be found. Thus, although one may readily produce a long series of random bits, it is impossible to prove that the series is actually random.

The definition and quantification of randomness provided by complexity theory may be all well and good, but not particularly helpful to the physicist who is trying to determine whether radioactive nuclei decay randomly. It is simply not practicable to search for algorithms, if indeed any exist, by which to compress long strings of data into short formulas whose information content in bits (i.e., "complexity") can be measured. I have found an alternative and, from the standpoint of physics, more useful way to quantify the randomness of the nuclear decay processes (and, by extension, any quantum decay process). It is based on the concept of a Markov chain.

Suppose, for example, that the outcome (H or T) of a coin toss influenced the outcome of the next toss in a prescribed way. If the outcome is H, then the probability of obtaining H on the next toss is enhanced; if the outcome is T, then the probability of obtaining T on the next toss is enhanced. The degree of enhancement can be gauged quantitatively by an adjustable bias parameter. If the bias parameter is the same for both outcomes, then the resulting numbers of Hs and Ts should be approximately the same. However, the outcomes of such a stochastic process are clearly history dependent, and the distribution of runs of Hs and Ts will, in principle, deviate from that of a sequence of Bernoulli trials. A chain of events in which each event is influenced at most only by the event immediately preceding it is known as a Markov chain.

In contrast to the theory of runs, where the question at issue was basically "How many times in succession are the outcomes of a process the same?," the essential question dealt with by a Markov chain is this: Given that a system is initially in a particular state (e.g., outcome H or T), what is the probability of finding the system n time intervals later in either the same state (retention probability p_{ii}) or in another state (transition probability p_{ij})? If the retention and transition probabilities differ, then the outcome in the nth time interval will depend on n. The difference $\Delta = p_{ii} - p_{ij}$ is therefore a useful measure of the extent to which the outcome of a stochastic process tends to persist, rather than change to the opposite outcome.

According to quantum theory (QT), Δ should be zero; that is, the probability of nuclear decay within a given time interval should not be influenced by the decay occurring in any preceding interval. Experimentally, that was precisely what the data showed. Evaluation of the requisite conditional probabilities from the numbers of counts in bins removed from any given bin by one unit, two units, three units, and so forth, led to a value of Δ that was zero to within approximately one part in a thousand for sets of data comprising approximately one million bins. The experiments were performed under two very different conditions. In the first, the mean number of counts per bin μ was high ($\mu \gg 1$), since it was conceivable that a high rate of nuclear disintegration in one time interval might lead to a diminished or enhanced rate in the following interval if QT-violating correlations were somehow dependent on sample size (i.e., number of decaying particles). In the second, the mean number of counts per bin was low ($\mu \ll 1$), since it was conceivable that a rare occurrence of a disintegration after a long period of nuclear quiescence might modify the decay probability of a subsequent particle if QT-violating correlations were somehow sensitive to proximity. In such a case, a violation of quantum theory would be more noticeable within a counting interval containing at most one particle than a hundred particles. Experimentally, the zero value of Δ under both circumstances lent strong support to the prediction that such correlations did not exist.

The investigations of nuclear decay by means of the theory of runs and the theory of Markov chains are actually complementary. One can ask, for example, how small the bias parameter (in the above coin-toss example) must be such that deviations from run theory are detectable? The answer, established by computer simulations, is that the longer the sequence of data, the smaller is the bias parameter. With a string of the order of one million bits, comparable to the number of bins of nuclear decay counts, a bias of approximately one part in a thousand—the same measure provided by the Markov chain analyis—could be discerned in the distribution of runs with virtual certainty. To detect smaller biases, the number of bins must be increased by shortening the counting interval or lengthening the total counting time or both.

8.7. Exponential Decay, Correlation, and Randomness: The Quantum Perspective

I have shown in Section 8.5 how von Schweidler's assumption—made long before the discovery of the atomic nucleus or creation of quantum mechanics—that a radioactive atom decays with constant probability independent of its past history [Eq. (8.13)], inexorably led to the exponential decay law [Eq. (8.14b)] and the statistics of nuclear disintegration [Eq. (8.15)]. No experiment of which I am aware refutes this assumption, and I have—somewhat carelessly—given the impression that this is exactly what quantum theory predicts. Strictly speaking, however, this is *not* what quantum theory predicts, and in concluding this chapter, I would like to disentangle several separate strands of inferences relating to von Schweidler's assumption—which is really two assumptions—and the quantum perspective.

First, according to quantum mechanics, the simple proportionality between probability of decay and the width of a time interval expressed in Eq. (8.13) is not valid for arbitrarily short time intervals. Rather, approximate solutions of Schrödinger's equation (employing Fermi's "Golden Rule") show that the decay "constant" λ is zero at the instant the unstable state is formed and, depending on the energy of the state and the interactions coupling it to lower states, reaches a constant value for a time interval beyond some approximate threshold. In the case of nuclear decay with emission of a gamma ray (as in the case of ^{137}Cs), λ is predicted to vary in an oscillatory way, becoming constant after a time interval of about 10^{-21} s.[31] No experiment to date has the time resolution to detect such a variation of λ.

Because a constant λ leads directly to an exponential decay law, the implication of the preceding paragraph is that unstable quantum systems—radioactive nuclei, excited atoms and molecules, whatever —cannot undergo transitions to lower-energy states in accordance with a strictly exponential decay law over arbitrary lengths of time. The law, in fact, should not be valid for times very short or very long compared with the half-life. There are other circumstances as well, such as the decay of a quantum system from a coherent linear superposition of quantum states (discussed in Chapter 4), which lead to nonexponential decay—in this case to the phenomenon of quantum beats.

Second, when quantum statistics are taken into account for a collection of unstable particles whose wave functions in some sense overlap, the conditional probability that a decay occurs after an earlier one depends on the time interval between the two events. In other words, there is a history dependence to quantum decay for time intervals short compared with a calculable correlation time. Consider, again, the example of the Hanbury Brown–Twiss (HBT) experiment discussed in Chapter 3. For a chaotic light source, the conditional probability of detecting a second photon of the same polarization as one already detected is strikingly higher for time intervals short compared to the longitudinal coherence time of the source than for longer intervals. The probability that two identical photons arrive at a detector simultaneously (zero time interval) is *twice* the probability of the coincidental arrival of two classical particles (Figure 3.10).

The nuclear experiments that I have described in this chapter examined radioactive processes under conditions in which these known quantum effects—nonexponential decay and quantum statistical bunching or antibunching—were not expected to occur. Had any of these nuclear processes led to correlated decays at variance with the theory of runs, the results could have been interpreted as incompatible with quantum theory. That would have been exciting to see—but for better or for worse, it did not happen.

Notes

1. Wheeler's remark is not so cryptic to a quantum physicist. It refers to the description of a quantum system by a wave function comprising a linear superposition of states, each state representing a potentially different experimental outcome. When a measurement is made, the system is forced randomly into one of these states, a process often referred to as "collapse" of the wave function.
2. Harvey B. Lemon, from the foreword to A. A. Michelson, *Studies in Optics*, University of Chicago Press, Chicago, 1962, p. xxi.
3. David Good, Maths for Softies, *New Scientist* (12 April 1997) 42–43. The reviewed book is *Goodbye, Descartes* by Keith Devlin (Wiley, New York, 1997).
4. Ask Marilyn®, *Parade Magazine* (9 September 1990).
5. In the book, *My Brain Is Open* (Simon & Schuster, New York, 1998), about mathematician Paul Erdös, a master of probability theory, author Bruce Schechter reports (p. 109) that the problem perplexed and upset Paul Erdös for several days.
6. The general result $P(A \text{ and } B) = P(A)P(B|A) = P(B)P(A|B)$ is known as Bayes' theorem.
7. The assumption of independence means that the probabilities for the various outcomes (H or T in this case) of a given event are uninfluenced by the outcomes of prior events. Thus, the probability of drawing an ace (of which there are initially 4) from a deck of 52 randomly mixed cards is $4/52 = 1/13$ each time one draws a card, provided that the drawn card is always returned to the deck. If drawn cards are discarded, then the prob-

ability of drawing an ace changes from draw to draw, and the outcomes of two successive draws are in that case not independent.

8. In general, if P is the probability of a particular configuration of tosses, then the number of times the configuration is expected to occur in n tosses is nP. For example, the probability of a single H is $\frac{1}{2}$, and, therefore, we would expect, on average, $\frac{1}{2} \times 256 = 128$ heads in 256 tosses of a fair coin.

9. For the general case of n_H heads and n_T tails in a total of n tosses, the exact expressions giving the statistics of runs are

$$\overline{R}_{kH} = \frac{n_H!(n_T + 1)(n - k)!}{(n_H - k)!\, n!}, \qquad (8.5')$$

$$\overline{r}_{kH} = \frac{n_H!\, n_T(n_T + 1)(n - k - 1)!}{(n_H - k)!\, n!}, \qquad (8.6')$$

$$\overline{R}_n = \overline{R}_{1H} + \overline{R}_{1T} = \frac{n + 2n_H n_T}{n}. \qquad (8.7')$$

10. For example, suppose the outcome of three experiments is the set of numbers $x_1 = 8$, $x_2 = 10$, and $x_3 = 6$. The mean value is $\bar{x} = \frac{1}{3}(8 + 10 + 6)$ $= 8$. The standard deviation of the data is $\sigma = \sqrt{\frac{1}{3}\{(8-8)^2 + (10-8)^2 + (6-8)^2\}}$ $= \sqrt{8/3} = 1.63$.

11. A good discussion of elementary statistics and their application to physics is given by F. Reif in *Fundamentals of Statistical and Thermal Physics* (McGraw-Hill, New York, 1965), Chapter 1, in particular pp. 15–16.

12. W. Feller, *An Introduction to Probability Theory and its Applications*, 2nd ed., Wiley, New York, 1957, Vol 1, p. 65.

13. The probability that a player has a positive accumulation for k out of n tosses is to good approximation given by $P(k, n) \sim 2/\pi\sqrt{k(n-k)}$. This expression is smallest for $k = \frac{1}{2}n$ and becomes unboundedly large for $k = 0$ (player never leads) or $k = n$ (player always leads). The exact formula, which is too cumbersome to be given here, remains finite for all k.

14. The expected gain at any toss is the value of the gain (+1 or −1) times the probability of the corresponding outcome ($\frac{1}{2}$ in both cases). Thus, $\overline{\text{Gain}} = (+1)(\frac{1}{2}) + (-1)(\frac{1}{2}) = 0$.

15. Supose, for example, that $A = 100$, $m = 16$, and $x_0 = 1$. Then $(100 \times 1)/16$ $= 6.25$ and $x_1 = 0.25$. The second iteration leads to $(100 \times 0.25)/16 = 1.5625$ and $x_2 = 0.5625$.

16. This result can be shown to follow from Eqs. (8.6) and (8.8).

17. In a χ^2 test, one divides the data into a number (n) of bins and calculates for each bin k the square of the difference between the observed (O_k) and theoretically expected (E_k) values, divided by the expected value i.e., $(O_k - E_k)^2/E_k$. The sum of these results over all the bins yields the single statistic, χ^2. Knowing n (referred to as the number of degrees of freedom) and χ^2, one can calculate (or look up in tables) the probability that subsequent experiments would yield greater or equal values of chi square.

18. I discuss the strange attributes of quantum mechanics in more detail in *More Than One Mystery: Explorations in Quantum Mechanics* (Springer-Verlag, New York, 1995).

19. Letter to James Franck cited in *Einstein, A Centenary Volume*, edited by A. P. French, Harvard University Press, Cambridge, MA, 1979, p. 6.

20. See M. P. Silverman, *Waves and Grains: Reflections on Light and Learning* (Princeton University Press, Princeton, NJ, 1998) for a description of the video recording of this experiment, which I frequently show when lecturing on quantum mechanics.

21. The designation "atomic weight," by which A is also known, is technically a misnomer. A represents the number of atomic mass units (amu) in which 1 amu has been defined as $\frac{1}{12}$ the mass of the carbon-12 nucleus ($^{12}_6C$). The mass of each proton and neutron is very close to 1 amu. The energy equivalence ($E = mc^2$) of 1 amu is approximately 931 MeV.

22. G. Gamow, *Zeitschrift für Physik* **51** (1928) 204; R. W. Gurney and E. U. Condon, *Nature* **122** (1928) 439.

23. See, for example, H. A. Bethe's classic, *Elementary Nuclear Theory* (Wiley, New York, 1947, p. 6). According to quantum theory, the probability for a particle of mass M and energy E to tunnel from a radial position a to radial position b in a spherically symmetric potential $U(r)$ is given by the approximate relation

$$P = \exp\left(-\frac{4\pi}{h} \int_a^b \sqrt{2M(U(r) - E)}\ dr \right).$$

The frequency with which an alpha particle of speed ν bounces back and forth across the diameter $2R$ of the nucleus is $f = \nu/2R$, and the rate at which the particle tunnels through the barrier is correspondingly equal to fP. Equating the de Broglie wavelength $h/M\nu$ of the alpha particle inside the nucleus with the nuclear radius R leads to $f = h/2MR^2$. Thus, the alpha decay rate of a nucleus should depend on the barrier height, barrier width, nuclear size, and alpha mass and energy.

24. For an alpha particle whose mass is some three orders of magnitude larger than the electron mass, the energy uncertainty due to confinement to a spherical region of radius $\sim 10^{-12}$ cm is roughly 8 MeV, and therefore lower than the nuclear barrier.

25. Recall from Chapter 2 that the exponential function e^x can be defined by the operation $\lim_{n\to\infty}(1 + x/n)^n$.

26. Expansion of the binomial expression $(p + q)^n$ yields the sum $\sum_{k=0}^n \binom{n}{k} p^k q^{n-k}$. If $p + q = 1$, so that p is the probability that an event occurs in a single trial and q is the probability that the event does not occur, then the term $\binom{n}{k} p^k q^{n-k}$ represents the probability that the event occurs k times out of n trials.

27. The demonstration that the binomial distribution reduces to the Poisson distribution under conditions of low probability p and large number of trials n is ordinarily accomplished by the tedious method of applying the Stirling approximation to the various factorial expressions occurring in the binomial coefficient. There is a much simpler and more elegant way to proceed. The expression $(q + pz)^n$, with $q = 1 - p$, results in a power series expansion of the variable z with coefficients of z^k ($0 \le k \le n$) given

by the binomial distribution. One can transform this into the expression $e^{\mu(z-1)}$ for which the coefficients of the power-series expansion are given by the Poisson distribution. Substitute $q = 1 - p$ into $(q + pz)^n$ to obtain $(1 + np(z - 1)/n)^n$ and take the limit as $p \to 0$ and $n \to \infty$ with np remaining finite (the mean value μ). The limit yields the exponential function $e^{\mu(z-1)}$. (See Note 25.)

28. The symbol β is reserved for an energetic electron or positron created by the decay process. A pre-existing orbital electron is designated by the symbol e.

29. Readable accounts are given in G. J. Chaitin, Randomness and Mathematical Proof, *Scientific American* (May 1975) 47 and J. Ford, How Random Is a Coin Toss, *Physics Today* (April 1983) 40.

30. There are 2^k algorithms of length k bits that might generate a series of $n \geq k$ bits. Thus, there are $2^1 + 2^2 + \ldots + 2^{n-r-1} = 2^{n-r} - 2$ algorithms of length shorter than $n - r$. Each algorithm specifies one series of n bits. The fraction of series of length n that can be generated by algorithms of length less than $n - r$ is then approximately $2^{n-r}/2^n = 2^{-r}$. For $r = 10$ and 20, the fractions are respectively 1/1024 and 1/1,048,576.

31. F. T. Avignone, III, Comment on "Tests of the Exponential Decay Law at Short and Long Times," *Physical Review Letters* **61** (1988) 2624.

CHAPTER 9

A Universe of Atoms: Symmetry, Unity, Gravity, and the Problem of "Missing Mass"

The paradox is now fully established that the utmost abstractions are the true weapons with which to control our thoughts of concrete fact.

Alfred North Whitehead

9.1. Keep It Together! Keep It Together! Keep It Together!

In the zany film *Bowfinger*, actor Eddie Murphy, in the role of a neurotic film star whose mental state is becoming increasingly unhinged by encounters with strange people who keep following him, tries to regain self-control by frantically repeating over and over again his psychiatrist's mantra: "Keep it together!" Mr. Murphy did not realize it, of course, but those three words summarize in the tersest way possible the principal task of theoretical physicists who study either the smallest building blocks of matter or the largest structures of the cosmos.

For the first group, the task is to find a single self-consistent mathematical framework for understanding the fundamental forces in nature: gravity, electromagnetism, and the weak and strong nuclear interactions. For the the second group, the task is to explain what "missing" matter or energy keeps the contents of galaxies and clusters of galaxies from flying apart. Surprisingly, the two problems have much in common. Rather than being boundary points at opposite ends of a linear spectrum of phenomena, the physics of the ultra small and the physics of the astronomically large join like the snake in Kekulé's dream, grasping its tail to form a circle.[1]

The origin and destiny of the universe have interested me for as long as I can remember. As a child, popular writings of Eddington, Jeans, Hoyle, Gamow, Hubble, Shapley, and other astronomers and cosmologists opened up grand vistas of the imagination that helped draw me into physics. Indeed, it may well have been so for most physicists of my generation, for not many children, I believe, would have been first attracted to physics by the prospect, however lucrative, of designing better transistors. Nevertheless, I did not become a cosmologist. Bearing in mind Lev Landau's caustic remark that "cosmologists are often wrong, but never in doubt," I set about to make my mark in other areas of physics which rested, I believed, upon more solid ground.

Over the years, however, the experimental foundations of the science of the universe have become increasingly firmer—and, in the past few years especially, the capacity to assess accurately cosmological quantities (galactic distances, rate of universal expansion, primordial abundances of light elements, fluctuations in the cosmic microwave radiation background, etc.) has become truly breathtaking. Hardly a day passes without a report in a major newspaper of one cosmological breakthrough or another. (*The New York Times* is perhaps the *de facto* premier periodical for publication of new astrophysics!) Although I am still not a cosmologist, I became interested again—this time more seriously—in its unsolved problems.

Ironically (in light of what I have written in Chapter 1), my attention to problems of space, time, and the universe was sparked at least in part by science fiction, a genre of entertainment that ordinarily interests me very little. However, living with children who for many years were avid "Star Trek" fans, I had on numerous occasions heard about the crew of one episode or another zipping back and forth in time with no more ado than taking a lift between decks of the Enterprise. Was this sheer nonsense? From the standpoint of physics, *is* time travel conceivable?

In case the reader is expecting a definitive answer to that question in this chapter, I hasten to state that I do not have one. For what it is worth, however, my professional opinion is that it is *not* possible. With certain qualifications, I believe that the past is irretrievable and the future is unknowable. Yes, I am aware of various well understood effects of special or general relativity by which motion or gravity can affect the passage of time. Yes, I know that looking up at the sky at night is tantamount to looking back in time. Yes, I also realize that there are solutions to Einstein's equations of gravity that describe universes with so-called "closed time lines." All the same, as for any realistic prospect of shuttling back and forth between past and future, my advice is don't bet on it.

Nothwithstanding a decidedly negative outlook on one of science fiction's most cherished literary devices, the question of time travel still intrigued me, but I approached the subject from what I believed to be an original perspective hitherto untried by either screenwriters or physicists. What, I wondered, would a universe be like in which there were two independent dimensions of time? In analogy to the optical phenomenon of birefringence,[2] I referred to one dimension, along which light propagates at the familiar speed c in vacuum, as "ordinary" time and the other dimension, along which the speed of light need not necessarily be c, as "extraordinary" time. This was, consequently, a universe with five dimensions—three spatial and two temporal—in contrast to the four-dimensional space–time upon which Einstein's special and general theories of relativity are based.

To those readers to whom the idea of a universe with five dimensions may seem overtly preposterous, I would point out that, by the standards of contemporary theoretical physics, this is actually a relatively modest stretch of the imagination. Among theoretical attempts to unify all of the interactions in nature, it is presently common to find theories comprising eleven or more spatial dimensions with all but the familiar three wrapped up ("compactified") into hyperdimensional cylinders. Such theories, however, still contained a single time dimension. In a theory with two time dimensions, it is conceivable—at least mathematically—to make a kind of rotation from one time axis to the other. Perhaps one could rotate from ordinary time into extraordinary time, move forward in extraordinary time, and then make another temporal rotation into either the ordinary future or ordinary past. Would this be possible physically? I strongly doubt it. Mathematics may admit of many potentialities that nature has chosen to ignore.

All the same, the idea of mathematically modeling a five-dimensional universe with two time dimensions in such a way as to avoid conflicts with presently known physical laws was a conceptually interesting challenge, and I and a colleague (R. L. Mallett) decided to work on the problem together. Our primary goal, however, was considerably more sober than carrying out an exercise motivated by science fiction. We hoped to develop a theory of gravity by appeal to certain principles of wide applicability in quantum physics. The theory of space–time that we put together, combining various key ingredients (field theory, gauge invariance, and spontaneous symmetry breaking) that have characterized nearly all major advances in theoretical physics over the past half century—and which will be discussed in the sections to follow—was a mathematical model stripped to its bare essentials and explored for ideas and insights rather than with an expectation that it would yield a realistic description of the world as we know it.

The theory, however, led unexpectedly to a bizarre result with striking implications. It predicted the existence of a particle with a rest mass lower than that of any known particle except the photon, which is believed to have a rest mass of precisely zero. Investigating the physical properties of these particles, I found that in the aggregate they condensed, like steam to water, into a form of matter that physicists have been trying to produce ever since Einstein first predicted its existence in 1924. By applying to a material gas, the novel statistics proposed that same year by Indian physicist S. N. Bose to account for the Planck blackbody radiation law, Einstein concluded that[3]

From a certain temperature on, the molecules "condense" without attractive forces, that is, they accumulate at zero velocity. The theory is pretty, but is there also some truth to it?

The "truth to it" was not revealed until over seventy years later. In 1995, researchers at the Joint Institutes of Laboratory Astrophysics (JILA) in Boulder, Colorado managed to achieve the lowest temperatures ever produced in a laboratory (170 billionths of a degree above absolute zero) and thereby were able to prepare approximately 2000 rubidium atoms in a so-called Bose–Einstein condensate.[4] The result was hailed as an experimental *tour de force* and has since opened up new fields of study in atomic physics and optics.[5]

Although it is rather unlikely that the actual physical universe has two dimensions of time, my colleague and I immediately realized, upon examining the various steps in the development of our theory, that the prediction and properties of the new particle did not depend at all on the unusual space–time geometry that we hypothesized, but emerged intact from broader considerations firmly grounded in established physical principles and the astrophysical data upon which we relied. Reworking the basic ideas of our first model specifically for the actual four-dimensional universe in which we live, we were pleased, but not surprised, to find that ultra low-mass particles emerged again together with a set of equations that described their gravitational interactions. The outcome was no longer merely an amusing diversion, but a mathematical model with far-reaching astrophysical implications whose validity could ultimately be decided by experiment and observation.

Ecstatic over the JILA group's success in creating a Bose–Einstein condensate, Carl Wieman, one of the project directors, waxed rhetorically "It really is a new state of matter. It has completely different properties from any other kind of matter," to which co-director Eric Cornell added "This state could never have existed naturally anywhere in the universe. So the sample in our lab is the only chunk of this stuff in the universe, unless it is in a lab in some other solar system."[6]

I smiled, when I first read those remarks, at the ironical twists of fate of which nature is capable. For if the strange, but ineluctable, conclusion of our theory is sustained, then the "new state of matter" whose creation took physicists seven decades to accomplish and which even today can be created on Earth only in relatively small quantities under conditions of extraordinarily low temperatures, may well be the *most abundant* form of matter in the cosmos, filling the voids of interstellar space and keeping galaxies together.

9.2. Symmetries for the Mind's Eye

To those who appreciate the beauty of symmetry, nature abounds in shapes and patterns to delight the eye. Crystals, for example, afford one of the richest sources of visual symmetry in science. Who could fail to marvel at comely golden cubes of iron pyrite, vitreous rhombohedrons of calcite with their intriguing image-doubling properties, hexagonal prisms of quartz, or the infinite variety of six-sided dendritic crystals of ice that come to us gratuitously as snowflakes on a cold wintry day?

It is at the atomic or molecular level however—beyond the gross morphology accessible to our senses—that the symmetries of crystals reveal fundamental aspects of the interactions that hold matter together. There, the bewildering variety of natural structures reduce to manageably small sets of symmetry elements, classifiable into different symmetry groups. Space groups, for example, characterize the various ways the lattice of an ideal crystal (i.e., one without edges or defects) can be displaced without altering the original pattern. In general, a symmetry element is a transformation that leaves a pattern invariant. To make the abstraction of a space group concrete, take a look, if at all possible, at some of the marvelous designs of M. C. Escher—in particular the numbered patterns designated as Symmetry Works.[7]

The lesson, if I may construe it so, of translational symmetry in crystallography is that there is no special or preferred point of origin in a crystal; the full lattice can be reconstructed from knowledge of a certain minimal neighborhood about any point within the crystal. This may seem obvious, and perhaps in the case of crystals it is. However, in addition to those geometric symmetries that please the eye, there are more abstract mathematical symmetries in nature discernable only by the "mind's eye" (to borrow Richard Feynman's colorful imagery). The objects of interest are not physical objects like crystals, but physical laws expressed, for example, as equations of motion. In these cases, invariance under transformations analogous to translations or rotations are not trivial, but have profound implications.

Reflecting upon such symmetries in his Nobel lecture, C. N. Yang wrote:

... Nature seems to take advantage of the simple mathematical representations of the symmetry laws. When one pauses to consider the elegance and the beautiful perfection of the mathematical reasoning involved and contrast it with the complex and far-reaching physical consequences, a deep sense of respect for the power of the symmetry laws never fails to develop.

Among simple symmetry laws displaying "elegance and beautiful perfection," Lorentz invariance and gauge invariance are two of the most consequential symmetries in physics. Together, they pose severe constraints on any candidate in the physicist's search for an ultimate theory of everything.

Undetectability of Uniform Motion: The Principle of Relativity

Every symmetry in physics is a statement about something that *cannot* be observed. The restriction is not an instrumental one; it is one of principle. It is a statement that the quantity in question is physically meaningless. This is a point that has not been fully appreciated before the creation of relativity theory and quantum mechanics.

Seated on a Boeing 747 cruising seven miles above ground at 900 km/hr, passengers around me either look idly out the windows or read their books and magazines. Even though I have flown countless times, I cannot refrain from observing each time the cup of tea on the tray beside me. What is it doing? In the absence of turbulence, it does nothing at all! Like Sherlock Holmes' reference to the dog in the night,[8] it is the *non*occurrence of any event that is remarkable. That cup of tea could just as well have been resting on my kitchen table at home. Uniform motion (through field-free space), no matter how fast, is undetectable.

According to Newton's second law of motion, the net force on a system is proportional to its total mass and acceleration:

$$\boldsymbol{F} = m\boldsymbol{a} = m\frac{d\boldsymbol{v}}{dt} = m\frac{d^2\boldsymbol{r}}{dt^2}. \tag{9.1}$$

Once a mathematical expression for the force $\boldsymbol{F}$ is supplied and boundary conditions stipulated, Eq. (9.1) can be solved, at least in principle, to yield the velocity $\boldsymbol{v}$ and coordinate vector $\boldsymbol{r}$ as a function of time. Now, supposing that Eq. (9.1) applies to a marble rolling on my kitchen table, then a transformation (called a "boost") of that table and everything on it to the constant velocity $\boldsymbol{V}$ of the Boeing 747 would transform the force on the marble as follows:

$$\boldsymbol{F}' = m\boldsymbol{a}' = m\frac{d\boldsymbol{v}'}{dt} = m\frac{d(\boldsymbol{v}+\boldsymbol{V})}{dt} = m\frac{d\boldsymbol{v}}{dt} = \boldsymbol{F}. \tag{9.2}$$

There is no change. The invariance (termed Galilean invariance) of the Newtonian force law under a boost is illustrative of a deep principle of physics. There is no single preferred frame of reference (e.g., that of the "ether" at rest) in which to express the laws of classical mechanics.[9]

In 1905, Einstein extended this observation into a more general principle of relativity by requiring the laws of physics, and not just mechanics, to take the same form in all inertial (i.e., nonaccelerating) reference frames. In addition, he assumed the speed of light in vacuum c to be a universal constant, the same for an observer in any inertial reference frame. As a consequence of these two postulates, there followed a set of transformation relations by which observers in different inertial reference frames could compare spatial measurements and time intervals—and therefore every other physical quantity (velocity, acceleration, force, field, etc.) that involved space and time. This set of relations is known as the Lorentz transformation (recall Section 5.2) because the Dutch physicist H. A. Lorentz had derived it a year earlier in an attempt to reconcile the null result of the Michelson–Morley experiment with the existence of an ether as a medium through which electromagnetic waves propagate. The true significance of the Lorentz transformation, however, reflecting the nature of measurement of space and time independent of any specific model of matter, is due solely to Einstein.

In the simplest case of a reference frame ("primed" frame) moving at speed V along the $+x$ axis of our own rest frame (the "unprimed" frame), the Lorentz transformation takes the form

$$x' = \gamma(x - Vt),$$
$$y' = y, \qquad z' = z,$$
$$t' = \gamma\left(t - \frac{Vx}{c^2}\right), \tag{9.3}$$

in which $\gamma = 1/\sqrt{1-(V/c)^2}$ is a factor that appears ubiquitously in relativistic formulas.[10] From Eq. (9.3) follow all the apparently counterintuitive effects—length contraction, time dilation, the non-Euclidian velocity addition law,[11] the twin paradox, among others—associated with special relativity. The word "relativity" reflects the fact that spatial intervals Δx (the difference between two spatial coordinates) and time intervals Δt (the difference between two time coordinates) are relative, that is, depend on the reference frame of the observer. For example, I would measure the length of a meter stick moving past my line of sight at speed V to be shorter than 1 m by the factor γ; only an observer at rest with respect to the meter stick would measure its length to be exactly 1 m. Likewise, I would find a clock moving past me at a speed V to run more slowly than a clock at rest beside me again by the factor γ. The factor γ, however, does not depart substantially from 1 until the relative velocity V is close to the speed of light.

Although the theory of relativity has removed from physics longstanding incorrect suppositions of absolute space and absolute time, this theory is, for all that, still very much a theory of absolutes, or invariances, as a physicist would more likely say. I have already discussed the Lorentz invariance of charge in Chapter 5 and pointed out in the preceding paragraphs that Einstein's derivation of the Lorentz transformation assumed the invariance of the speed of light in vacuum. Moreover, the requirement that the laws of physics take the same form in any inertial reference frame leads to the recognition that certain ostensibly different physical quantities like space and time, energy and linear momentum, or electric and magnetic fields are actually representable by a single geometrical structure, as, for example, by a four-dimensional vector or a four-dimensional tensor. The scalar magnitudes of 4-vectors and 4-tensors provide additional invariant quantities.

The formalism for expressing special relativity in a manifestly Lorentz invariant way was developed by one of Einstein's former mathematics teachers, Hermann Minkowski. It is reported that Einstein did not particularly appreciate at first the geometrization of his theory, but regarded it as "überflüssige Gelehrsamkeit" or superfluous learnedness.[12] (It might also be added that Minkowski did not think particularly highly of Einstein as a student.) Subsequently, the tensor formalism of Minkowski was to be indispensable to Einstein in his creation of the general theory of relativity.

The transformation of coordinates expressed in Eq. (9.3) can be looked upon as the transformation of a single coordinate vector with four components $x^\mu = (x^0, x^1, x^2, x^3) \equiv (ct, x, y, z)$, the Greek superscript μ taking values 0, 1, 2, 3. (Note carefully that the superscript μ is an index or label, and not an exponent or power.) With this modest modification of notation and the definition $\beta \equiv V/c$ standard in special relativity, the transformation of the coordinate 4-vector takes the simpler form

$$x'^0 = \gamma(x^0 - \beta x^1),$$
$$x'^1 = \gamma(x^1 - \beta x^0),$$
$$x'^2 = x^2, \quad x'^3 = x^3, \tag{9.4}$$

displaying the symmetry between temporal (x^0) and spatial (x^1) coordinates. It is important to emphasize that coordinates have no physical significance in themselves; only coordinate differences—lengths and time intervals—are measurable. In Minkowski's geometry, the magnitude of a space–time interval between two events, defined by

$$(\Delta s)^2 = (c\Delta t)^2 - (\Delta x)^2 - (\Delta y)^2 - (\Delta z)^2, \tag{9.5}$$

in which each component Δx^μ is a coordinate difference, is a Lorentz invariant, i.e., it has the same numerical value for any inertial observer of the same two events. Relation (9.5) resembles somewhat the Pythagorean theorem, except that the spatial terms are subtracted from (not added to) the temporal term. This distinction is crucial. In Euclidian geometry, to which the Pythagorean theorem applies, the magnitudes of vectors are invariant under rotations; in Minkowski geometry, the magnitudes of vectors are invariant under boosts.

A general way of writing geometrical intervals is by means of a structure known as a metric (or metric tensor). Thus, in the case of differentially small coordinate displacements, Eq. (9.5) takes the form

$$(ds)^2 = \sum_{\mu=0}^{3} \sum_{\nu=0}^{3} \eta_{\mu\nu} dx^\mu dx^\nu, \tag{9.6a}$$

in which the so-called Minkowski metric $\eta_{\mu\nu}$ is a diagonal 4×4 matrix with elements $(1, -1, -1, -1)$. It may seem like notational overkill to take a simple sum of four terms like Eq. (9.5) and write it as the double sum over the components of a matrix and two vectors like Eq. (9.6a), but the latter form provides a compact way to show explicitly that the expression is a scalar quantity—i.e., unchanged by a coordinate transformation—even when the metric may not be diagonal and the sum involves cross-terms. Lorentz invariance is referred to as a "global" invariance because the elements $\eta_{\mu\nu}$ are the same for all inertial observers no matter where they are.

To help avoid cumbersome mathematical expressions, physicists often employ the Einstein convention in which summation symbols are omitted and repeated indices (one raised, one lowered) are automatically summed. With the Einstein convention, Eq. (9.6a) becomes

$$(ds)^2 = \eta_{\mu\nu} dx^\mu dx^\nu = dx_\mu dx^\mu. \tag{9.6b}$$

The coordinate vector with the lowered index in Eq. (9.6b) is defined by the sum $dx_\mu = \eta_{\mu\nu} dx^\nu$. The lowering of an index is not merely idle notation. Although a discussion of the matter would take us beyond the intended scope of this chapter, it is worth noting here that there is, in general, a significant geometrical distinction between a contravariant vector (raised index) and a covariant vector (lowered index). For our purposes, however, the notation provides a quick way to recognize scalar quantities i.e., those of the form $A_\mu A^\mu$ (with sum over μ).

Any set of quantities that transform in the same way [Eq. (9.3) or (9.4)] as the components of the coordinate 4-vector are also 4-vectors and, consequently, give rise to an invariant expression like Eq. (9.5). For example, the energy-momentum 4-vector $p^\mu = (p^0, p^1, p^2, p^3) \equiv (E/c, \boldsymbol{p})$ of a particle of mass m, energy E, and (3-vector) linear momentum $\boldsymbol{p}$ leads immediately to the invariance relation $p_\mu p^\mu = E^2/c^2 - |\boldsymbol{p}|^2 = m^2 c^2$. The scalar invariant may be readily verified by evaluating $p_\mu p^\mu$ in the rest frame of the particle in which $\boldsymbol{p} = 0$ (the particle is obviously at rest in its rest frame) and $E = mc^2$.

It is of historical interest to point out that Einstein did not create the special theory of relativity in response to any perceived experimental violation of Newton's laws; none was known at the time. Rather, he sought a description of motion consistent with the laws of electromagnetism deduced by James Clerk Maxwell in the 1850s. It is from these laws that another symmetry for the mind's eye emerges— one that was to be a critical signpost along the path to uncovering the common structure to all (or nearly all) known fundamental interactions.

Undetectability of Potential and Phase: The Gauge Principle

Formulated in words, the laws of electromagnetism comprise four disconnected empirical statements:

(1) The electric force between two point charges is proportional to each charge and falls off as the square of their separation. [Coulomb's law]
(2) All magnets have both "north" and "south" ends. [Nonexistence of magnetic monopoles]
(3) A change in magnetic flux gives rise to an electrical potential difference (also called an electromotive force). [Faraday's law]
(4) Magnetism is generated by electric currents or by a change in electric flux. [Ampere–Maxwell law]

It is only when these statements are expressed mathematically in terms of electric (E) and magnetic (B) fields, as first worked out by James Clerk Maxwell, that they cease to be disconnected and lead to a self-consistent theory of all classical electromagnetic and optical phenomena.

At first thought, the introduction of magnetism may present a disturbing paradox. Consider the passage of a charged particle through the field of a stationary magnet and recall that the magnetic force on the particle (termed the Lorentz force) is proportional to the magnitude of the charge, the strength of the field, and the speed of the particle. A charged particle moving with constant velocity V into a magnetic field B will be deflected (as long as V is not parallel to B). Suppose, however, that the particle is at rest in the laboratory and a magnet moves past it at the constant velocity $-V$. Although, from the perspective of relativity, the two situations should be equivalent, the Lorentz force on the particle is now theoretically null. Would the stationary particle be deflected by a moving magnet? Can the existence or nonexistence of a physical force depend on which of two purportedly equivalent inertial reference frames is chosen to describe the phenomenon? For that matter, is there not a violation of the principle of relativity when a particle moving at a *constant* velocity experiences a force? (Remember the cup of tea in the Boeing 747.)

Last things first. A charged particle moving through a uniform magnetic field may have a constant speed but not a constant velocity; the magnetic force continuously changes the direction of particle motion and, therefore, accelerates the particle. The situation is not the same as that of a Lorentz-boosted cup of tea, which always remains in an inertial frame. Not only does the Lorentz force not violate the principle of special relativity, but, using special relativity, one can derive the Lorentz force from Coulomb's law.

The resolution of the preceding paradox is that the initially stationary charged particle is indeed deflected, but as a result of an *electric* force, not a magnetic force. The Lorentz transformation that relates the space and time coordinates of two inertial frames also relates the electric and magnetic fields in these frames in the following way. Components of E and B along the direction of the relative velocity V are unchanged. Components of E and B perpendicular to the relative velocity are intermixed. For the specific Lorentz transformation of Eq. (9.3) or (9.4), in which the boost is along the x axis, the transformation of electric and magnetic fields takes the form

$$E'_x = E_x, \qquad\qquad B'_x = B_x,$$

$$E'_y = \gamma\left(E_y - \frac{V}{c} B_z \right), \qquad B'_y = \gamma\left(B_y + \frac{V}{c} E_z \right),$$

$$E'_z = \gamma\left(E_z + \frac{V}{c} B_y \right), \qquad B'_z = \gamma\left(B_z - \frac{V}{c} E_y \right). \qquad (9.7)$$

V is the velocity of the "primed" frame (the rest frame of the charged particle in the preceding example) relative to the "unprimed" frame (the rest frame of the magnet, or laboratory frame). In the case of the stationary magnet and moving charged particle, B is the only field presumed present in the laboratory frame. In the rest frame of the particle, however, there exist both an electric field $E' = \gamma(V/c) \times B$ and a magnetic field $B' = \gamma B$. The latter has no effect because the particle is stationary, but the former exerts an electric force (charge × electric field) of the same form as that exerted by the magnetic field in the laboratory frame.

The laws of electromagnetism—specifically Maxwell's equations and the combined electric and magnetic force law—are invariant under the Lorentz transformations (9.3) and (9.7) of both coordinates and fields. From the perspective of relativity developed by Minkowski, therefore, electric and magnetic fields constitute a single geometrical entity, an electromagnetic field tensor $F^{\mu\nu}$ (with both sets of indices assuming values 0, 1, 2, 3), whose elements are intermixed under a Lorentz transformation. The field tensor is antisymmetric, i.e., $F^{\mu\nu} = -F^{\nu\mu}$, with elements $F^{01} = E_x$, $F^{02} = E_y$, $F^{03} = E_z$, $F^{12} = B_z$, $F^{23} = B_x$, $F^{13} = -B_y$.[13]

Diagonal elements must vanish because the numerical relation $F^{\mu\mu} = -F^{\mu\mu}$ is satisfied only by zero.

Written in tensor notation, the four laws of electromagnetism reduce to two independent equations, of which one relates the field tensor to charges and currents [i.e., combines statements (1) and (4)], and the other expresses a symmetry relation among field components [i.e., combines statements (2) and (3)]. The relativistic formalism provides not only an economy of expression but also insights into the geometrical and physical meaning of the laws, as well as leads to another Lorentz invariant, $F_{\mu\nu}F^{\mu\nu} = B^2 - E^2$, which, in analogy to Eq. (9.5), is interpretable as the square of the magnitude of the field tensor.[14]

Maxwell's equations, however, incorporate another kind of symmetry that complements those deriving from special relativity, and this symmetry, referred to as gauge invariance, has turned out to be of profound importance. The geometrical embodiment of the first statement (Coulomb's law) is that electric lines of force diverge radially and isotropically from each point electric charge. The analytical expression of this fact takes the form of an equation in which the density of electric charge (i.e., amount of charge per volume), which is the source of the electric field, is related to the spatial variation of the field as represented by a particular sum of derivatives called the "divergence."[15] Since the second statement denies the existence of point magnetic charges (isolated north or south poles), there can be no corresponding lines of magnetic force diverging radially and isotropically from any region of space. Instead, magnetic lines of force form loops about their current sources. This fact is succinctly expressed by the statement that the divergence of the magnetic field is zero.

The algebraic consequence of the fact that the divergence of a magnetic field vanishes is that every magnetic field, no matter what its origin or configuration, can be expressed in terms of a vector of spatial derivatives [the curl—see Eq. (3.4b)] of a function designated the vector potential $\boldsymbol{A}$. I have introduced the vector potential in Chapter 3 in conjunction with the uniquely quantum mechanical phenomenon known as the Aharonov–Bohm effect. In revisiting the subject at this point, however, my focus of attention is not on electron interference, but on the symmetries of electromagnetism and their consequences.

When the equation $\boldsymbol{B} = \text{curl } \boldsymbol{A}$ is substituted into the mathematical representation of Faraday's law [statement (3)], there results a general expression for the electric field $\boldsymbol{E}$ in terms of the time variation of $\boldsymbol{A}$ and the spatial variation (the gradient) of another function called the scalar potential ϕ. The exact relation is given by Eq. (3.4a), but the essential point here is that by replacing the electric and magnetic fields in Maxwell's equations by the preceding relations, the laws of electromagnetism can be reformulated entirely in terms of potentials, rather than fields. Because the electric and magnetic forces exerted on matter are directly proportional to the electric and magnetic fields, these fields must be (and are) uniquely determined in any well-posed system of charges and currents—otherwise classical electromagnetism would lead to inconsistencies and paradoxes. The potentials A and ϕ, however, are not unique.

As discussed briefly in Chapter 3 (in Note 13), a set of potentials (ϕ, A) that generate fields E and B can be transformed into another set of potentials (ϕ', A') that generate the same electromagnetic fields by means of a so-called gauge transfomation:

$$\phi' = \phi - \frac{\partial \Lambda}{c\partial t},$$

$$A' = A + \mathrm{grad}\,\Lambda, \tag{9.8a}$$

in which Λ is, to a large extent, an arbitrary function although it cannot be a mathematically "pathological" function or one that corresponds to physically unacceptable charge and current sources. Because Maxwell's equations and the Lorentz force law are unchanged under the transformation (9.8a), electromagnetism is said to be gauge invariant.

The word "gauge"—or "Eich" in the original German expression of mathematician Hermann Weyl—conveys the notion of size or scale, as in the gauge of a railroad track. In a certain sense (although not actually the sense in which Weyl meant it), the expression "gauge invariance" is an apt one, for it signifies the invariance of Maxwell's equations under a change in the origin (or zero setting) of the scale that records the scalar and vector potentials. That the electromagnetic field equations remain unchanged under a "shift" of potentials as specified by Eq. (9.8a) should not be surprising; after all, the scalar potential corresponds to the "voltage" in an electrical circuit, and only differences in voltage are physically meaningful and measurable.

The gauge invariance of electromagnetism appears explicitly when the theory is again formulated in relativistic notation. The scalar and vector potentials are then found to constitute the components of a 4-vector, $A^\mu = (\phi, A)$ or $A_\mu = (\phi, -A)$, and the electromagnetic field tensor (with lowered indices) is simply expressed by the antisymmetric combination of derivatives $F_{\mu\nu} = \partial A_\mu/\partial x^\nu - \partial A_\nu/\partial x^\mu$. For spatial indices, the expression yields precisely the components of the curl of the vector potential and, therefore, the components of the magnetic field. For mixed spatial and temporal indices, the expression is the same (to within a minus sign) as Eq. (3.4a) and generates the components of the electric field. In relativistic notation, the gauge transformation (9.8a) is expressible by the single relation

$$A'_\mu = A_\mu - \frac{\partial \Lambda}{\partial x^\mu}, \tag{9.8b}$$

which is readily seen to have no effect on the field tensor because the extra term $\partial^2\Lambda/\partial x^\nu\partial x^\mu$ is canceled by the extra term $\partial^2\Lambda/\partial x^\mu\partial x^\nu$. (Partial derivatives can be taken in any order.)

Were one to consider nothing else but electromagnetism, the gauge invariance just described might be regarded as little more than an interesting curiosity. I am not sure whether Maxwell himself paid any more attention to this feature of his theory than as a mathematical device for facilitating calculation.[16] It is from the perspective of quantum mechanics, however, that a deeper significance of gauge symmetry is revealed.

In contrast to classical mechanics, it is not the particle location r or velocity v that quantum mechanics provides, but an abstract complex-valued wave function $\psi(r, t)$ by aid of which the probability of finding a particle within a specified region of space can be calculated. The quantum "wave," unlike waves of sound or light, does not represent the undulation of any physical medium, either material like air or immaterial like an electromagnetic field. Rather, it is a purely mathematical function conveying statistical information about particles. As emphasized in Chapter 3, wavelike phenomena are manifested collectively in the behavior of many quantum particles or in repeated observations of one or a few particles, but never in the single observation of a single particle. The strangeness of the quantum world lies in the observable fact that undulatory behavior like diffraction and interference can emerge even when the experiment is performed one particle at a time and, therefore, in the absence of any cooperative particle interactions.

The quantum equation of motion of a free particle of mass m and charge q moving at a speed low compared to the speed of light is the Schrödinger equation

$$\left(\frac{p^2}{2m}\right)\psi(r, t) = i\hbar\frac{\partial\psi(r, t)}{\partial t}, \tag{9.9}$$

in which the linear momentum p is a mathematical vector operator with components $p_x = (\hbar/i)(\partial/\partial x)$, etc. The "square," $p^2 = p \cdot p$, is proportional to a sum of second derivatives referred to as the Laplacian, $\nabla^2 = \partial^2/\partial x^2 + \partial^2/\partial y^2 + \partial^2/\partial z^2$. Since the particle is presumed free (i.e., not subject to any external potential), the electric charge q, which couples the particle to electromagnetic potentials, is entirely absent from Eq. (9.9). The fact that the state of a quantum particle is represented by a wave function rather than by a trajectory leads to a symmetry requirement on the equation of motion that may, at first glance, seem useless, but which has extensive ramifications. Like the potentials of electromagnetic theory, the wave ψ is not unique, but can be multiplied by any phase factor of the form $e^{i\Lambda}$, where Λ is a constant signifying a global phase adjustment. Because only relative phase (like relative voltage) is physically meaningful, both ψ and $e^{i\Lambda}\psi$ contain, in their spatial and temporal evolution, the same dynamical information about a physical system.

The preceding statement is obviously true for constant Λ, since $e^{i\Lambda}$ can be dropped from both sides of the Schrödinger equation (9.9). However, if an admissible quantum equation of motion is *required* to be invariant under a *local* phase transformation (the gauge principle)—that is, a transformation in which the phase $\Lambda(r, t)$ can be reset at each and every point of space and time through which the particle moves—then a cursory inspection of Eq. (9.9) shows immediately that the spatially varying function $e^{i\Lambda(r,\ t)}$ does not drop out; the operator p^2 gives rise to first and second derivatives of Λ on the left side with no compensating terms on the right. The gauge principle cannot be implemented for free particles.

It is at this point that the conceptually separate paths of quantum physics and classical electrodynamics fruitfully merge. If the electromagnetic potentials are inserted into Eq. (9.9) according to the long-established procedure known as "minimal coupling"

$$\left(\frac{1}{2m}\left(\boldsymbol{p}-\frac{q}{c}\boldsymbol{A}\right)^2 +q\phi\right)\psi(\boldsymbol{r},t)=i\hbar\frac{\partial\psi(\boldsymbol{r},t)}{\partial t}, \tag{9.10}$$

then it is not difficult (albeit a bit tedious) to confirm that the new Schrödinger equation (9.10) is unchanged in form under a gauge transformation (9.8a) of the potentials *and* a phase transformation

$$\psi' = e^{iq\Lambda(r,t)/\hbar c}\psi \tag{9.11}$$

of the wave function, both transformations being effected by the same arbitrary function $\Lambda(\boldsymbol{r}, t)$. From now on, it is to be understood that the two transformations together comprise a (local) gauge transformation. Originally, Weyl believed (incorrectly) that Einstein's theory of gravity, which will be discussed shortly, should be invariant under a *scale* change e^Λ—hence, his adoption of the word "gauge." Quantum mechanics reveals that the appropriate invariance for a particle subject to electromagnetic interactions involves phase, not scale.

From Eq. (9.10) follow all the interactions of a single nonrelativistic charged particle with electric and magnetic fields, including the Zeeman effect, Stark effect, paramagnetism, diamagnetism, the emission, absorption, and scattering of radiation, and the Aharonov–Bohm effect, to cite but some of the most familiar examples. The relativistic (i.e., Lorentz-invariant) extension of Eq. (9.10) takes different forms, depending on whether the charged particle in question is a fermion or a boson. In the former case, the Dirac equation is the appropriate equation of motion, and in the latter, the equation of motion is known as the Klein–Gordon equation. In both cases, the electromagnetic interactions of the charged particle derive from expressions that contain scalar contractions of the 4-vector $p_\mu - (q/c)A_\mu$ in which $p_\mu = i\hbar(\partial/\partial x^\mu)$ is a differential operator.[17]

In the historical development of electromagnetism, the force between elementary charges (Coulomb's law) was first inferred from experiment and eventually led, in conjunction with other empirically determined laws, to Maxwell's equations, the recognition of light as an electromagnetic wave, and the minimal coupling $p_\mu - (q/c)A_\mu$ in the equation of motion of a charged particle subjected to electromagnetic fields (as mediated by the potentials). The imposition of local gauge invariance and Lorentz invariance is so restrictive, however, that the chain of reasoning could have preceded in the opposite direction—that is, by the insertion of "light" (i.e., the gauge potential A^μ) into the appropriate quantum equation of motion of a free charged particle to make the equation gauge invariant, with subsequent *deduction* of the electrostatic force law and other Maxwell equations. Exactly how this program of deriving a fundamental interaction from the "thin air" of gauge symmetry works involves mathematical considerations that go beyond the objectives of this chapter. It will be useful for the following sections, however, to examine the procedure, at least in broad outline.

At the outset, the model builder seeking to account for the interactions of particles endowed with a general property of charge (not necessarily electric charge) specifies the internal degrees of freedom of the particles and the group of symmetry elements that locally reset internal variables as the particles move from one space–time point to another. In electromagnetism, the internal degree of freedom (or internal quantum number) of the particle is its electrical charge, to which the corresponding local variable is the phase of the wave function. However, there are other possibilities. In the first extension of gauge theory beyond electromagnetism, C. N. Yang and R. L. Mills attempted to model the strong nuclear interactions by constructing a gauge theory of the "nucleon" whose internal space had two components called "isospin." Isospin is a kind of "up" and "down" directionality (like the two components of electron spin) whereby the "up" state corresponded to the proton and the "down" state to the neutron. In this case, gauge invariance was invoked to ensure that the resulting equations of motion remained unchanged in form under an arbitrary rotation of the isospin direction, a transformation with three adjustable parameters.

In general, the construction of a gauge theory works more or less in the following way:

1. Write down in a manifestly Lorentz-invariant form a certain scalar function, called the Lagrangian, for the free particles. In classical dynamics, the Lagrangian expresses the difference in kinetic and potential energies of the particles. The Lagrangian for a single free particle takes the form of a kinetic energy term (proportional to $p^\mu p_\mu$ for a boson and $\gamma^\mu p_\mu$ for a fermion, in which γ^μ is a set of constant 4×4 matrices known as Dirac matrices) minus a term proportional to the particle mass, which is a kind of potential energy.[18]

2. Require that the Lagrangian be invariant under transformations of the internal variables by introducing, as needed, an appropriate number of gauge potentials A^μ and subsequently replacing each ordinary derivative $\partial/\partial x^\mu$ that comes from the momentum p_μ with a "gauge-covariant" derivative D/Dx^μ constructed from the gauge potentials. [In electromagnetism, the two derivatives are related by $D/Dx^\mu = \partial/\partial x^\mu + i(q/\hbar c)A_\mu$.[19]]

3. Construct the fields (analogous to E and B) from the gauge potentials via the antisymmetric relation $F_{\mu\nu} = DA_\mu/Dx^\nu - DA_\nu/Dx^\mu$. In the case of electromagnetism, the preceding equation gives precisely the field tensor defined previously, since the terms $A_\mu A_\nu$ and $A_\nu A_\mu$ cancel. However, in the case of more complex interactions, such as in Yang–Mills theory, the potentials are noncommutative (i.e., the order of their appearance in a product matters) and additional terms appear in $F_{\mu\nu}$ that have no analog in electromagnetism. Add to the Lagrangian the sum of terms $-\frac{1}{4}F_{\mu\nu}F^{\mu\nu}$ representing the energy of the gauge fields $F_{\mu\nu}$.

4. Finally, deduce from the Lagrangian (by the methods of the calculus of variations) a set of relations known as the Euler–Lagrange equations that prescribe the interaction of the particles with the potentials and fields, as well as the generation of the fields by particle currents. From the antisymmetry of the field tensor $F_{\mu\nu}$ follow other relations among the fields themselves. In the case of electromagnetism, this step leads to the Dirac or Klein–Gordon equation for the particles and to Maxwell's equations for the fields.

Although the procedure as just outlined may read like a cookbook recipe and, in general, is not too difficult for a mathematical physicist to implement, the resulting equations may be of daunting complexity. For example, in stark contrast to electromagnetism in which the fields themselves are uncharged although they are produced by charged particles, the components of Yang–Mills fields are subject to the same force as the particles and, therefore, interact with one another. As a consequence, the principle of superposition—vital to the solution of many electromagnetic problems—does not apply, except under some exceptional circumstances. However, complexity aside, the program itself, following ineluctably from the imposition of Lorentz and gauge invariance, provides a tractable, self-consistent scheme for deriving the form of fundamental interactions from the supposition of symmetry. Prior to this, one would have had to guess equations of motion from a profusion of experimental data.

9.3. Spontaneous Symmetry Breaking

From the perspective of quantum theory, the interactions between particles are mediated by the exchange of bosons. In electromagnetism, for example, the Coulomb repulsion of two electrons results from the exchange of a photon between them. Experiments strongly support the belief that photons have zero rest mass, a property that accounts for the infinitely long range of the electromagnetic interaction. We get a sense of this long range (compared to atomic or nuclear dimensions) whenever we move bits of paper about with a statically charged comb or paper clips with a magnet.

As a rough guide, the range of an interaction mediated by a particle of nonzero rest mass m is the Compton wavelength of the particle $\lambda_C = h/mc$, previously introduced in Chapter 7. The gauge theory of electromagnetism accounts perfectly for the masslessness of the photon, for the particles corresponding to gauge potentials introduced into a Lagrangian in the manner outlined in the foregoing section must have zero rest mass. If this were not the case, the Lagrangian would have to contain an additional scalar term quadratic in the potentials—a term proportional to $m^2 A_\mu A^\mu$—which is not invariant under a gauge transformation.

Were the gauge principle applicable only to electromagnetism, it would be of rather limited interest. On the contrary, two of the most significant achievements of the past half-century of physics are the elucidation of the strong and weak nuclear interactions in terms of gauge-invariant quantum field theories. This may seem at first glance paradoxical because the range of the nuclear interactions does not extend much beyond the size of a nucleon, or approximately 10^{-15} m, and the mediators of these interactions might, therefore, be expected to be massive particles.

The expectation is false in the case of the strong nuclear interaction—the interaction that binds protons and neutrons together in atomic nuclei. Called quantum chromodynamics (QCD) from the Greek root for color, the theory describes the interactions of subnuclear massive fermions, whimsically named quarks (from a fanciful passage in James Joyce's novel *Finnegan's Wake*), exchanging massless gauge bosons termed gluons. In contrast to electromagnetism, in which electrons and other fermions, but not photons, are endowed with electrical charge, the quarks and gluons of QCD are both endowed with the corresponding quantum property of color. Color, of which there are three varieties (often designated red, blue, and green) and their anti versions, have no relation at all to optical color; rather, it is another whimsical term to characterize the attribute of particles that makes them subject to the strong nuclear force. The motivation for the term color is that a mixture of the three primary colors produces the senation of white (i.e., of no color). Similarly, according to QCD, all strongly interacting particles must contain colored quarks in an appropriate linear superposition resulting in a net color attribute of zero.[20]

Because gluons, as well as quarks, possess color, the theory gives rise to nonlinear interactions, such as the coupling of gluons to gluons, for which there are no analogs in electromagnetism. It is from these nonlinear interactions and the requirement that physically observable particles display no net color charge that there arises one of the extraordinary features of QCD known as confinement. The effective potential energy of quarks *increases* with their separation. (By contrast, the potential energy of two electrons decreases with the first power of the distance between the particles.) Ultimately, an infinite amount of energy would be required to break the two quarks free of one another. Before this occurs, however, the energy put into the system (e.g., through bombardment of a nucleus with externally accelerated particles) goes into creating other composite physical particles within which quarks and antiquarks again remain bound. There are, according to the predictions of QCD, no free quarks. QCD is an exact gauge theory; the gluons are presumed to have no rest mass, and the short range of the interaction derives from dynamics of the gauge fields leading to quark confinement.

The situation is quite different, however, with the weak nuclear interactions—those responsible for a variety of nuclear transmutation processes such as beta decay. Here, the carriers of the weak interaction are particles about 80–90 times more massive than a proton (i.e., of a mass comparable to a bromine or rubidium atom). How can one construct a theory that is invariant to gauge transformations, yet at the same time has massive gauge particles? The answer to this question was found in an ingenious application to elementary particle physics of a process long recognized in the physics of condensed matter. It is referred to as "spontaneous symmetry breaking" because the theory is gauge invariant at the outset, but gives rise to particle masses in a natural way by a kind of phase change, without theorists having to insert these masses by hand into the Lagrangian (which would have violated gauge invariance). From a mathematical point of view, spontaneous symmetry breaking reflects the fact that the equations of motion of a theory can possess certain symmetries, yet give rise to solutions that are less symmetric. As this process is to play an important role in the sections to follow, it is worth examining in more detail.

Consider a thin cylindrical metal rod pressed vertically against a tabletop by your hand. Assuming that you are pressing straight down, there is no reason to expect the bar to deform in any special direction; all directions around the rod are equivalent. Yet, when the pressure becomes sufficiently great, the bar will buckle in some particular direction, thereby breaking the cylindrical symmetry. Physically, the bar has undergone a transformation from a state with cylindrical symmetry to a lower-energy state with less symmetry as a consequence of the pressure reaching some threshold value. Were the experiment to be repeated numerous times with different, but equivalent, bars, buckling would occur with equal probability in any direction—and, in this sense, the cylindrical symmetry of the physical configuration is preserved. The deformation of the bar in a given direction, however, represents one possible solution of an infinity of solutions to the cylindrically symmetric equation describing the bar under pressure.

There is another physical example more specifically within the realms of electromagnetism and quantum physics that brings out the seemingly magical appearance of mass in what is initially a strictly massless field. One of the characteristics of superconductivity, mentioned briefly in Chapter 3 for its utility in demonstrating the Aharonov–Bohm effect, is the Meissner effect, the expulsion of a magnetic field from the interior of a (Type I) superconductor. Ordinary electric currents passing through nonmagnetic wires do not expel magnetic fields. Thus, the short penetration depth[21] of a magnetic field into a superconductor may be likened to the Compton wavelength of a massive photon. (Recall that the Compton wavelength of a particle is a measure of the range of the interaction that it mediates.) How does an initially massless magnetic field acquire the semblance of a mass?

Although the interaction between the magnetic field and the charge carriers within the superconductor is still governed by the "minimal coupling" characteristic of gauge invariance, the charge carriers themselves are not individual electrons, but pairs of electrons (Cooper pairs) that constitute a coherent flow of charge. The entire system of Cooper pairs is described by a macroscopic wave function with a phase determined only by the dynamics of the Cooper pairs and independent of external gauge fields; this wave function therefore breaks the gauge symmetry of the Schrödinger equation. It is the interaction of the magnetic field with a self-coherent field of particles that generates a short-range interaction without violating the gauge invariance of Maxwell's equations. The breaking of gauge symmetry occurs spontaneously with the "condensation" of the electrons into the Cooper pairs of the superconducting phase when some external parameter, in this case temperature, is lowered through a critical value.

The essential ideas outlined above can be expressed more quantitatively by a simple phenomenological model based on a model first proposed by Russian physicists Ginzburg and Landau before a fully quantum theory of superconductivity was developed by Bardeen, Cooper, and Schrieffer. In the simplified model outlined here, the free energy per volume of the system $V(\phi)$ depends on a field ϕ according to the relation

$$V(\phi) = a\phi^2 + b\phi^4,   \text{(9.12)}$$

which is symmetric about the axis $\phi = 0$. The parameters a and b are taken to be real-valued with b positive. [If b were negative, the function $V(\phi)$ would have no finite minimum value.] Thermodynamically, the stable state of the system corresponds to the value of ϕ that minimizes the energy $V(\phi)$. Setting the first derivative $dV(\phi)/d\phi$ to zero, one finds that there are three distinct solutions: $\phi = 0$, $\pm\phi_0$, where $\phi_0 = \sqrt{-a/2b}$.

The behavior of the system depends on the sign of the parameter a. If a is positive, then the shape of $V(\phi)$ is parabolic (see Figure 9.1), and the minimum is clearly $V_0 = 0$ at $\phi = 0$, representing the nondegenerate ground state of the normal conductor. If however, as a consequence of the phase change to the superconducting state, the parameter a changes sign, then $V(\phi)$ has the shape of a double well with the minimum $V_0 = -a^2/4b$ at each of the two values of the field, $\phi = \pm\phi_0$. These two values correspond to two different values of the phase θ of the field, which can be written as $\phi = \phi_0 e^{i\theta}$, where $\theta = 0$ or π. [Note that the word "phase" in the preceding sentence refers to the argument of a mathematical function, not the physical state (solid, liquid, superconductive, normally conductive, etc.) of a medium.] As the tempera-

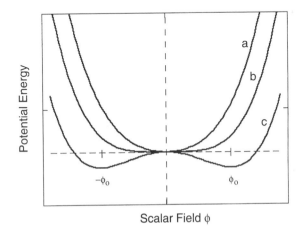

Figure 9.1. Self-interaction potential energy $V(\phi) = a\phi^2 + b\phi^4$ as a function of the scalar field ϕ for $b > 0$ and (a) $a > 0$, (b) $a = 0$, (c) $a < 0$.

ture is lowered below the critical temperature for the onset of super-conductivity, the state $\phi = 0$ becomes an unstable equilibrium point (since it is now a local maximum), and the system settles unpredictably into one of the two minima, thereby spontaneously breaking the original reflection symmetry (across the axis $\phi = 0$) of the equations of motion that govern this system.

The preceding heuristic model has all the essential ingredients of the mechanism—known as the Higgs mechanism—that was proposed to account for the masses of the gauge particles which carry the weak interactions. These particles, two of which are electrically charged (W^+, W^-) and one of which is neutral (Z^0), together with the photon, derive from the four gauge fields which, when inserted (via the appropriate covariant derivatives) into the Lagrangian of free fermions (electrons, muons, neutrinos, and quarks), lead to a theory of "electroweak" interactions that has so far satisfactorily passed all experimental tests. At the outset, however, all the gauge fields represent massless particles, for the Lagrangian cannot contain any terms quadratic in the individual gauge fields, or it would no longer be gauge invariant.

According to the Higgs mechanism, however, one assumes that, in addition to the fundamental fermions and gauge bosons, there is a universally present scalar field—the Higgs field ϕ—which, like the free energy ϕ of the superconductor, has a self-interaction potential of the form given by Eq. (9.12). One must then add to the Lagrangian terms that represent the difference in kinetic energy and potential energies of the Higgs field. At a sufficiently low temperature, such as is presumed to have occurred after the earliest moments of the origin of the universe, the Higgs field condensed into a self-coherent state, thereby randomly selecting one of an infinite number of possible stable equilibrium values ϕ_0. Mathematically, one accounts for this phase change by re-expressing the Higgs field in the Lagrangian as an excitation $\bar{\phi}$ relative to this new minimum ($\phi = \phi_0 + \bar{\phi}$), rather than relative to the value $\phi = 0$, which, after condensation, became an unstable equilibrium state. Subsequent regrouping of terms in the Lagrangian to sort out the interactions of the physical particles leads to quadratic products of individual gauge fields with coefficients that are functions of ϕ_0. As a consequence of their interaction with the Higgs field, the gauge particles carrying the weak interaction, but not the photon which carries the electromagnetic interaction, thereby acquire mass.

The masses of the gauge particles, however, are not directly calculable from ϕ_0, as this value is not known. Indeed, at the time of writing this chapter, experiments have already confirmed the existence of all the fermions and gauge bosons required by the gauge theories of the weak, electromagnetic, and strong interactions—with the exception of the Higgs particle. This particle has not yet been detected, presumably because the energy equivalent to its rest mass is greater than, or at the upper limit of, the energy that present particle accelerators, operating within their designed ranges, can furnish. Researchers using the Large Electron Positron (LEP) collider at CERN, the European high-energy physics laboratory in Geneva, Switzerland, believe that they might have caught a glimpse of the Higgs particle in a final series of experiments just before the LEP was shut down permanently for the construction of the higher-energy Large Hadron Collider (LHC), a project expected to take about 5 years.[22]

If the Higgs particle is found, it would provide the final element required for complete confirmation of the so-called Standard Model, which has given a common gauge-theoretical structure to the electromagnetic, weak, and strong interactions. So complex, in particular, are the equations of motion of the weak and strong interactions that, were it not for the stringent constraints posed on the Lagrangian by Lorentz and gauge invariance, it is highly doubtful that physicists would ever have been able to deduce these equations from empirical results alone.[23]

All the same, the hypothesis of a Higgs field and the mechanism of spontaneous symmetry breaking may strike a reader as the theoretical equivalent of splitting nuts with a sledgehammer. Why *not* simply construct a gauge-invariant Lagrangian and then break the symmetry by inserting the masses of gauge particles by hand as needed, thereby avoiding the extra theoretical baggage of an all pervasive "ether" of presently undetectable and possibly nonexistent Higgs particles? The reason quite simply (to state, not to prove) is that spontaneously broken gauge field theories are demonstrably "renormalizable," a word that refers to their calculability.

Because of an infinite number of virtual interactions that can occur in a quantum field theory (e.g., the emission and reabsorption of a photon by a particle, or the creation and immediate annihilation of a particle–antiparticle pair), the "bare" theoretical parameters like particle charge or mass that are put into a Lagrangian at the outset do not correspond to the experimental values of these quantitities. As a consequence, calculation (usually by perturbation theory) of the transition probabilities or cross sections of various physical processes frequently result in divergent integrals. In a renormalizable theory, however, the infinite terms can be collected together as a sum of terms multiplying a bare charge or bare mass, the resulting divergent product then being redefined as the experimental charge or mass. Many physicists regard this procedure as merely sweeping the intrinsic difficulties of the theory under the rug, but from the standpoint of practicality, the infinities are eliminated and the theory can be made to yield finite calculations. For the Standard Model, these calculations have so far been in agreement with experiment.[24]

In addition to the three fundamental interactions—weak, strong, and electromagnetic—for which the Standard Model provides a unifying group-theoretic framework, there remains gravity, the intrinsically weakest of the four basic interactions and the one whose most elementary features have been known the longest. Yet, for all its ostensible familiarity, gravity is still an interaction apart. It remains outside the purview of the Standard Model or any other quantum field theory. From the perspective of Einstein's theory of general relativity—the most successful explanation of gravity to date—the universal attraction between masses results not from the exchange of gauge bosons but from the "warping" of space and time.

9.4. What Is the Matter with Gravity?

Looking up at the sky at night, one cannot help but sense a certain emptiness. Despite the numerous pinpoints of light shimmering in the blackness, the overall impression of the cosmos as seen by naked eye from the Earth is nevertheless much more one of space than of matter. A view of the universe, however, through the lens of the Hubble Space Telescope gives a very different impression. Like a drop of pond water teeming with small organisms, a Hubble deep-field image (Figure 9.2) reveals a cosmos teeming with galaxies in every imaginable orientation.

Astronomers estimate that the visible universe contains about 100 billion (10^{11}) galaxies, each with about 100–1000 billion stars. It is hard to talk about the contents of the Universe without having to say "billions and billions" (although the late astronomer and media celebrity Carl Sagan is certain he never said it![25]). If an average star has the mass of the Sun (about 2×10^{30} kg), then it would seem that there is an immense amount of matter in the Universe—somewhere in the vicinity of 10^{52} kg—bound up in stars alone, with perhaps an equal amount in the form of predominantly hydrogen interstellar gas. But enormous as this number is, it represents only a small fraction of the total matter that astronomers believe the Universe must contain.

The discrepancy between what is "out there" somewhere and what can be seen through telescopes at all electromagnetic wavelengths (not just visible light) has been called the "missing mass" and may well constitute over 90% of all the mass (or energy equivalent) in the cosmos. The mass is not missing, however; it is just not luminous. The necessity of the existence of dark matter is inferred through the effects of its gravity.

Despite the fact that gravity is the weakest of the four basic physical interactions, it is the force that shapes the cosmos. Electromagnetism, although stronger, has little influence because most of the

Figure 9.2. Hubble Space Telescope deep-field image of galaxies. The arcs of light are distortions of the images of distant galaxies as a result of gravitational lensing by closer objects.

matter of the Universe is believed to be electrically neutral, and the strong and weak nuclear interactions have far too short a range. From the perspective of Newtonian physics, the gravitational force between any two elementary bits of matter is an instantaneous mutual attraction proportional to the mass of each bit and the inverse square of their separation. Expressed quantitatively, Newton's law for the attractive force between two point masses M_1 and M_2 a distance r apart takes the form

$$F = G \frac{M_1 M_2}{r^2}, \qquad (9.13)$$

in which G is the universal constant of gravity ($6.7 \times 10^{-11}\,\mathrm{N\,m^2/kg^2}$). The law applies not only to elementary point masses, but, more importantly, as Newton first demonstrated, to the attraction between any two spherical masses, provided the spheres do not overlap.

Equation (9.13), used in conjunction with Newton's second law and a modicum of empirical numbers whose values were determinable even in Newton's time, makes it possible to deduce facts about celestial objects that would otherwise have remained forever inaccessible. For example, together with the acceleration of free fall ($g = 9.8\,\mathrm{m/s^2}$) and the radius of the Earth ($R_E = 6.4 \times 10^6$ m), the law of gravity yields the mass of the Earth ($M_E = 6.0 \times 10^{24}$ kg), an object one could never hope to weigh directly.[26]

The law of gravity can be used again to weigh the Earth in an entirely independent way by utilizing the period of the Moon (1 month) and the center-to-center distance between the Moon and the Earth ($r \sim 60R_E$). Any object in uniform circular motion with speed v about a central mass M at a distance r experiences a centripetal acceleration v^2/r.[27] Use of Eq. (9.13) together with Newton's second law for the inertial force on the orbiting object leads to the very useful relation

$$M = \frac{v^2 r}{G} \tag{9.14a}$$

for determining the total mass around which the object orbits. Since the satellite covers a total distance of $2\pi r$ in one rotational period T, one can substitute $2\pi r/T$ for the velocity in Eq. (9.14a) to obtain the equivalent expression

$$M = \frac{4\pi^2 r^3}{GT^2}, \tag{9.14b}$$

which is a special case of Kepler's third law of planetary motion. [In Chapter 4, we applied this law to atomic orbits; see Eq. (4.6a).] This same mode of reasoning, employing the period of the Earth about the Sun (1 year) and the distance of the Earth from the Sun (1.5×10^{11} m), leads to the mass of the Sun ($M_S = 2.0 \times 10^{30}$ kg). Indeed, the law of gravity and second law of motion enable us to weigh any distant star with an observable orbiting companion, or even the mass of entire galaxies.

It was in the measurement of the mass of galaxies that astronomers became acutely aware of a serious problem: There is, in general, insufficient luminous matter to account for the rotational velocity of the outlying galactic matter (in the form of individual stars or hydrogen gas). One does not have to look too far to encounter this problem; it occurs close to home in the features of our own galaxy, the Milky Way. The Sun, a typical star in the outskirts of one of the spiral arms, orbits the galactic center, some 28,000 light-years[28] (ly) distant, at a speed of about 220 km/s. From Eq. (9.14a), it readily follows that for the Sun to be gravitationally bound with this velocity, it must be attracted by a total central mass of approximately 10^{11} solar masses. Astronomers, however, can actually detect only about one-tenth of the requisite matter from observations throughout the electromagnetic spectrum. There is apparently much less luminous matter than gravitational matter.

Observations of other galaxies show even more graphically the seriousness of the dark-matter problem. From Eq. (9.14a), one would

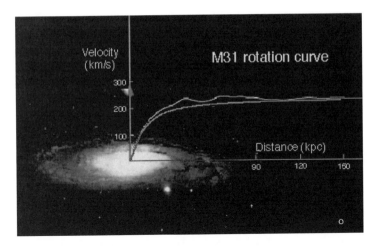

Figure 9.3. Rotation curve of the Andromeda Galaxy (M31), a spiral galaxy 2.3×10^6 lightyears distant, and comparable in size to the Milky Way. Distance is plotted in kiloparsecs (kpc) and velocity in km/s. (1 kpc is approximately 3300 ly.) The smooth curve is the theoretical prediction based on Eq. (9.23).

expect that the velocity of matter rotating in the plane of a spiral galaxy outside the central bulge in which most of the visible mass M is concentrated to diminish inversely as the square root of the orbital distance r according to the relation

$$v = \sqrt{\frac{GM}{r}}. \tag{9.15}$$

Equation (9.15) defines what is called the Keplerian rotation curve (v plotted against r), because it characterizes the motions of planets obeying Kepler's laws. Figure 9.3 shows the rotation curve of the Andromeda Galaxy (labeled M31 in the Messier catalog), a huge "island universe" approximately 2.3×10^6 ly from the Milky Way with a total mass of more than 4×10^{11} solar masses. Within the central bulge, matter rotates more or less like a solid body, the velocity increasing roughly linearly with radial distance. Beyond the observable galactic edge, however—even more than 100,000 ly beyond—the speed of rotating stars or gaseous hydrogen clouds circulating with the galaxy does not fall off as predicted by Eq. (9.15), but remains flat. In other cases, such as the smaller spiral Triangulum Galaxy (M33), which is approximately 2.4×10^6 ly distant and may be a satellite galaxy of M31, the rotation curve (not shown here) continues to rise at thousands of light-years beyond the luminous disk.

The most reasonable interpretation of a flat or rising rotation curve is that there must be more matter in the galaxy than meets the eye—

that is, the eye as enhanced by detectors covering the entire electromagnetic spectrum. That about 90% of galactic matter cannot be seen is perhaps worrisome enough, but it is only when a fuller tally of the matter of the universe is made that the problem takes on a truly alarming proportion. For one then realizes, if recent measurements of the cosmic radiation background and of the luminosities of distant supernovae hold true, that luminous matter may constitute a mere 1% of whatever it is that holds the universe together—or, as the case may be, pushes it apart.

Astonomers and astrophysicists frequently address the question of dark matter in terms of a density parameter omega, Ω, the ratio of the actual mean density ρ of mass in the Universe to the critical density ρ_c required to slow the expansion of the Universe eventually to zero. The idea that the Universe is expanding is admittedly rather difficult to visualize. The evidence for this, first adduced by Edwin Hubble in the late 1920s, comes from the Doppler shift of spectral lines[29] from distant galaxies. (It is by measurement of Doppler shifts that galactic rotation curves can be determined.) No matter the direction of observation, the spectral lines are displaced toward the red end of the spectrum ("red shift"), signifying that the radiating matter is receding from Earth. The extent of red-shifting, represented by the symbol z, is defined as the ratio of the spectral line displacement $(\lambda_{observed} - \lambda_{emitted})$ to the wavelength emitted by the source in its rest frame $(\lambda_{emitted})$.

To make sense of an isotropic recession of distant galaxies, one must conclude that either the Earth is at the center of the Universe, which is highly implausible (and, in fact, makes no sense within the framework of general relativity), or that the Universe as a whole is expanding. To visualize the expansion of the Universe, one is frequently asked to imagine blowing up a balloon upon which many small dots (the "galaxies") are inked. From the perspective of any one dot on the expanding surface, all other dots are receding. Moreover, it is straightforward to show that the apparent recession velocity of any dot relative to a given reference dot increases in proportion to its distance from that reference.[30]

This picture is adequate up to a point, as long as one keeps in mind that the analogy is between the two-dimensional *surface* of the balloon and the entire Universe; the Universe, in contrast to the balloon interior, has no center. Also, although the balloon is expanding into space previously occupied by air, the Universe is not expanding *into* anything; the expansion itself creates space. And last, it would be better to think of the dots as small rigid disks affixed to the balloon with a bit of adhesive, rather than as spots inked directly to the surface. Ink spots would get larger as a balloon expands, but a galaxy (or even a cluster of galaxies) in an expanding space is essentially unchanged in size because of the gravitational binding of its constituents.

The mass density of the Universe, and hence the parameter Ω, are intimately connected to the geometry and fate of the Universe, provided that the Universe contains more-or-less familiar kinds of matter and energy, a point to which I will return in the next section. If $\Omega > 1$, the Universe has a positive geometric curvature (like that of a sphere) and is said to be closed. The gravitational attraction between matter is sufficient to halt the expansion, and the contents of the Universe will eventually collapse into a singularity ("the big crunch"), a suitably dramatic conclusion, perhaps, to its formation in a "big bang". Conversely, if $\Omega < 1$, the Universe has a negative curvature (like that of a saddle surface) and is said to be open. There is insufficient matter to halt the expansion which continues indefinitely, resulting in a universe immeasurably dilute of matter. The threshold condition, $\Omega = 1$, characterizes a flat universe with zero curvature (like that of a plane). This condition represents a delicate balance between kinetic and potential energies in which the universal expansion asymptotically slows to zero. However, a universe described by $\Omega = 1$ is in an unstable equilibrium, the slightest perturbation in density rapidly (on an astronomical timescale) leading to either a hot, dense collapse or cold, sparse expansion. Yet, as implausible as it might first appear, both theory and experiment presently point to "1" as the likely value for the mass parameter of the Universe.

Although the full significance of an expanding universe can be understood only with Einstein's theory of general relativity, Newton's laws can be used again to deduce the value of the critical density ρ_c and provide at least a rudimentary insight into the magnitude and nature of the dark-matter problem.

For any compact distribution of matter of total mass M, there is a minimum or threshold speed v_{esc} at which a small object gravitationally bound to M must be launched in order to escape forever this gravitational attraction and just coast to rest at an infinite distance away. Consider, for example, a baseball on the surface of the Earth. With what minimum speed must that ball be thrown "up" so that it never comes "down"? For each unit of its mass, the ball at the instant of launch has a kinetic energy $\frac{1}{2}v_{\mathrm{esc}}^2$ and a potential energy $-GM/R$, in which M and R are respectively the mass and radius of the Earth. If the ball is to come to rest (kinetic energy = 0) at an infinite distance from the Earth (potential energy = 0), then its total energy (kinetic + potential) must be zero. Provided that energy is conserved, the total energy per unit of mass at the outset ($\frac{1}{2}v_{\mathrm{esc}}^2 - GM/R$) must also be zero, in which case it follows straightforwardly that $v_{\mathrm{esc}} = \sqrt{2GM/R}$. For the mass and radius of the Earth given previously, the escape velocity of the ball, or any other object at the Earth's surface, would be approximately 11.2 km/s (24,000 miles/hour). The best baseball pitcher (throwing the ball at a speed under 100 miles/hour) could never even come close.

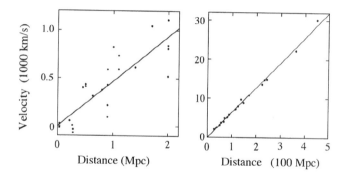

Figure 9.4. Hubble plots of galactic recession velocity against distance as determined originally by Hubble (~1929) (left) and from recent (~1996) measurements of Type Ia supernovae (right).

If the mass M is uniformly distributed with density ρ within a spherical volume of radius R, then substitution of $M = (4\pi/3)R^3\rho$ into the preceding expression leads to an escape velocity $v_{esc} = \sqrt{8\pi G\rho/3}\,R$, which is directly proportional to the radial distance of the ball at its launch from the center of the Earth.

Now, one of the consequences of an expanding universe discovered empirically by Hubble through his red-shift measurements is that the relative velocity of recession of two distant unbound galaxies is proportional to the spatial separation of the galaxies (as in the balloon analogy). This proportionality constant, which theoretically is not constant but varies with the age of the Universe, is known as the Hubble parameter H. Because of uncertainties in the determination of galactic distances, Hubble's original plot (~1929) of recession velocity against distance showed a wide dispersion about the best-fit line (left frame of Figure 9.4), but recent data (~1996), obtained from red shifts ($z < 1$) of distant Type Ia supernovae, show a linear relationship (right frame of Figure 9.4) to such perfection as one can only marvel at. Assuming that the Universe "coasts to rest" (i.e., to zero expansion rate) at spatial infinity, we can apply the expression derived for the escape velocity of a baseball from Earth to the expansion of the Universe by setting the mass density ρ equal to the critical density ρ_c and then equating the factor $\sqrt{8\pi G\rho_c/3}$ with the present value of the Hubble parameter. The resulting critical mass density of the Universe is then

$$\rho_c = \frac{3H^2}{8\pi G}. \tag{9.16}$$

The numerical value of the Hubble parameter is perhaps one of the most contentious issues in astronomy and cosmology and has led to complex acrimonious debates[31] extending over at least six decades since Hubble first reported a linear relation between recession velocities and distances. A number that is perhaps not too far from what various factions are converging upon is $H \sim 60\,\text{km/s/Mpc}$ (megaparsec). The astronomical distance unit of 1 pc is effectively the altitude of an isoceles triangle of apex angle of 1 arc-second (1/3600 of a degree) and base length equal to the distance of the Earth from the Sun. This amounts to approximately 3.3 ly. Thus, conversion of H into standard metric units results in $H \sim 2 \times 10^{-18}\,\text{s}^{-1}$. The Hubble constant has the dimension of inverse time, and it is reasonable to interpret the reciprocal of H (i.e., $5 \times 10^{17}\,\text{s}$ or 15–16 billion years) as an approximate measure of the age of the Universe, the amount of time that has passed since the singular event ("big bang") giving rise to space, time, matter, and energy. Adopting this value of H in Eq. (9.16) leads to a critical mass density $\rho_c \sim 7.1 \times 10^{-27}\,\text{kg/m}^3$, or about the mass of four protons per cubic meter of space.

The idea that the Universe began as a singular explosion of particles and radiation that subsequently evolved over billions of years into the elements and structures observable today was proposed by George Gamow in the late 1940s. Derisively termed the "big bang" theory of cosmology by adherents of an alternative and then prevailing "steady-state" cosmology, Gamow's theory (and the more recent variations incorporating a brief period of exponential expansion referred to as "inflation"[32]) successfully accounted for the relative abundances of primordial light elements—principally hydrogen, deuterium, helium, and lithium—and predicted the existence of an all pervasive cosmic background radiation (CBR). The CBR was detected (unknowingly) by Bell scientists Arno Penzias and Robert Wilson around 1965 and has since been measured extraordinarily precisely over a wide range of frequencies by balloon-borne and satellite-based instrumentation. The spectral distribution of the radiation follows nearly perfectly the Planck blackbody radiation curve[33] for a cosmic background temperature of $2.728 \pm 0.004\,\text{K}$. At this temperature, the peak radiation intensity, deducible from Wien's displacement law,[34] occurs at a wavelength of approximately 1 mm. Since this peak is squarely in the microwave portion of the electromagnetic spectrum, the radiation has also been referred to as the cosmic microwave background (although the high-frequency tail of the curve extends into the infrared and beyond).

If there is anything in contemporary physics that corresponds to a universal reference frame, such as embodied in the long-discarded notion of an "ether," it is the CBR. The radiation is isotropic,[35] bathing the Earth from all directions in space, and uniform in temperature to about one part in 10^5. Remarkable as this uniformity is, it is not perfect—the angular distribution of the radiation across the sky reveals minuscule intensity, and therefore temperature, fluctuations—

and therein lies an experimental fact quite literally of cosmic significance.

The CBR, a relic dating back to a mere 300,000 years after the big bang (a blink of an eye in cosmological history), is the most ancient signal that has yet been detected and, indeed, the oldest that *can* be detected until newly developed gravitational wave detectors are put into service. Before this time, according to the standard cosmological model (i.e., the inflationary big-bang cosmology), the temperature of the Universe was too high to permit the combination of charged particles into neutral atoms. Instead, ionized matter and radiant energy coexisted in a hot, dense, opaque plasma. With continued expansion, the temperature of the plasma fell, neutral atoms formed (at approximately 3000 K), and the resulting hot gas became transparent to electromagnetic radiation. The photons that constitute the CBR decoupled from matter and have traveled undisturbed ever since, increasing in wavelength (and decreasing in frequency and energy) as the Universe expanded. To observe this radiation is to look back in time and see a part of the cosmos as it was many billions of years ago.

Although the hot gas from which the CBR originated was highly uniform in density, acoustic waves through the gas produced a spectrum of density fluctuations—regions of greater and lesser concentrations of matter—which, in turn, generated fluctuations in the energy density and temperature of the escaping radiation.[36] Given the distance over which CBR photons have propagated to Earth and knowledge of the wavelengths of the acoustic resonances in the hot gas, the angular variation of these ripples across the sky is predictable. The pattern of ripples depends, however, on the geometry of the space through which the photons have traveled (Figure 9.5)—and this geometry, according to general relativity, in turn depends on the density parameter Ω.

Imagine looking at the squares of a chessboard through a large convex lens (with the lens less than a focal length from your eye). The curvature of the lens results in a magnification of the squares so that each square subtends a larger angular width in the observer's field of view than if viewed through a parallel-sided slab of window glass. By contrast, the squares would be minified and subtend a smaller angular width if examined (at an appropriate distance) through a concave lens. A universe with $\Omega > 1$ is like a convex lens; the larger the value of Ω, the larger is the angular width of the patches of CBR temperature variations across the sky. Conversely, a universe with $\Omega < 1$ is like a concave lens; the CBR patches appear smaller. A spatially flat universe is like a lens with no curvature, i.e., a flat windowpane, in which the rays of light follow the straight lines of Euclidian geometry. The observed angular size of the principal contribution to CBR fluctuations, independently measured and nearly simultaneously reported in

Curvature and Angular Size

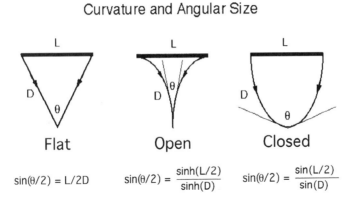

Flat	Open	Closed
$\sin(\theta/2) = L/2D$	$\sin(\theta/2) = \dfrac{\sinh(L/2)}{\sinh(D)}$	$\sin(\theta/2) = \dfrac{\sin(L/2)}{\sin(D)}$

Figure 9.5. Relation between spatial curvature and the angular size (θ) of a structure of known length (L) and distance (D). Photons reaching the observer by traveling along the geodesics of an open (closed) space generate an angle smaller (larger) that the angle generated in a flat space. [In the formulas for $\sin(\theta/2)$, distances are expressed in units of the radius of curvature and are therefore dimensionless quantities.]

1999 by three different teams of researchers,[37] has consistently turned out to be about 1 degree on the sky, or approximately twice the apparent angular size of the full Moon seen from Earth (Figure 9.6). This is precisely what inflationary cosmology predicts for $\Omega = 1$. The Universe seems to be flat.

For the Universe to be flat, it must contain an average mass density nearly coincident with the critical density ρ_c, provided (as we have been assuming so far) that gravity is the only relevant long-range interaction affecting the distribution of matter and the expansion of the Universe. However, the best estimates of the total mass of all the matter that can be observed with telescopes yield a density parameter $\Omega \leq 0.05$. In other words, if the chain of reasoning connecting the geometry of the universe and the matter within it is sound, then over 95% of the mass in the Unverse is not visible. What is this matter and where is it?

A substantial part of the dark matter, as inferred from galactic rotation curves, must undoubtedly lie in the halos encompassing individual galaxies. Such halos are expected to contain relatively cold stars, like white dwarfs, which have ceased to generate energy through nuclear fusion reactions, or "failed" stars, like brown dwarfs, which from the outset were not massive enough to initiate or sustain nuclear fusion. (Brown dwarfs might fuse deuterium, but the supply would not last very long, astronomically speaking.) These stars would at best glow in the infrared with a low undetectable luminosity until eventu-

-300 μK ▬▬▬▬▬▬▬▬▬▬▬▬▬▬▬ 300 μK
                    0
                    ●

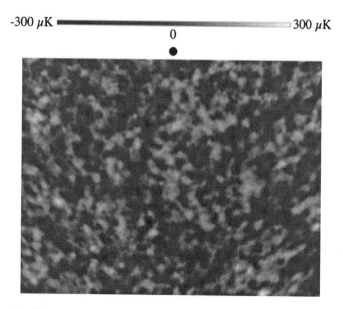

Figure 9.6. Fluctuations (spanning a range from –300 μK to +300 μK) in the cosmic background radiation temperature measured in a portion of the sky over Antarctica in 1998–1999 by the BOOMERANG project (Balloon Observation of Millimetric Extragalactic Radiation and Geomagnetics). The black circle above the figure shows the angular size (approximately 0.5 degree) of the full Moon subtended at the Earth.

ally becoming burnt-out stellar cinders. Indeed, why not solve the dark-matter problem in its entirety simply by assuming the existence of an immense population of these cold dark objects both bound in galaxies and wandering alone through the cosmos? Regrettably, the assumption is untenable for at least two reasons.

First, although dark matter by definition cannot be seen directly through its own radiant emissions, it can, if concentrated in compact objects like white and brown dwarf stars, *block* the light coming from luminous objects (like galaxies or quasars) behind it that lie in the line of sight of a terrestrial telescope. Such luminous objects might, therefore, be expected to "wink out" as the invisible foreground object crosses the line of sight. Actually, the effect looked for by astronomers is even more interesting, for a dark compact object can behave more like a lens than like a shutter. One remarkable consequence of Einstein's theory of general relativity is that light rays are deflected by gravity. To use an optical analogy again, imagine holding a convex lens within a distance of a focal length from your eye and moving the lens horizontally across your view of a small incandescent light bulb across the room. The image of the light is momentarily larger and brighter as the lens passes by. Similarly, the light rays from a distant luminous object, e.g., a quasar, skirting the periphery of a compact dark foreground object, are gravitationally diffracted toward the observer, giving rise to a variation in luminosity of the background object over the course of passage. This variation in luminosity can take a wide range of forms, from a short-term brightening as a result of "microlensing" by a dark object of low mass (e.g., 0.1 solar mass) to multiple images and rings engendered by a dark foreground object of galactic size. Numerous gravitational lensing events have been observed both within and outside the Milky Way (see Figure 9.2), but not enough to account for the vast preponderance of dark matter. A small sample of microlensing events suggests at present that less than 20% of the dark matter in the Milky Way may consist of compact halo objects.

Second, the standard cosmological model puts a rather stringent limit on the amount of baryonic matter (principally neutrons and protons) that could be produced in the aftermath of the big bang. In minutes following the initial fireball (and long before the decoupling of radiation from matter), while the temperature of the Universe remained above 10^{10} K, neutrons and protons in the cosmic plasma could transmute into one another by means of nuclear weak interactions. Neutrons, however, are slightly more massive than protons by nearly 1.3 MeV/c^2. Thus, when the temperature of the expanding universe cooled to about 7×10^9 K, the available thermal energy per particle[38] was insufficient to make up the neutron–proton mass difference, and the ratio of protons to neutrons froze out at a value of about 7 to 1. As the Universe expanded and cooled further, all remaining neutrons eventually underwent beta decay to protons or combination reactions to form deuterium ($^2_1$H) and isotopes of helium ($^3_2$He, $^4_2$He). In a universe with sufficiently high mass density, virtually all the undecayed neutrons would have ended up in $^4_2$He, which, in fact, makes up the preponderance (~23% by mass) of primordial matter apart from hydrogen. However, if the density of the Universe is low enough, the conversion of $^2_1$H and $^3_2$He into $^4_2$He would be incomplete. The relative abundances of these elements compared to hydrogen as observed today depend sensitively on the mass parameter Ω, and the approximate value which best reproduces these proportions is $\Omega_B \sim 0.1$, where B explicitly denotes baryonic matter.

There is a deep mystery in that simple constraint $\Omega_B \sim 0.1$, for it signifies that some 90% of the mass in the universe contains matter different from that with which we are familiar. It is the ultimate embodiment of the Copernican principle; not only is the Earth not the center of the Solar System, nor the Solar System the center of the Galaxy, nor the Galaxy the center of the Universe, but we are not even made of the dominant material of the cosmos. The problem of dark matter, however, is yet stranger and more subtle than depicted so far, for, if conclusions drawn from recent observations of Type Ia supernovae are sustained, then much of the "missing mass" of the Universe may not even be mass.

A supernova is the cataclysmic death of a star in which, within the span of less than a second, the luminosity can increase a billionfold to rival that of an entire galaxy. The mechanisms and end products of these spectacular stellar explosions generally fall into two basic categories. Type II supernovae represent the gravitational collapse of a star of perhaps 10–100 solar masses (see Figure 9.7), leaving behind a neutron star, a highly compact stellar remnant containing one or more solar masses within a diminutive volume only 10 km in diameter, or a black hole, a star with gravitational attraction so strong that not even light can escape. Pulsars, whose train of radio pulses constitute the most precise chronometers in the universe, are known to be rapidly rotating, highly magnetized neutron stars. In a Type II supernova, the source of the explosion is gravitational potential energy. By contrast, a Type I supernova (see Fig. 9.8) is believed to be a thermonuclear explosion of a white dwarf star in a binary star system. Over time, the dwarf accretes more and more matter from the normal companion, until its mass reaches a critical value, the Chandrasekhar

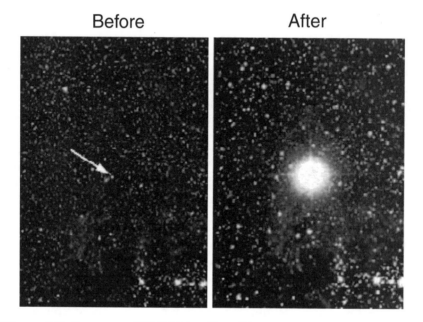

Before   **After**

Figure 9.7. Supernova SN1987A is a Type II supernova observed in 1987; it is about 170,000 ly away.

Figure 9.8. Supernova SN1994D, the bright star at the lower left of the figure, is a Type I supernova observed in 1994; it is at a distance of 45,000,000 ly, which, by the Hubble relation, corresponds to a relatively small red shift $z \sim 3 \times 10^{-4}$.

limit (approximately 1.4 solar masses), at which point the star detonates and is totally disrupted. No stellar remnant is left behind.

Because all Type Ia supernovae are believed to detonate at approximately the same theshold mass and size, and therefore radiate energy isotropically at a rate that should be the same for all such explosions, astronomers can use Type Ia supernovae as a sort of standard candle for gauging distances to other galaxies, especially those that have high red shifts and therefore (by Hubble's relation) are very far away. The intrinsic luminosity is determined by first examining nearby Type Ia supernovae whose distances can be measured by other means. Once the intrinsic luminosity is known, the distance to supernovae in the far depths of the cosmos can be deduced from measurement of the radiant flux (power per square meter) that reaches the Earth.[39]

Actually, the situation is not quite so simple (it never is!), for different Type Ia supernovae are found to have somewhat different intrinsic luminosities correlated with their light curves, i.e., the variation in brightness of the event with time. In general, the expanding fireball reaches peak brightness in a few weeks and then declines over the course of a few months. However, the duration of larger, brighter events is longer than that of less energetic, fainter events. From measurements of these light curves, astronomers were able to produce a highly linear calibration curve, such as shown in Figure 9.4 for distances up to 500 Mpc (corresponding to red shifts up to $z \sim 0.1$), in which nearby Type Ia supernovae of the same red shift have the same distance. In this way, two separate groups[40] independently measured the distances and corresponding red shifts of about 50 supernovae, including over 30 with high red shifts.

The results for the highest red-shifted supernovae were unexpected and startling in their implications. From the red-shift data and use of Hubble's relation, the distance to each supernova could be estimated and compared with the distance deduced from the observed radiant flux. They were not the same, the luminosity-based distances being on average about 10–15% larger. Stated differently, the supernovae appeared to be too faint to be at the distances yielded by Hubble's relation if the intrinsic luminosities were actually as large as astronomers believed them to be. If it is indeed the case that the Type Ia supernovae are farther than the Hubble relation predicts, then one interpretation is that the rate of expansion of the Universe is greater now than it had been in the distant past when these stars actually exploded.[41]

That the Hubble parameter is different in the present epoch than it was billions of years ago is not in itself a surprising revelation, because general relativistic models of an isotropically expanding universe containing (at the largest scales) a homogeneous distribution of matter predict exactly how the cosmic scale factor should vary in time. However, a universe containing only gravitationally attracting matter and radiation should expand more *slowly* with the passage of time, not accelerate.

The initial reports of this accelerated cosmic expansion created something of a sensation. Highlighted in television broadcasts, newspapers, and scientific journals, the discovery was selected by *Science* as the "Top Research Advance" of 1998. Figure 9.9 summarizes the situation at the time by plotting, as a function of red shift z, the difference between the observed supernova brightness and the brightness expected from Hubble's relation with the present value of the parameter H_0. If the unexpected faintness of distant Type Ia sypernovae is interpreted as a cosmological effect, then the ordinate of the plot is a measure of the variation in the value of the Hubble constant from the present value. Superposed on the data are theoretical curves for accelerating and decelerating cosmic expansions.

Readers can decide for themselves the extent to which the dispersed data points at the right (i.e., high-z) side of Figure 9.9 discriminate among the three possibilities: acceleration, deceleration, no change. On occasions like this, I think of Mark Twain's perceptive comment in *Life on the Mississippi* that

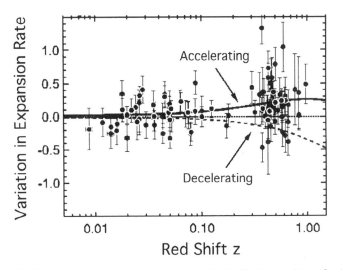

Figure 9.9. Variation in the expansion rate of the Universe (i.e., the Hubble parameter) as a function of red shift as determined by observation of the brightness and red shift of Type Ia supernovae. The horizontal baseline (at 0.0) signifies the present value of the Hubble parameter H_0. The higher-lying solid theoretical curve shows an accelerating cosmic expansion; the lower-lying dashed theoretical curve shows a decelerating cosmic expansion. Data points from both the Supernova Cosmology Project and the High-Z Supernova Search Team represent the difference between observed brightness and brightness inferred by Hubble's relation for specified z and constant H_0. (Adapted from http://cfa-www.harvard.edu/cfa/oir/Research/supernova/HighZ.html.)

There is something fascinating about science. One gets such wholesale returns of conjecture out of such a trifling investment of fact.

(although it must be said that the investment in equipment and time to observe and analyze Type Ia supernovae is far from trifling).

Other explanations for the faintness of distant supernovae are conceivable. For example, the supernovae may have appeared dimmer than expected because of scattering by interstellar dust or because the intrinsic luminosity of Type Ia supernovae many billions of years ago was lower than that of more recent supernovae.

That the cosmological inference may be justified, however, is supported by a serendipitous rediscovery in 2001 of a Type Ia supernova (SN1997ff) photographed four years earlier with the exceptionally high red shift $z \sim 1.7$.[42] SN1997ff is believed to have detonated over 10 billion years ago at an epoch in the evolution of the Universe when the high density and, therefore, stronger gravitational attraction of matter *retarded* cosmic expansion. Under such circumstances, a supernova should appear *brighter*, not fainter, than it would for the same z

and uniform expansion. Analysis of the luminosity of SN1997ff over several spectral regions purportedly confirms this brightening. Presuming that the event has been correctly identified as Type Ia, this finding is inconsistent with what one would expect from supernova evolution or from light scattering by dust. At the present time, it would seem that accelerated cosmic expansion provides a viable explanation of Type Ia supernova data.

Statistical fits of various cosmological parameters to the supernova data constrain the mass parameter to $\Omega_M \sim 0.3$, where now a subscript M explicitly denotes that this is a measure of the total density of *mass* in the Universe, luminous and dark combined. Moreover, the data are consistent with a flat universe and, therefore, a total omega parameter of 1. If the expansion of the universe is indeed accelerating, there must be something other than matter and radiation to drive it. Because both CBR and supernova data are presently consistent with $\Omega \sim 1$, the unknown agency is believed to contribute approximately 0.7 to omega, a contribution denoted by the symbol Ω_Λ, where the subscript Λ refers to a term that Einstein once inserted and subsequently removed from his equations of general relativity. Designated equivalently as the "cosmological constant" (from the perspective of general relativity) or the "energy density of the vacuum" (from the perspective of quantum field theory), the origin and nature of this bizarre and dominant component is one of the outstanding problems of contemporary astrophysics and cosmology.

9.5. Shedding Light on Dark Matter

If the universe is to be filled with some kind of nonbaryonic matter, theorists have had little difficulty in coming up with possibilities; Twain's observation applies widely in particle physics and cosmology. The numerous "returns of conjecture" fall roughly into two categories: hot dark matter (HDM), comprising fast-moving relativistic light particles principally in the form of neutrinos, or cold dark matter (CDM) in the form of sluggish nonrelativistic **weakly interacting massive particles** (designated generically by the acronym WIMP).

One advantage of HDM models is that neutrinos, neutral spin-$\frac{1}{2}$ fermions associated with each of the three leptons (electron, muon, and tau particle) and their antiparticles, are known to exist. Although originally considered to be massless, there is mounting evidence, based principally on the phenomenon of neutrino oscillations—the periodic transmutation of one neutrino "flavor" into another—that at least one (and most likely all) types of neutrinos have a nonzero rest mass. Neutrino oscillation experiments do not determine the absolute masses of the various neutrinos, only differences in mass (actually differences in the squares of the masses). Because there are numerous neutrinos in the universe,[43] even a small neutrino mass (e.g., a few eV, as compared with an electron mass of one-half million eV) could lead collectively to a substantial contribution to Ω. The difficulty with neutrinos, however, is that such high-velocity particles would form structures on scales larger than those observed and that the time for fragmentation into galaxy-sized structures would take an appreciable fraction of the age of the Universe. Thus, in contrast to prevailing evidence, galaxies would have formed only recently.

In view of these deficiencies, cosmologists turned instead to CDM models with WIMP masses tens to thousands of times the mass of a proton. Cosmological models based on WIMPs have been highly successful in accounting for galaxy-sized structures, but at smaller scales, they have led to overly dense galactic cores, dense substructures in galactic halos, and too many galactic halos within a galactic cluster.

Another worrisome attribute of CDM models is the inability of increasingly sensitive experiments to find any WIMPs. If WIMPs exist as part of a halo of nonluminous matter permeating the Milky Way and other galaxies, then, from time to time, one of them should collide with a target nucleus in a suitable detector, leading to either an elastic recoil or a nuclear excitation. Ongoing nuclear experiments to search for some manifestation of these interactions have led to intriguing contradictory results. In one series of experiments (DAMA) carried out at the Gran Sasso laboratory in Italy, an annual modulation was found in the presumed elastic scattering of WIMPs from the nuclei in a scintillating target material (thallium-activated NaI, the same material used in the nuclear decay experiments discussed in the previous chapter). Such an annual variation is precisely what one would expect, according to the DAMA researchers, if the Earth experiences a WIMP "wind" as it rotates about the Sun. In June, when the velocity of the Earth about the Sun is parallel to the velocity of the Solar System through the Galaxy, the flux of WIMPs should be larger and the number of elastic scattering events more numerous than in December, when the two velocities are directed oppositely. The DAMA group claim to have seen indications of this phenomenon. The claim, however, is refuted by a second group, the Cryogenic Dark Matter Search (CDMS), employing an entirely different detector (a large germanium crystal) at an underground facility on the campus of Stanford University. According to the CDMS team, events observed by DAMA were most probably due to stray neutrons produced by cosmic ray collisions in the atmosphere. Subsequent experiments seem to confirm this interpretation. Further research with more sensitive detectors will eventually resolve the issue, but for the present, evidence for the existence of WIMPs is tenuous at best.

If neither neutrinos nor WIMPs are the principal constituents of dark matter, then of what is the preponderance of nonbaryonic dark matter made? It is at this point that the idea of a Bose–Einstein condensate (BEC) of ultra light bosons, with which I began this chapter, becomes very appealing.

Suppose that nonbaryonic dark matter consisted of extremely low-mass neutral bosons subject only to gravity. Above a certain transition temperature, a gas of these particles would behave more or less like a photon gas (i.e., relativistic hot dark matter), but *below* the transition temperature, the particles would drop into the lowest-energy states to form a degenerate quantum gas or BEC. Because the particles in the condensed phase are in a coherent quantum state of fairly sharp momentum centered about zero, the gas is nonrelativistic and behaves like cold dark matter. Individual bosons are likely to be found anywhere within a spatial region defined by the coherence length of the condensate, which, according to the Heisenberg uncertainty principle, varies inversely with the sharpness of the momentum distribution.[44] If it should turn out that the coherence length is of astronomical size, then a cosmic BEC, despite—in fact, because of—the low mass of its constituent particles, would not have the undesirable property of collapsing into dense nuggetlike structures at the centers of galaxies. The coherence length of the condensate, as I shall show shortly, is larger, the smaller the mass of the boson—and would, indeed, assume an astronomical size for a particle of sufficiently low mass. Moreover, the corresponding transition temperature for condensation turns out to be so high that, except for a brief period following the origin of the Universe, BEC dark matter would have been present throughout the important period of galaxy formation and up to the present time.

Because a pervasive condensate of very light neutral bosons constitutes a sort of **w**eakly **i**nteracting **deg**enerate **et**her, I designated the particles by the acryonym WIDGET. The word "weak" here takes on its standard English meaning and does not refer specifically to weak nuclear interactions (although such interactions may be incorporated into some future model).

The chain of reasoning leading to WIDGETs brings into play all the basic ingredients of theoretical physics described in the preceding sections of this chapter. Let us assume that a scalar field ϕ permeates all space. Within the framework of modern physics, this is not an unreasonable assumption; indeed, similar assumptions involving other scalar fields underlie the Higgs mechanism for generating the masses of elementary particles in the Standard Model and the exponential cosmic expansion in the inflationary big-bang model of the Universe. It is assumed further that the potential energy of the scalar field is given by Eq. (9.12), a form widely employed in condensed matter physics to account phenomenologically for attributes of superconductivity and superfluidity. This is not a coincidence. Both types of "super" phenomena are manifestations of a Bose–Einstein condensation involving phase changes from a high-temperature incoherent normal state to a low-temperature coherent quantum state as a result of spontaneous symmetry breaking.

Look again at Figure 9.1, which shows the general shape of the potential energy curve represented by Eq. (9.12). As applied to the problem of dark matter, the state of minimum energy of the scalar field in a sufficiently high-temperature universe is $\phi = 0$, the minimum of the parabolic curve that corresponds to the potential parameter $a > 0$. As the Universe cools, however, the potential energy curve evolves into the double-well shape with parameter $a < 0$, and the state $\phi = 0$ is no longer a global energy minimum, but a local maximum. The system then undergoes a transition into one of the two states of minimum energy $\phi = \pm\phi_0$ (where $\phi_0 = \sqrt{-a/2b}$), thereby breaking the reflection symmetry (across the axis $\phi = 0$) of the equations of motion.

The equations of motion describing the dynamics of the scalar field interacting with gravity are obtained by following a procedure very similar to that outlined at the end of Section 9.2 and in the description of spontaneous symmetry breaking in Section 9.3. One starts by adding the Lagrangians for the scalar field and for gravity, i.e., for Einstein's general relativity *without* a cosmological constant.[45] The principle of gauge invariance determines how the (spinless) particles of this theory must couple to gravity; in other words, it specifies the form of the covariant derivative in the kinetic energy term. Next, one accounts for spontaneous symmetry breaking by re-expressing the full Lagrangian in terms of the excitation $\bar{\phi}$ defined with respect to the broken-symmetry field $+\phi_0$ or $-\phi_0$. The dynamics of the scalar field interacting with gravity then follow by calculating, from the Lagrangian, the Euler–Lagrange equations for the scalar field $\bar{\phi}$ and the gravitational fields.

According to general relativity, gravity is a manifestation of the geometry of a four-dimensional space–time. Like a deformable landscape, space–time is warped by the presence of matter and energy, and these ethereal "hills" and "valleys" define the paths, referred to as geodesics, along which matter and radiation travel. A short pneumonic (due to J. A. Wheeler, I believe) that succinctly expresses this dual dependence of space–time and mass–energy is this: Matter tells space–time how to bend; space–time tells matter how to move. Mathematically, the geometry of space–time is reflected in a differential line segment of the form of Eq. (9.6a) or Eq. (9.6b), but in which the constant elements $\eta_{\mu\nu}$ of the Minkowski metric are now replaced by functions $g_{\mu\nu}$ that can vary spatially and temporally. The metric elements $g_{\mu\nu}$ (or the elements $g^{\mu\nu}$ of the inverse metric tensor) are the gravitational potentials of Einstein's theory. In Newtonian gravity, there is only one potential, but in general relativity, there are, as a result of the symmetry $g_{\mu\nu} = g_{\nu\mu}$, ten independent metric elements. It is through insertion of these elements in the kinetic energy term of the Lagrangian that the scalar field experiences gravity.[46]

To find the elements $g_{\mu\nu}$ that describe space–time in the presence of a distribution of matter and energy, one must solve Einstein's field equations, which take an ostensibly compact form

$$R_{\mu\nu} - \frac{1}{2} g_{\mu\nu} R = -\left(\frac{8\pi G}{c^4}\right) T_{\mu\nu} \qquad (9.17)$$

that belies its formidable complexity as a set of 10 coupled nonlinear differential equations. The interpretation of Eq. (9.17), however, is straightforward and elegant. The left-hand side, constructed from the so-called contracted Riemann tensor $R_{\mu\nu}$ and the Riemann curvature scalar R (both of which depend nonlinearly on the set of gravitational potentials $g_{\mu\nu}$), is a measure of the curvature (i.e., geometry) of space–time. The right-hand side, which is constructed from the stress-energy tensor $T_{\mu\nu}$ of matter and radiation, is pure physics.[47] It has been remarked by Einstein that the left side is like marble (the firm beautiful rigor of mathematics) and the right side is like wood (the soft messy details of the physical world).

As a consequence of the spontaneous symmetry breaking of the scalar field, two important things occur. First, as in the case of the electroweak interactions discussed in Section 9.2, the quanta of the scalar field (i.e., the WIDGETs) acquire a mass m (expressed through the particle Compton wavelength λ_C)

$$\lambda_C \equiv \frac{h}{mc} = \frac{2\pi}{\sqrt{-2a}} \qquad (9.18)$$

that depends on the quadratic parameter a of the scalar-field potential energy. Second, the Lagrangian for gravity acquires a cosmological constant Λ

$$\Lambda = \frac{2\pi G}{c^4}\left(\frac{a^2}{b}\right) \qquad (9.19)$$

dependent on both parameters of the potential energy.

Shortly after formulating general relativity,[48] Einstein realized to his chagrin that the field equations for a universe with matter did not lead to a static universe, but to a universe that collapsed, unless it was expanding. To circumvent this problem, he added a term $\Lambda g_{\mu\nu}$ to the left-hand side of Eq. (9.17), which had the effect of pushing space outward everywhere. To sustain the Universe from collapse, the cosmological constant Λ need assume only a very small value and would not have affected the otherwise successful applications of general relativity to local systems as, for example, the precession of the perihelion of Mercury about the Sun. However, upon learning of Hubble's discovery that the Universe was in fact expanding, Einstein removed the cosmological term, expressing his regret (as recounted by Gamow) for having made so bad a "blunder."

Inclusion of the cosmological term, however, is not generally regarded as a blunder today, for a positive value of Λ results in an accelerating cosmic expansion in keeping with the recent observations of Type Ia supernovae. However, there is no need to insert a cosmological constant into Einstein's field equations "by hand"; spontaneous symmetry breaking of the scalar field does this naturally.

The equilibrium size of a cloud of BEC dark matter is the outcome of two competing forces. On the one hand, the gravitational attraction among all the particles compresses the gas as much as possible so as to lower the potential energy. On the other hand, the confinement of quantum particles to smaller regions raises their kinetic energy.[49] A balance between inward gravitational attraction and outward quantum pressure is reached when the energy of the condensate is minimized, i.e., when the derivative of the total energy of the gas with respect to its radius vanishes. The equilibrium radius of the gas is the condensate coherence length, which up to an unimportant numerical factor of order unity takes the form

$$\xi_c = \frac{h^2}{GMm^2} = \left(\frac{3h^2}{4\pi Gm^2 \bar{\rho}} \right)^{1/4}, \qquad (9.20)$$

where m is the mass of the scalar boson (WIDGET) and M and $\bar{\rho}$ are respectively the total mass and mean density of the condensate.

A simple heuristic argument can be given for the existence of a scale, referred to as the Jeans length λ_J, separating gravitationally stable and gravitationally unstable density fluctuations in matter. Perturbations in the density of matter of a size $\lambda < \lambda_J$ are gravitationally stable and propagate through the matter as acoustic waves. (The term "acoustic" here refers to longitudinal waves, like sound waves, and does not imply that the corresponding frequencies are audible to humans.) By contrast, perturbations of a size $\lambda > \lambda_J$ are gravitationally unstable and can grow or decay exponentially. Exponential growth results in the condensation of gravitationally bound clouds of matter such as is believed to have occurred in the formation of galaxies. The timescale over which a cloud of matter collapses under its own weight is roughly $\tau_g \sim 1/\sqrt{G\bar{\rho}}$.[50] Acting against this collapse is the gas pressure within the cloud. The timescale for gas pressure to respond is $\tau_p \sim \lambda/v$, where λ is the wavelength of a density fluctuation and—in the case of a classical gas—v is the velocity of sound v_s. At the threshold size, where gas pressure just counterbalances gravitational collapse, the equality $\tau_p = \tau_g$ leads to the Jeans scale $\lambda_J \sim v_s/\sqrt{G\bar{\rho}}$.

In a classical gas, the velocity of sound depends on the compressibility of the gas—that is, on the change in gas pressure with volume. However, the compressibility (and therefore the classical sound velocity) of an ideal BEC vanish, because the pressure of the gas depends only on temperature. One might be tempted to conclude that $\lambda_J = 0$ and, hence, density perturbations at *all* wavelength scales would be gravitationally unstable, but this is not the case. In a BEC, which is a quantum gas, the pertinent velocity v of wavelike perturbations through the medium is obtained from the (nonrelativistic) de Broglie wavelength $\lambda = h/mv$ of the bosons. Identifying λ with λ_J and substituting $v = h/m\lambda_J$ into the expression for τ_p lead to a Jeans length which corresponds very closely to the coherence length given by Eq. (9.20). Thus, density perturbations of a size smaller than the coherence length are gravitationally stable, as inferred previously from an argument based on the uncertainty principle.

To estimate the mass of a WIDGET in the halo of dark matter about a galaxy, let us consider again the Andromeda Galaxy (M31), which, as shown in Figure 9.3, has a luminous core extending about 30 kpc ($\sim$98,000 ly) from the center. The mean mass density of the galaxy (out to about 150 kpc) is $\bar{\rho} \sim 2.0 \times 10^{-24}$ kg/m^3. If the bulk of Andromeda is made up of dark matter assumed to be in the form of a BEC, then there follows from Eq. (9.20) a boson mass m of the order of 10^{-59} kg or about 10^{-23} eV/c^2. For comparative purposes, recall that the mass of an electron is 9.11×10^{-31} kg or 5.11×10^{5} eV/c^2 and that the putative mass of the lightest neutrinos, inferred from neutrino oscillation experiments, is only a few eV/c^2. If they exist—and there is no evidence at present to rule out the possibility—WIDGETs would constitute by a wide margin the lightest of all particles with a nonvanishing rest mass.

Presuming that such bosons do exist, how can one be certain that they form a BEC rather than a relativistic gas like the cosmic background radiation? The theory of the transition from a hot gas of bosons (within which particle momenta are distributed broadly according to the classical Maxwell–Boltzmann distribution) to a BEC (within which the narrow momentum distribution uniquely reflects the quantum attrributes of Bose–Einstein statistics) is well understood. From the quantum expression for the distribution of bosons over energy states, one can calculate rigorously the transition temperature for formation of a BEC. However, the same result can be estimated accurately from a simple heuristic argument.

The classical (i.e., Maxwell–Boltzmann) regime of a gas corresponds to conditions in which the mean distance between particles is very large compared with the thermal de Broglie wavelength λ of a particle. Under these circumstances, the particles are essentially noninteracting (except for direct "billiard-ball" collisions); any one particle is largely unaware of the presence of the others. At the other extreme, when the mean distance between particles is small compared with a de Broglie wavelength, the single-particle wave functions overlap and the motion of many particles is highly correlated. In the quantum (i.e., Bose–Einstein) regime, it is not even physically meaningful to speak of "different" particles within the system because all particles are identical and therefore indistinguishable (as I discussed in Chapter 3). The threshold for the transition to a BEC, which marks the onset of the quantum regime, is therefore determined by the mean particle density $\bar{n}$.

The transition temperature T_c may be defined as the temperature at which there is, on average, one particle within a volume of a cubic de Broglie wavelength λ^3, or $\bar{n}\lambda^3 = 1$. At temperatures close to and above T_c, the bosons, being of very low mass, move with relativistic speeds; like photons whose momentum and energy satisfy $p = \varepsilon/c$, the bosons have a de Broglie wavelength $\lambda = h/p = hc/\varepsilon$ (and *not* h/mv), where $\varepsilon = kT_c$ is the mean thermal energy per particle and k is Boltzmann's constant (see Note 13 of Chapter 1). From the preceding considerations, it then follows that the transition temperature and particle density are related by

$$T_c \sim \left(\frac{hc}{k}\right)\bar{n}^{1/3}. \qquad (9.21)$$

From the mass of a WIDGET and the mean mass density $\bar{\rho} = m\,\bar{n}$ of the Andromeda halo, one calculates a particle density $\bar{n} \sim 5 \times 10^{34}\,\mathrm{m^{-3}}$ and, from Eq. (9.21), a transition temperature $T_c \sim 5 \times 10^9\,\mathrm{K}$. T_c corresponds to the temperature of primordial nucleosynthesis at about a few seconds after the big bang and is much higher than the temperature ($\sim 3000\,\mathrm{K}$) at which matter and radiation decoupled (thereby leading to galaxy formation) or the present temperature ($\sim 2.7\,\mathrm{K}$) of the cosmic background radiation. If galactic dark matter is composed of scalar bosons, one can be reasonably sure that these particles are in a BEC.

9.6. A Galactic Superfluid?

I have deduced so far by heuristic arguments some of the principal attributes of nonbaryonic BEC dark matter in galactic halos. However, as with any quantum system, a more complete and rigorous understanding can be acquired only by solving the appropriate equations of motion, which, for a nonrelativistic self-gravitating gas of neutral spinless particles, is the Schrödinger equation. However, in contrast to usual systems of interest in which quantum particles are subjected to external potentials, the particles of BEC dark matter are subject to a gravitational interaction that arises from their own collective masses.[51] Because the gravitational potential energy depends on the particle density and the particle density is proportional to the square of the magnitude of the wave function $|\Psi|^2$, the Schrödinger equation for Ψ in the present case turns out to be an intractable nonlinear equation.

Symmetry helps somewhat. If, in keeping with astrophysical evidence, it is assumed that dark matter forms essentially spherical halos, then Ψ can be expressed as the product of a radial wave function $\psi(r)$ and an angular wave function (known as a spherical harmonic) that depends on the polar and azimuthal angles. There then results an equation in the radial coordinate only which, upon further approximation, can be brought to a form known as a cubic Schrödinger equation. This equation is also nonlinear, but can be solved exactly to yield an analytical solution

$$\psi(r) = \frac{A \tanh(r/r_c)}{r}. \tag{9.22}$$

The two constants in the wave function are a multiplicative factor A, which depends on the number, mass, and energy of the bosons in the condensate, and a characteristic length r_c, which may be interpreted as the de Broglie wavelength. In the approximation that the density within a BEC of mass M is uniform, the magnitude of the ground state energy is $G^2 M^2 m^3 / 2\hbar^2$ (the result one would obtain by solving the "gravitational Bohr atom") and the characteristic length reduces to the coherence length of Eq. (9.20) within a factor of order unity. Note that the factor A is not a normalization constant arbitrarily imposed to make $\psi(r)$ interpretable as a probability amplitude. Because the Schrödinger equation is nonlinear in $\psi(r)$, one does not have this freedom; the equation itself leads to a unique prefactor A.

If $\psi(r)$ is interpretable as a probability amplitude, then $\rho = M |\psi|^2$ (which has the dimension of mass per volume) is the condensate mass density. To obtain the radial mass distribution $M(r)$ within the halo, one has only to integrate the amount of matter ($4\pi r^2 \rho \, dr$) contributed by each concentric spherical shell of thickness dr between the center of the halo and radial distance r. The resulting expression for $M(r)$, upon substitution into the velocity formula [Eq. (9.15)], yields the theoretical prediction

$$v(r) = v_\infty \sqrt{1 - \frac{\tanh(r/r_c)}{(r/r_c)}} \tag{9.23}$$

for the rotation curve of luminous matter orbiting the galactic center. The velocity of matter infinitely far from the center, v_∞, depends on the ratio of boson energy to mass. One might expect the velocity to drop eventually to zero for a halo of finite size, but the solution is derived, after all, from an approximate equation, and the wave function (9.22) is, in fact, not "square-integrable" (i.e., it does not yield a finite value when $|\psi|^2$ is integrated over all r). Nevertheless, because the rotation curves of many galaxies are either flat or rising for as far from the center as they have been measured, it is of interest to see how well Eq. (9.23) accounts for these observations.

The smooth curve in Figure 9.3 shows a fit of Eq. (9.23) to the rotation curve of the Andromeda Galaxy (M31). The fit, made visually by means of computer simulation, yielded the expression, $v_{M31} \sim 249.2\sqrt{1-\tanh{(0.11r)}/0.11r}$, where v is in km/s and r is in kpc. The empirical constants r_c and v_∞ together with theoretical relations from the model allow one to deduce the boson mass and energy for a galactic halo of given total mass. The result, $m \sim 10^{-24}\,\mathrm{eV}/c^2$, is very close to the mass deduced previously and independently by assuming a coherence length of the size of the luminous core. A comparable mass was also obtained when the model was applied to the Triangulum Galaxy (M33), whose total mass is tenfold smaller.

Because the scalar bosons constituting BEC dark matter interact only through gravity, their direct experimental detection would be difficult and require detection schemes quite different from those employed to search for highly massive WIMPs. Such light bosons would not show up in accelerator-based experiments or searches in ambient particle fluxes for characteristic decays or nuclear excitations. One intriguing possibility, however, by which the gravitational presence of degenerate dark matter might be discerned is by its superfluid vorticity.

Superfluidity, like superconductivity, is a macroscopic quantum phenomenon entailing the dissipationless flow of matter. The first discovered and most widely studied superfluid is that of the helium isotope ^{4}He which condenses to a normal liquid (He I) at 4.2 K and becomes a superfluid (He II) below the so-called lamda point at 2.2 K. The term "lambda" refers to the shape of the plot of specific heat against temperature, which resembles the lowercase Greek letter. In the vicinity of the lambda temperature, the specific heat of ^{4}He shows a discontinuity that is strikingly similar to the cusp in the theoretical specific-heat curve that characterizes Bose–Einstein condensation. Indeed, most of the properties of He II can be understood, at least qualitatively, as the properties of a BEC of interacting bosons. With two protons and two neutrons, the ^{4}He nucleus has spin 0 and constitutes a composite boson.

The behavior of superfluid ^{4}He is utterly unlike that of any ordinary liquid. It flows freely through the finest capillary tubes that would obstruct the flow of He I or even of gaseous helium. It runs spontaneously up and over the walls of an open container. In a tube packed with fine powder (e.g., emory powder), submerged in a He II bath, and heated at the upper end by the light of a small lamp, superfluid will flow from the low-temperature end to the high-temperature end (in contrast to the normal direction of convective flow of matter). If the upper end of the tube is attached to a capillary extending out of the bulk liquid, the force of the flow can produce a superfluid jet rising to heights of 30 or 40 cm.

Two of the most extraordinary properties of superfluid ^{4}He, however, relate to its behavior under rotation. An ordinary liquid in a container at rest will, shortly after the container is made to rotate (e.g., by being mounted on the turntable of a record player), acquire a rotational motion in the same sense and for as long as the container is turning. If a sample of He I is rotated in the same way at a sufficiently low angular frequency (i.e., below a certain critical frequency ω_c), it too will rotate with the container. However, should the rotating fluid be subsequently cooled below the lambda point, the resulting superfluid will come to rest in the laboratory (actually with respect to the fixed stars) even though the container continues to turn.[52] This signifies that the true thermodynamic equilibrium state of superfluid ^{4}He is the nonrotating state.

A related, but distinct, effect concerns the behavior of the superfluid when the rotating container is brought to rest. Because of friction with the walls of the container, an ordinary liquid will eventually come to rest too. In striking contrast, however, He II will continue to circulate in the container for a seemingly indefinite duration, provided that the initial angular frequency exceeded ω_c. This state of fluid rotation within a stationary container is not an equilibrium thermodynamic state, but represents, instead, a condition of long-lasting metastability.

Both effects can be understood if it is assumed that a rotating atom in the superfluid state, like an electron in an atomic orbit, has an orbital angular momentum quantized in integral values of $\hbar$. ^{4}He atoms in the superfluid are not rotating independently of one another, however, but are part of a BEC in which all atoms are in the same quantum state, thereby giving rise to a total angular momentum of macroscopic size. The angular momentum can change only in discrete units of $\hbar$, but, because all atoms of the superfluid are in the same quantum state, such a change entails a change in the correlated motion of an astronomical number of atoms. Thus, the transition of the condensate from one angular momentum quantum state to another is impeded by a high energy barrier.

When the angular frequency of the superfluid is below the critical frequency necessary for each atom in the circulation to have the minimum unit of angular momentum, the superfluid cannot rotate. When rotating at a higher angular frequency, the superfluid cannot come to rest after the container is brought to rest because to do so would require surmounting the energy barrier. Instead, the fluid rotates in a quantized angular momentum state of such quantum number that the corresponding angular frequency most closely matches the rotation frequency of the container. The critical frequency for the onset of rotation can be estimated roughly from the classical expression $L = mR^2\omega$ for the angular momentum of a ^{4}He atom of mass m at the radius R of the container. The lowest angular frequency ($\omega = \omega_{cr}$) occurs for the lowest nonvanishing angular momentum ($L = \hbar$), leading to $\omega_{cr} \approx \hbar/mR^2$.

The manner by which superfluid helium acquires rotational motion is quite extraordinary and fundamental to understanding its behavior. As the rotational angular frequency of the container is increased beyond ω_c, the superfluid becomes threaded by a symmetrical array of parallel vortices with quantized circulation. Although the superfluid appears to rotate with uniform angular frequency, the bulk of the superfluid is actually at rest; only matter within a relatively small cylindrical region of radial extent a_0 (approximately the size of the de Broglie wavelength of a ^{4}He atom in He II) about a vortex axis is circulating. In fluid dynamics, the term "circulation" has a technical meaning; it is the integral of velocity over a closed contour about the rotation axis. Such an integral is zero over any contour within a superfluid that does not enclose a vortex and has an integer value of h/m for each vortex that is enclosed. It can be shown that a system is more stable with two vortices each of unit circulation than with one vortex of two units. The quantization of circulation is precisely equivalent to the quantization of atomic angular momentum.[53]

Thermodynamically, the condition for equilibrium of a rotating superfluid is that the energy (technically the Helmholtz free energy) of the superfluid in the reference frame rotating with the container be a minimum.[54] This means that a vortex will form in the rotating fluid, provided that the energy is lower than what it would be without the vortex. Implementation of this condition for a container of radius R leads to the critical angular frequency

$$\omega_{cr} = \frac{\hbar}{mR^2} \ln\left(\frac{R}{a_0}\right) \tag{9.24}$$

for formation of a single vortex of minimum circulation (one unit of h/m). Equation (9.24) differs from the rough estimate above only by a logarithmic factor. The vorticity of a fluid is defined as the total circulation within unit area. Thus, the line density or number of vortices of circulation h/m per unit area in a superfluid rotating at angular frequency ω is[55]

$$n_v = \frac{2m\omega}{h} \tag{9.25}$$

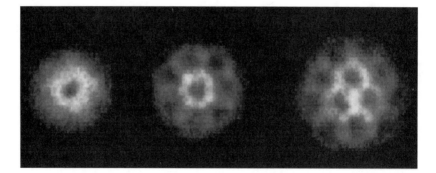

Figure 9.10. Vortex formation in a rotating BEC of ^{87}Rb atoms "stirred" by a laser beam. The higher the angular frequency, the greater is the number of resulting vortices of unit circulation. (Adapted from http://www.lkb.ens.fr/recherche/atfroids/vortex.html.)

and the total number of vortices in a sample of radius R is

$$N = \pi R^2 n_v. \tag{9.26}$$

Although there is little, if any, doubt that superfluid He II is a BEC, not all such condensates will necessarily be superfluids. Nevertheless, recent investigations confirming the formation of superfluid vortices in rotating BECs of low-density alkali metal vapors[56] (such as those of rubidium and lithium) encourage the belief that a galactic BEC may likewise manifest superfluidity. If that is the case, then one might expect the dark-matter halo of a galaxy to be threaded with vortex lines such as those shown in Figure 9.10 for a spherical condensate of ^{87}Rb atoms. (The atoms were driven into rotation at different angular frequencies by a laser beam whose position was controlled by an acoustic-optic modulator.)

The implications of the preceding considerations for the Andromeda Galaxy (M31) are striking. If a minimal halo radius of 150 kpc and a coherence length of 30 kpc are assumed, then Eq. (9.24) predicts a critical frequency of approximately 2×10^{-19} radians/s. However, the observed rotation curve shows that the velocity of matter at 150 kpc is approximately 250 km/s, corresponding to a rotation rate of $\omega \sim 5 \times 10^{-17}$ radians/s. Since $\omega \gg \omega_{cr}$, it would seem that it would actually be difficult to keep vortices from forming in a dark-matter superfluid comprising the Andromeda halo. The vortex line density estimated for M31 from Eq. (9.25) is about 1 vortex per 208 kpc^2, which would indicate that nearly 340 vortices could have formed within the M31 halo.

If this were the case, how might such vortices be observed? Since dark matter is, after all, dark and, consequently, does not emit or scatter light, rotational motion of BEC vortices would not show up as red- and blue-shifted subgalactic regions. Nevertheless, according to general relativity, rotating matter affects the geometry of space–time differently than stationary matter. The existence of dark matter vortices could be sought in gravitational effects on the imaging or polarization of light from distant background sources transmitted through the halos of foreground galaxies.

9.7. And So . . .

Some day, the dark matter–dark energy problem will be solved. Increasingly comprehensive surveys of the Universe by techniques of greater scope and greater sensitivity will provide the necessary data, and the model presented here, as well as others, will be tested. When this problem is laid to rest (at least temporarily, because every "solved" problem in physics usually generates new questions), we will have a better idea of what comprises the 95% or so of the cosmos beyond the minuscule fraction now familiar to us. By that time, there will, I suspect, still remain the fundamental question: "What is gravity?"

My investigation of the nature of dark matter as a BEC actually began—as the beginning of this chapter relates—as a study of that question. To a certain extent, Einstein has told us what gravity is. It is an apparently attractive interaction between particles, including massless ones, arising from the curvature of space–time. From a quantum mechanical perspective, however, all interactions between particles are mediated by the exchange of bosons, the range of the interaction varying inversely with the boson mass. Eectromagnetism, for example, is an infinitely long-range interaction because the exchanged particle, the photon, has zero rest mass. Gravity is also believed to be an infinitely long-range interaction, for which the mediator, the graviton, is a spin-2 massless boson. This belief rests on the fact that a quantum field theory of spin-2 particles leads to equations of motion that coincide with a linearized approximation to Einstein's gravitational field equations. No experiment to my knowledge has ever detected a graviton, nor do I see such a prospect on the distant horizon.

As discussed previously, all the fundamental interactions except gravity—i.e., the electromagnetic and the weak and strong nuclear interactions—are describable by gauge field theories. Among the attractive features of gauge field theories is that they are renormalizable; in effect, one can calculate with them. Although such theories have a geometrical structure analogous in some ways to that of general relativity, it has not yet proven possible to construct a gauge field theory of gravity that reproduces Einstein's field equations in their entirety and not just in a linearized approximation. The pursuit of such a theory has long been a sort of holy grail for many theorists. When I and my colleague initially examined a scalar field in a five-dimensional universe, it was with that goal in mind—to arrive at a theory in which the full equations of general relativity (augmented, perhaps, by additional interactions) emerged naturally as a consequence of imposing gauge invariance upon some free boson field with specified internal degrees of freedom. However, the procedure yielded only a part of these equations and it did not seem likely that further elaboration would yield the rest. Nevertheless, it was an exploration that produced a novel conception of dark matter.

Interestingly, in the aftermath of this first attempt to uncover gravity, there is reason to believe that the procedure, which did not work with bosons, may work with fermions. The critical distinction lies in the nature of the wave function which represents these fields. As fields of particles with integer spin, boson wave functions comprise scalar (spin-0), vector (spin-1), or tensor (spin-2 or higher) functions. Fermions, by contrast, are particles with odd half-integer spins represented by spinor wave functions. Spinor fields couple to gravity in a different way than do scalar, vector, or tensor fields. Although further discussion would take us too far afield, I will conclude with an intriguing and thoroughly speculative possibility.

If it should prove true that gravity as we presently understand it (via general relativity) can be shown to arise through the imposition of gauge invariance on some spinor field, then what particle corresponds to the associated fermion? Furthermore, since identical fermions cannot occupy the same quantum state, would the lovely picture of dark matter as a BEC be all for naught? Not necessarily. Perhaps these fermions pair up to form composite bosons. We already know of at least one example: The phenomenon of superconductivity in metals arises from the Bose–Einstein condensation of Cooper pairs of electrons. Of course, the cosmos is not a superconductor. But, if the development of physics over the past century has illustrated anything at all, it is the enormous explanatory power of a few key unifying ideas to reveal the nature of phenomena that span unimaginably broad ranges of size, energy, and area of application.

Notes

1. The chemist, Auguste Kekulé, after having pondered deeply the structure of the benzene molecule, reportedly realized, upon awakening from a dream of a snake grasping its tail, that the atoms formed a closed six-carbon ring.

2. See Chapter 4. Also, I discuss birefringence comprehensively in *Waves and Grains: Reflections of Light and Learning* (Princeton University Press, Princeton, NJ, 1998).

3. The quote is taken from a letter from Einstein to Ehresfest, cited in A. Pais, *"Subtle is the Lord . . .": The Science and the Life of Albert Einstein* (Clarendon Press, Oxford, 1982, p. 432).

4. M. H. Anderson, J. Ensher, M. Matthews, C. Wieman, and E. Cornell. Observations of Bose-Einstein Condensation in a Dilute Atomic Vapor, *Science* **269** (14 July 1995) 198–201.

5. The 2001 Nobel Prize in Physics was awarded to three researchers (Eric Cornell, Carl Wieman, and Wolfgang Ketterle) for their creation of Bose–Einstein condensates.

6. Joint news release by The National Institute of Standards and Technology and the University of Colorado on 13 July 1995, Physicists Create New State of Matter at Record Low Temperature, http://jilawww.colorado.edu/www/press/bose-ein.html.

7. An excellent discussion is given by the author himself in M. C. Escher, *Escher on Escher: Exploring the Infinite* (H. N. Abrams, New York, 1986). Also, all the one- and two-dimensional space-group symmetries have been reproduced in lovely Hungarian folk needlework; see I. Hargittal and G. Lengyel, The Seven One-Dimensional Space-Group Symmetries Illustrated by Hungarian Folk Needlework, *Journal of Chemical Education* **61** (1984) 1033–1034.

8. A. Conan Doyle, The Silver Blaze, in *The Complete Sherlock Holmes* (Doubleday, New York, 1930, pp. 383–401). *Inspector Gregory:* "Is there any other point to which you would wish to draw my attention?"; *Holmes:* "To the curious incident of the dog in the nighttime."; *Gregory:* "The dog did nothing in the nighttime."; *Holmes:* "That was the curious incident." [p. 397]

9. In anticipation of discussion of the cosmic microwave background radiation (CBR), I note that this thermal radiation (at a temperature of about $2.7\,\mathrm{K}$) permeates all space and therefore constitutes a sort of ether with respect to which one's motion can be measured by means of the Doppler effect. Thus, at any point in space, the frame in which the CBR appears isotropic may be singled out as a preferred frame of reference. This does not violate the postulate of special relativity, however, which asserts that all inertial reference frames are equivalent for the description of *local* physics phenomena.

10. This is the same factor that led to the time-dilation operator in Chapter 5; see Eqs. (5.4b) and (5.8b).

11. It would seem almost a theorem of geometry, rather than a law of physics, that the velocity of a marble rolling at speed v' down the aisle of a bus traveling in the same direction with speed V should be $v' + V$ to an observer at rest by the side of the road. This intuitive velocity addition law, however, is not generally valid because of the different rates at which clocks in the two frames (bus and curbside) keep time.

12. A. Pais, Note 3, p. 152.

13. If the components of the electric and magnetic fields are designated by numbers (1, 2, 3) rather than by letters, then the elements of $F^{\mu\nu}$ follow

the simple pattern: $F^{0k} = E_k$ and $F^{ij} = \varepsilon_{ijk}B_k$. The completely antisymmetric tensor ε_{ijk} is equal to +1 if (i, j, k) is an even permutation of $(1, 2, 3)$, −1 if (i, j, k) is an odd permutation of $(1, 2, 3)$, and 0 if any two indices are equal. I have labeled the components of **E** and **B** with letters rather than numbers, however, because they are not components of a Lorentz 4-vector.

14. The field tensor with lowered indices is obtained from the original field tensor by $F_{\mu\nu} = \eta_{\mu\alpha}\eta_{\nu\beta}F^{\alpha\beta}$. From the form of the Minkowski metric, one sees that, in $F_{\mu\nu}$, the sign of each component of the electric field is opposite that of $F^{\mu\nu}$, whereas each component of the magnetic field is unchanged.

15. The divergence of the field is expressed by $\nabla \cdot \mathbf{E} = \partial E_x/\partial x + \partial E_y/\partial y + \partial E_z/\partial z$.

16. In his *Treatise on Electricity and Magnetism*, Maxwell uses the gauge transformation of the vector potential [second equation of relations (9.8a)] to eliminate a certain scalar function. He does not introduce the corresponding transformation for the scalar potential nor comment on the invariance of the field equations under such a combined transformation. To my knowledge H. A. Lorentz was the first to recognize gauge invariance as a general principle in electromagnetism in lectures delivered at Columbia University in 1906. See H. A. Lorentz, *The Theory Of Electrons*, Second Edition (Dover Publications, New York, 1952) 239.

17. A scalar contraction of two 4-vectors a^μ and b^μ is the quantity $a_\mu b^\mu$. The Dirac equation takes the form

$$\left[\gamma^\mu\left(p_\mu - \frac{q}{c}A_\mu\right) - mc\right]\psi = 0$$

in which $p_\mu = i\hbar\partial/\partial x^\mu$ and the factors γ^μ constitute a set of sixteen constant 4×4 matrices (that make up what is called a Clifford algebra). The Klein–Gordon equation takes the form

$$\left[\left(p_\mu - \frac{q}{c}A_\mu\right)\left(p^\mu - \frac{q}{c}A^\mu\right) + m^2c^2\right]\psi = 0.$$

By recalling that $p^\mu = (E/c, \mathbf{p})$ and therefore $p_\mu = (E/c, -\mathbf{p})$, one sees that the definition $p_\mu = i\hbar(\partial/\partial x^\mu)$ is consistent with the usual quantum mechanical operator representations $E = i\hbar(\partial/\partial t)$ and $\mathbf{p} = (\hbar/i)\nabla$.

18. The Lagrangian for a free electron takes the form $\bar{\psi}(\gamma^\mu p_\mu - mc)\psi$ where ψ and $\bar{\psi}$ are Dirac spinors with four components (two each to represent the spin of the electron and the spin of its antiparticle, the positron). The Lagrangian for a free spin-0 boson takes the form $(p_\mu\phi)(p^\mu\phi) - m^2c^2\phi^2$, where in the simplest case ϕ is a real-valued scalar wave function.

19. From the definitions of the canonical momentum $p_\mu = i\hbar(\partial/\partial x^\mu)$ and the kinetic momentum

$$P_\mu = i\hbar\frac{D}{Dx^\mu} = p_\mu - \frac{q}{c}A_\mu,$$

it follows that

$$\frac{D}{Dx^\mu} = \frac{\partial}{\partial x^\mu} + i\frac{q}{\hbar c}A_\mu.$$

20. The color neutrality of particles can occur two ways. Protons, neutrons, and other baryons (massive fermions subject to the strong interactions) comprise three quarks each of a different color, resulting in colorless quantum states. Pions and other mesons (massive bosons subject to the strong interactions) comprise pairs of quarks and antiquarks in which the anticolor of the antiquark neutralizes the color of the quark, much as positive and negative electric charges cancel each other.

21. The so-called London penetration depth is given by $\lambda_L = \sqrt{mc^2/4\pi Nq^2}$ in which m, q, and N are respectively the mass, charge, and number density of the charge carrier in a superconductor. The charge carrier is a Cooper pair with twice the mass and charge of a single electron.

22. P. F. Schewe and B. Stein, An Intriguing Hint of the Higgs Boson, *The American Institute of Physics Bulletin of Physics News*, N. 502 (September 14, 2000).

23. It is worth noting that the fundamental equations of electromagnetism, Maxwell's equations, were also not deduced from empirical results alone. Although Coulomb's law for static charges and Ampere's law for currents were based on experiment, Maxwell's modification of Ampere's law (addition of the "displacement current") was not required by any experimental result at the time.

24. Recent measurements of the precession of muon spin in a magnetic field gave results for the anomalous magnetic moment of the muon that differed from previous calculations based on the Standard Model by approximately two standard deviations. See H. N. Brown, *et al.*, Precise Measurement of the Positive Muon Anomalous Magnetic Moment, *Physical Review Letters* **86** (2001) 2227. If sustained by further experiments with better statistics and by confirmations of the theoretical prediction, this disagreement would indicate the observation of a physical effect beyond the Standard Model. However, subsequent recalculation of the higher-order terms in the expression for the muon magnetic moment have revealed an earlier error in sign which now makes the claim of a violation of the standard model much less likely. See B. Schwarzschild, Correcting a Correction Weakens a Whiff of Supersymmetry, *Physics Today* **55** (February 2002) 18.

25. Carl Sagan, *Billions & Billions*, Random House, New York, 1997.

26. Newton's law of gravity together with Newton's second law $F = mg$ applied to a small mass m falling to the Earth leads to $M_E = gR_E^2/G$.

27. The term "centripetal" derives from Latin for "seeking the center." In Chapter 5, the same acceleration was termed "centrifugal," from Latin for "fleeing the center." The difference is one of reference frame. An object in uniform circular motion undergoes centripetal acceleration to an observer in an inertial reference frame, but centrifugal acceleration to an observer at rest with respect to the object.

28. One light-year is the distance traveled by light in a year, or approximately 9.5×10^{15} m.

29. For light propagating in vacuum, the ratio of the observed frequency ν_o to the emitted frequency ν_e, or the ratio of the observed wavelength λ_o to the emitted wavelength λ_e, depends on the relative velocity of the source and receiver according to the expression

$$\frac{\nu_0}{\nu_e} = \frac{\lambda_e}{\lambda_o} = \gamma\left(1 - \frac{V}{c}\cos\theta\right)^{-1}$$

in which V is the relative speed, $\gamma = 1/\sqrt{1-(V/c)^2}$, and θ is the angle between the light ray from the source and the velocity of the source. If the source recedes from the observer, $\theta = 180°$; the observed frequency is, therefore, lower than the emitted frequency, and the observed wavelength is longer than the emitted wavelength. The wave is said to be red-shifted.

30. As a simple example, consider three collinear dots located at coordinates 0, 1, and 2 in some arbitrary unit. The dot at 0 is the reference dot. Suppose that the linear space expands to three times its length in 1 s, at which time the coordinates of the dots become 0, 3, and 6. The distance of the third dot from the first has increased by $6 - 2 = 4$ units in the same span of time that the distance of the second dot from the first has increased by $3 - 1 = 2$ units. Thus, the third dot, whose distance to the reference dot is twice that of the second dot, is receding from the reference dot twice as fast.

31. An entertaining and informative account of the lives and researches of prominent astronomers and cosmologists can be found in Dennis Overbye, *Lonely Hearts of the Cosmos* (HarperCollins, New York, 1991).

32. The theory of inflation maintains that the Universe, very shortly after its creation, underwent an exponential expansion, vastly increasing in size well beyond the distance that light could have traveled during that same time interval. There is no violation of relativity theory so long as the speed of massive objects relative to the expanding space does exceed the vacuum speed of light c. After the brief period of inflation (concluded in under 10^{-30} s), the Universe evolved along the lines of the standard big-bang theory. See A. Guth, Inflationary Universe: A Possible Solution to the Horizon and Flatness Problems, *Physical Review D* **23** (1981) 347.

33. The Planck radiation law $u_\nu = (8\pi h\nu^3/c^2)(e^{h\nu/kT} - 1)^{-1}$ expresses the energy density of thermal radiation at absolute temperature T within a frequency interval $d\nu$ about frequency ν. Expressed in terms of wavelength $\lambda = c/\nu$, the radiation density within a wavelength interval $d\lambda$ about λ becomes $u_\lambda = (8\pi hc/\lambda^5)(e^{hc/\lambda kT} - 1)^{-1}$.

34. The Wien displacement law, $\lambda_{max}T = 0.029$ m-K, relates the absolute temperature T and the wavelength λ_{max} at which the thermal radiation intensity is greatest. It derives from the expression for u_λ in Note 33. The corresponding relation obtained from the expression for u_ν is $\frac{\nu_{max}}{T} = 62.5$ GHz K^{-1}. Note that the frequency at which u_ν is maximum does *not* correspond to the wavelength at which u_λ is maximum ($\nu_{max} \neq c/\lambda_{max}$).

35. The CBR is isotropic in the rest frame of the matter with which it was in equilibrium before radiation and matter decoupled. However, the Galaxy (Milky Way) is moving as part of the Local Group of galaxies toward the Virgo Cluster at a speed of some 600 km/s relative to the CBR. Thus, CBR photons received from the direction in which the Local Group is moving are blue-shifted; CBR photons received from the trailing direction are red-shifted. The result is a perceived anisotropy known as the cosmic

microwave background dipole. To detect the small temperature fluctuations in the CBR, the dipole anisotropy must first be subtracted from the data.

36. The effect of mass density on radiation temperature is an example of what is known as a gravitational red shift. Radiation propagating out of the gravitational potential well produced by surrounding matter loses energy and therefore incurs longer wavelengths depending on the depth of the well (which itself depends on the density of matter). The greater the mass density, the greater the gravitational red shift. Variations in the energy density u of thermal radiation result in variations in radiation temperature T via the Stefan–Boltzmann relation, $u \propto T^4$. The gravitational red shift was demonstrated terrestrially with gamma rays by Pound and Rebka using the Mössbauer effect; see R. V. Pound and G. A. Rebka, Jr., Gravitational Red-Shift in Nuclear Resonance, *Physical Review Letters* **3** (1959) 439.

37. See the news reports: (a) J. Glanz, Radiation Ripples from Big Bang Illuminate Geometry of Universe, *The New York Times on the Web* (26 November 1999)[http://www.nytimes.com/library/national/science/112699sci-big-bang.html]; (b) B. Schwarzschild, Balloon Measurements of the Cosmic Microwave Background Strongly Favor a Flat Cosmos, *Physics Today* **53** (July 2000) 17–19.

38. The mean thermal energy per particle is approximately kT, in which k is Boltzmann's constant. Equating this energy to the difference in rest-mass energy of the neutron and proton leads to a temperature of about 7 billion K.

39. If L is the intrinsic luminosity and F the observed flux, then the so-called luminosity distance to the supernova is $d_L = \sqrt{L/4\pi F}$. There are other ways, however, of determining distances. For example, the angular diameter distance $d_A = D/\delta$ is the ratio of the proper diameter D of a structure (e.g., a galaxy) to the observed angular diameter δ. In a static Euclidian geometry, the various definitions are equivalent, but this is not the case in an expanding and possibly curved space.

40. A. G. Riess et al., Observational Evidence from Supernovae for an Accelerating Universe and a Cosmological Constant, *Astrophysical Journal* **116** (1998) 1009–1038; S. Perlmutter et al., preprint astro-ph/9812133.

41. From the assumed luminosity and measured apparent brightness of a Type Ia supernova, one can infer a distance d_L, as described in Note 39. Similarly, from measurement of the red shift z and use of Hubble's relation (with present value of the Hubble constant H_0), one can calculate a distance

$$d_z = \frac{c}{H_0}\left[\frac{(z+1)^2-1}{(z+1)^2+1}\right] \propto \frac{1}{H_0}.$$

The discovery that $d_L > d_z$ suggests that $d_L \propto 1/H$ in which the Hubble constant H at a "look-back" time corresponding to z is smaller than the present value H_0. If this line of reasoning is correct, then the Universe is presently expanding at a greater rate than at the times when the observed high-z supernovae occurred.

42. (a) Blast from the Past: Farthest Supernova Ever Seen Sheds Light on Dark Universe, Press Release No. STScI-PR01-09 (Space Telescope Science Institute, Baltimore MD, 2 April 2001). See http://oposite.stsci.edu/pubinfo/PR/2001/09/pr.html. (b) A. G. Riess *et al.*, The Farthest Known Supernova: Support for an Accelerating Universe and a Glimpse of the Epoch of Deceleration, Astrophysics Journal **560** (2001) 49.

43. It is estimated on the basis of the big-bang cosmology that there should be about as many relic neutrinos and antineutrinos in the cosmic background radiation as there are photons, approximately 10^9 times the number of baryons.

44. In the 1994 "Ig Nobel Prize" award ceremony at MIT, the (real) Nobelist William Lipscomb (Chemistry, 1976) interpreted the Heisenberg uncertainty principle for the U.S. Congress: "If your position is everywhere, your momentum is zero."

45. The Lagrangian that leads to Einstein's equations of the gravitational field is $L_R = \left(\dfrac{16\pi G}{c^4}\right)^{-1} R$ where R, the Riemann curvature scalar, is a measure of the curvature of spacetime.

46. The kinetic energy of the scalar field is proportional to $g_{\mu\nu}\partial^\mu\phi\partial^\nu\phi$ or $g^{\mu\nu}\partial_\mu\phi\partial_\nu\phi$.

47. The stress–energy tensor $T_{\mu\nu}$, which is symmetric in the indices, incorporates the energy density of matter (element T_{00}), the pressure exerted by matter (elements T_{11}, T_{22}, T_{33}), and the stresses produced in matter (off-diagonal elements).

48. Einstein's path from completion of special relativity in 1905 to publication of the final form of the field equations of general relativity in 1915 was a rather tortuous one with numerous false starts. Einstein and mathematician David Hilbert independently deduced the field equations of gravity at almost the same time. Hilbert's manuscript was submitted for publication first, but recent historical sleuthing seems to indicate that Hilbert subsequently made a change in his equations and that Einstein's derivation therefore preceded Hilbert's.

49. A particle confined to a linear region of extent r has a momentum uncertainty $p \sim h/r$ and, therefore, a kinetic energy that may be as large as $p^2/2m \sim h^2/2mr^2$.

50. The free-fall acceleration of a particle of matter at distance r from the center of a uniform spherical cloud of mass M and radius R is given by Newton's laws of gravity and motion: $d^2r/dt^2 = GM(r)/r^2$. Recall that the particle is attracted only by the quantity of mass $M(r) = M(r^3/R^3)$ within the volume of radius r and experiences no force from the matter outside this radius. This equation can be rewritten as $d^2r/dt^2 + r/\tau_g^2 = 0$, in which $1/\tau_g^2 \equiv GM/R^3 \sim G\bar\rho$ defines the characteristic free-fall time τ_g.

51. The exact form of the gravitational interaction of the scalar bosons is derivable by solving the coupled equations that result from the total Lagrangian of the scalar field ($\bar\phi$) and gravitational field ($g_{\mu\nu}$). In general, this interaction will depend on the energy and pressure of the scalar field. Because the BEC comprising a galactic halo is essentially a nonrelativis-

tic gas of low density, it is reasonable to employ, as a first approximation, the Newtonian gravitational potential in the Schrödinger equation.

52. G. B. Hess and W. M. Fairbank, Measurements of Angular Momentum in Superfluid Helium, *Physical Review Letters* **19** (1967) 216.

53. For uniform circular motion, the circulation is defined by $\kappa \equiv \oint v\,ds = 2\pi r v$ $= 2\pi\ell/m$, in which $\ell = mvr = n\hbar$ is the angular momentum (with integer n). Thus, $\kappa = nh/m$.

54. The internal energy $U(S, V, N)$ of a system in thermodynamic equilibrium is a function of the entropy, volume, and the number of particles (or mols). The Helmholtz free energy $F(T, V, N)$ provides equivalent information about the system through a tranformation (the Legendre transform) of U that replaces the entropy by the temperature as the independent variable. For a system in contact with a heat reservoir, the equilibrium state leads to the lowest value of the Helmholtz free energy of any state with the same temperature as the reservoir. In a rotating reference frame, the Helmholtz free energy (F') takes the same form as the energy eigenvalues of an atom in a rotating rotating reference frame, Eq. (7.8)—that is, $F' = F - \omega L$, where ω is the angular frequency and L is the total angular momentum of the rotating fluid. A vortex of minimum circulation is formed when F' vanishes, leading to the critical frequency, $\omega_{cr} = F/L$, whose explicit evaluation yields Eq. (9.24).

55. For a rotating container of radius R, the total circulation around the periphery is $(2\pi R)(\omega R)$, which must equal $n_v(h/m)(\pi R^2)$ if the superfluid is to appear to rotate with uniform angular frequency ω. Equation (9.25) follows from this equality.

56. K. W. Madison, F. Chevy, W. Wohlleben, and J. Dalibard, Vortex Formation in a Stirred Bose-Einstein Condensate, *Physical Review Letters* **84** (2000) 806; F. Chevy, K. W. Madison, and J. Dalibard, Measurement of the Angular Momentum of a Rotating Bose-Einstein Condensate, *Physical Review Letters* **85** (2000) 2223. See also http://www.lkb.ens.fr/recherche/atfroids/.

CHAPTER 10

Science and Wonder

In all the years that I have been pursuing physics as a career, I have never found the subject dull, nor exhausted the multifarious store of interesting questions to investigate. Maybe it is lack of imagination on my part, but I have rarely felt the desire to change fields, for I could think of nothing else as deeply satisfying. Poincaré has aptly depicted the source of such feeling when he wrote (with perhaps only a modicum of exaggeration)[1]

The Scientist does not study nature because it is useful to do so. He studies it because he takes pleasure in it; and he takes pleasure in it because it is beautiful. If nature were not beautiful, it would not be worth knowing and life would not be worth living. . . .

The contrast between a working physicist's perception of physics and the attitude held by most other people is, in my experience, simply astonishing. I have often noted with amusement the reactions shown by strangers who, upon making my acquaintance, learn what I do for a living. Following almost ritual complimentary remarks, uttered ostensibly in admiration of an intellect that can master so difficult a subject, inevitably comes a confession that physics was the speaker's most difficult and least enjoyable subject in school. If the conversation continues long enough, however, it is usually disclosed that this discomfiture originated less in difficulty than in boredom—and I learn again, that the awe elicited by my occupation owes less to any presumed intellectual abilities that to the "Sitzfleisch" required to survive memorizing disconnected facts and formulas and working through countless hours of tedious exercises. For *that*, unfortunately, is the impression widely left by courses that introduce (and often enough terminate) the study of physics.

Yet, it is a fact that, throughout my entire career as a professional physicist, I have almost never had to memorize facts and formulas, but simply avail myself of the appropriate references, or, when necessary, derive the mathematical expressions I needed. What calculations I have performed—and many were indeed lengthy and time-consuming—were voluntarily undertaken to explore the topics that interested me. How can it be that a subject which provides continued intellectual challenge and pleasure to one person is at the same time the epitome of tedium and pointlessness to so many others? Anyone who teaches physics may have had occasion to think about this question. The answer, in fact, is obvious: The way in which one encounters physics in school usually bears little resemblance to the activities of a working physicist, and, to varying degrees, this may be true of other sciences as well.

*   *   *

I am in the main a physicist, not a philosopher. Yet, an important component of both occupations, I believe, is ably addressed in philosopher Alan Watt's self-characterization[2]:

A philosopher, which is what I am supposed to be, is a sort of intellectual yokel who gapes and stares at what sensible people take for granted, a person who cannot get rid of the feeling that the barest facts of everyday life are unbelievably odd. As Aristotle put it, the beginning of philosophy is wonder.

Generally speaking, one might assert as well that the beginning of science is wonder. In a way, all healthy children are naturally born scientists; they come into this world with an innate and intense desire to investigate everything around them. Parents of young children know only too well (and perhaps to their frequent frustration) how difficult it is to prevent a child from exercising this inborn drive. What has happened to so many people along their paths from infancy to adulthood to so dull or cripple the natural inclination to explore and understand? Educational theorist Jerome Bruner gets to the heart of the matter clearly[3]:

The will to learn is an intrinsic motive, one that finds both its source and its reward in its own exercise. The will to learn becomes a "problem" only under specialized circumstances like those of a school, where a curriculum is set, students confined, and a path fixed. The problem exists not so much in learning itself, but in the fact that what the school imposes often fails to enlist the natural energies that sustain spontaneous learning—curiosity, a desire for competence, aspiration to emulate a model, and a deep-sensed commitment to the web of social reciprocity.

Human curiosity lies at the root of all science and must be nourished, or at least not thwarted, if it is to thrive. Yet, science education, particularly at the introductory level, often amounts to conveying so much information that it is overwhelming, or so little that it is uninformative. The recipients of such an education can only conclude that science is mysterious and unapproachable, that scientific explanations either do not exist or cannot be understood. Then, the desire to learn science is lost, perhaps irretrievably. The paramount task of science education, as I see it, is to provide inspiration, and not merely information.

There is a common theme that echoes repeatedly in the responses of scientists whenever they are asked what it is that led them to their careers. What comes across, at least in the aggregate of many such enquiries,[4] is the sentiment that science is intellectually exciting, a challenge to one's mental and physical skills; that there is great beauty to science, whether in a bold and artful experimental solution to a seemingly insurmountable problem or in the aesthetic appeal and startling predictive powers of a set of equations; that there is a particular satisfaction in operating daily with universal laws and principles. Unfortunately, this does not seem to come across in the classroom. I have known many people with experience similar to those expressed by palaeontologist Kevin Padian[5]:

...I had a fascination with dinosaurs, but science teaching deadened that fascination by [stressing prematurely] the mindless details of plant cell structures and the genetic code. It was, after all, the Age of Sputnik.

My first semester in college I [nearly failed] a required science course. I was all set for another [low grade] the second semester when the professor paused in the middle of a particularly boring lecture and said, "You know, some of you may not be into this". (It was now the Age of Aquarius.) "If you'd like to do something else, see me after class." I did. He put me in touch with a professor who specialised in all the things I'd always wanted to learn about, who took the time to help me to study them independently and discuss them.

That led to my eventual career which I wouldn't trade for anything.

In a similar way, my own educational development was more often furthered by self-study or through the sympathetic counsel of those who shared their expertise with me, than through anything that occurred in the classroom.

I realize that the responsibility facing most science educators is not necessarily how to produce more scientists, but how to improve the quality of science instruction and to raise the level of scientific awareness of all those who would study science, even briefly. Nevertheless, it is my firm conviction that any person, whether interested in a science career or not, will be motivated to learn science for the same reasons that motivate scientists themselves—to satisfy their curiosity; to seek answers to personally meaningful questions.[6]

If my own experiences are in any way typical, then the tragedy confronting science teachers today is that past a certain stage of development too few students retain enough curiosity about nature to ask themselves any questions. And so conventional science instruction tells its beneficiaries what precisely they must know, by when they must know it, and how they must demonstrate on tests, homework, and laboratory work that they really know what they are supposed to. John Holt, an advocate of more self-directed education, has created a particularly appropriate simile to describe this type of teaching and learning: too often, teaching is like pouring liquid into bottles that come down a conveyor belt. Questions about education reduce principally to questions about what, and how much, to pour into the bottles— two semesters of introductory physics or four?—ignoring the lack of correspondence between inanimate receptacles waiting to be filled and real people who need to make sense of the world.

*   *   *

As a research physicist, I know that to perform my work well I must have a sound background of factual information. However, I also know why I need that information: to solve the problems that interest me. The key to successful science education must lie first in kindling curiosity where the spark has died and nurturing it where it lies latent. Only then will learning become a pleasure, and the mastery of theoretical concepts and empirical detail follow almost of their own accord.

Having taught for many years under the "specialized circumstances" referred to by Bruner, I understand only too well how deeply ingrained in traditional educational practice these circumstances are and how unlikely it is that they will be changed any time soon. How, then, are science teachers to sustain students' curiosity when they must do this under the restrictive conditions of parceled time and fixed curricula that contribute in the first place to its diminution?

Faced with an educational framework over which they may have but little control, educators can, nevertheless, exert considerable influence through their own attitudes toward their subject and their students. There are important ties between the perception of science and the teaching of science that affect whether a teacher will be a wellspring or a dry well of inspiration. Who can doubt, for instance, that a teacher for whom science is largely a technical discipline will provide a different type of instruction than one for whom science is more broadly construed as a cultural activity? Or that a teacher for whom the primary goal of science is the acquisition of accurate data will provide a different type of instruction that a teacher for whom the goal of science is the development of comprehensive theories? Attitudinal differences on the part of an educator may produce students who are interested in science for different reasons or whose working styles, should they become scientists, are different. Such differences can be important, yet educationally innocuous.

Far more serious in their educational consequences are perspectives that seriously misconstrue the essence of science. A teacher who sees science in terms of authority figures—the allegedly great and wise who pass down their knowledge to mere mortals—may well teach science in an authoritarian manner, emphasize and require memorization of material that scientists, themselves, would generally look up in reference books, assign problems involving needless repetition, discourage inquiries, and repel with indignation challenges of fact or interpretations. A teacher who sees science as a cut-and-dried impersonal subject, a repository of facts from which correct answers inevitably flow, may well communicate to students, implicitly or explicitly, that human attributes and human interactions do not matter, that scientific progress follows from slavish adherence to prescribed scientific methods and not from creative imagination and resourceful use of serendipity.

By contrast, a teacher who realizes that science is a multifaceted mode of enquiry and not a sepulcher of facts, that science involves personalities, and that personality and human distinctiveness affect discovery would likely have respect for his own students' individuality and regard that individuality as important and worth fostering. Such a teacher, even with the restrictions and inflexibility of an institutional environment, could create in the classroom an atmosphere in which students are encouraged to think, to experiment, to challenge—in short, to engage in the type of creative exploration of which science consists.

* * *

I have come to recognize three principal tasks that science educators need to accomplish if they are to motivate the study of science.

The first is to convey an accurate and sympathetic impression of science by revealing its humanistic ties to our general intellectual heritage. In the light of heightened public concern over the impact of science on the quality of life, the physical sciences in particular are too often perceived as cold, uninspiring activities pursued by people who are, at best, asocial and, at worst, dangerous. Norman Campbell's remarks of eighty years ago still seem apt today[7]:

It is certain that one of the chief reasons why science has not been a popular subject . . . and is scarcely recognized even yet as a necessary element of any complete education, is the impression that science is in some way less human than other studies.

If science is to be seen as a human endeavor, a quest by people for answers to significant questions, then science educators ought to provide some sense of historical perspective. Newton's laws of motion and law of gravity, for example, are among those enduring topics that will forever grace the introductory physics curriculum. Over the decades, many a student, having retired for the night glassy eyed from calculating the paths of falling projectiles, must certainly have wondered, "Why bother?" I have found, however, that when students understand better the circumstances of Newton's discoveries—that Newton addressed "the great unanswered question confronting natural philosophy" of his time;[8] that the answers did not fall to him with ease, but only after the intense labor of a "man transported outside himself"; that his answers had momentous impact on his contemporaries; that, being of a jealous and suspicious disposition, he had to be coaxed, flattered, and wheedled by Edmund Halley (of comet fame) into writing the *Principia*, and that even Newton had trouble with the concepts of circular motion (he was, after all, inventing these concepts, not reading them from a textbook)—they look with renewed interest upon the subject. Students can be helped to realize that the laws of motion and of gravity are not artificially concocted academic exercises to improve proficiency in calculation; rather, they are a precious part of our cultural legacy, a historical landmark in mankind's progress toward knowledge and truth and away from error and ignorance.

As part of humanizing science education, teachers need also to make their students aware of the aesthetic dimension that has long been a source of personal pleasure and intellectual stimulation to scientists. Sometimes, this beauty is explicitly visual, deriving from the color, shape, or transformation of physical systems. Sometimes, it is, as Feynman says, "a rhythm and a pattern between the phenomena of nature which is not apparent to the eye, but only to the eye of analysis."[9] And sometimes, it lies in the subtle intricacy or bold simplicity of an ingenious experimental stratagem to wrest from nature her closely held secrets. In whatever form it takes, the beauty of science is part of a vital feedback loop of learning: It provides motivation to explore and to comprehend while it increases with comprehension.

The second principal task is to help students appreciate that there is survival value to the acquisition of scientific knowledge, procedure, and attitude. In an age dominated by the fruits of science and technology, a person ignorant of the most basic scientific principles and experimental skills is at a severe disadvantage. Children, as I noted previously, are born with an innate sense of wonder; however, they experience not only intense curiosity but also a strong impression of the incomprehensible. The early years of childhood have been called the "magic years":

These are "magic" years because [the child's] earliest conception of the world is a magical one; he believes that his actions and thoughts can bring about events. Later he extends this magic system and finds human attributes in natural phenomena and sees human or super-human causes for ... ordinary occurrences of daily life ... But a magic world is an unstable world, at times a spooky world, and as the child gropes his way toward reason and an objective world he must wrestle with the dangerous creatures of his imagination. ...[10]

It is not exclusively children but also those without scientific knowledge and understanding who inhabit a "spooky world" where ordinary (and not so ordinary) occurrences of daily life may seem threatening. These people have no sound foundation upon which to rely to help distinguish plausible fact from wildly improbable speculation in the barrage of imminent calamities and breakthroughs that fill the news reports. They are often paralyzed in frustration by the failure or malfunction of the technological devices upon which they must depend. They are prey to the influence of occultism, mysticism, extreme religious fundamentalism, and bogus science. Like the world of early childhood, theirs, too, is an unstable one troubled by the dangerous creatures of their imagination.

The third principal task is to provide students an opportunity to pursue, to whatever extent possible given circumstances and resources, some form of scientific research. To participate in scientific activity directly, to have occasion to utilize the facts and techniques one is learning, is the greatest source of motivation to study science.

All science is, at root, an empirical activity involving the creative interaction between theory and the facts that emerge from observation and controlled experiment. Unfortunately, most students who take science courses will never understand the role, significance, and procedures of the experimental aspect of science, nor ever experience the exhilaration engendered by execution of a successful, self-motivated experiment following arduous and perhaps frustrating preparatory work. Yet, it is this experimental aspect of science—the planning, looking, touching, manipulating, controlling, measuring, recording, checking—this direct contact with the phenomena of nature for the purpose of satisfying one's own curiosity, that has provided many a scientist the strongest motivation and deepest satisfaction.

Scientific experimentation has almost nothing in common with instructional laboratories that provide practical exercises designed from the outset to yield clean, unambiguous data in a reasonably short time on previously well-studied phenomena with low probability of failure. Not only does such laboratory work *not* reflect what actually transpires in a research laboratory, but, worse still, it is ordinarily *assigned* work, rather than an activity that a person would willingly and enthusiastically undertake for the purpose of learning something. Even students with little scientific experience recognize the distinction between science and "cookbook" exercises that do not inspire—or perhaps even permit—innovation and that lead in the end to results of no interest to anyone outside the classroom.

An exploratory activity, on the other hand, does not have to be of momentous general significance to science as long as it is personally meaningful to the person doing it. At times, I think back to the reaction of British naturalist Alfred Wallace to the discovery of a butterfly[11]:

> None but a naturalist can understand the intense excitement I experienced when at last I captured it. My heart began to beat violently, the blood rushed to my head, and I felt much more like fainting than I have done when in apprehension of immediate death . . . so great was my excitement produced by what will appear to most people a very inadequate cause.

One of the most enjoyable scientific experiences I have had myself was when, as a child, I built a motor out of simple nails and wire. It was pleasurable in large part because I wanted to do it; had I been required to do it, the motivation would no longer have been my own, and the educational value of the project most likely would have been lost.

*   *   *

Although few people would willingly admit with any pride to ignorance of their culture's literature, music, and history, I have discerned over the years no such reluctance when it comes to lack of scientific knowledge. Indeed, among science educators, researchers, employers, and administrators in the United States and Great Britain in particular (and the problem is no doubt even more acute in technologically developing nations), the issue of scientific literacy—however that is interpreted—has been a subject of wide concern for years. The statistical details, if one believes them, are chilling, for they indicate massive indifference to science and gross misconceptions about the most basic facts of nature. Nor, if these reports are accurate, is there reason to expect significant favorable change in the half-century to come.

Solutions that I see proposed (depending on the level of instruction) call for such remedies as extensive standardized testing, mandatory science requirements, more classroom hours per day, longer school terms, and, of course, more money to pay for all this. With respect to colleges and universities, there seems to be a growing sentiment that faculty research is at variance with good instruction, for it means less time devoted to the classroom. The presumed remedy is to increase teaching duties and thereby reduce what critics perceive to be educationally unproductive free time.

If history is any guide, remedies predicated on the belief that science instruction and scientific research are incompatible, if not mutually adversarial, and that impose additional curricular requirements and testing as the foundation of better science teaching are bound to fail, just as they have in the past. For what the reports of scientific illiteracy dramatically show is quite simply that where there is no interest, science cannot be taught. One does not generate interest by increasing the very activities through which interest is lost. Real learning is, like science itself, a process of discovery, and, as educational reformers have often expressed, if one wants this process to occur in a school, then one must create the conditions under which discoveries can be made: leisure to think, freedom to explore.

Science educators whose idea of instruction goes no further than the textbook, whose notes have become fossilized from unvarying use, and whose concept of scientific activity is ritualized repetition of procedure cannot hope to motivate and inspire students. Teachers must, themselves, be motivated and inspired to read avidly and regularly in order to learn the lessons of the past and to keep abreast of the present and to undertake their own investigations, however modest in scope or means, in order to teach with confidence based on personal experience.

The keys to motivating science learning are all molded from the same metal: that science instruction is most efficacious and enduring when it reflects the intrinsic activities of science itself. To teach science well, one must have the philosophical attitudes of a scientist: to see science as culturally important, technically useful, and aesthetically moving; to understand that the pursuit and acquisition of scientific knowledge helps free the mind from the bondage of ignorance, superstition, and prejudice; to have a driving curiosity to comprehend the reason that manifests itself in nature and to enjoy sharing this curiosity with others.

Einstein's eloquent words say it all:

The fairest thing we can experience is the mysterious. It is the fundamental emotion which stands at the cradle of true art and true science. He who knows it not and can no longer wonder, no longer feel amazement, is as good as dead, a snuffed-out candle.[12]

Our task, as educators, is to light that candle.

Notes

1. Cited by W. N. Lipscomb, in *The Aesthetic Dimension of Science*, edited by D. W. Curtin, Philosophical Library, New York, 1980, p. 7.
2. A. Watts, *Does It Matter?*, Random House, New York, 1970, p. 25.
3. J. S. Bruner, *Toward a Theory of Instruction*, Harvard University Press, Cambridge, MA, 1966, p. 115.
4. See, for example, P. Weintraub, *The OMNI Interviews* (Ticknor and Fields, New York, 1984) and the article Seventy-Five Reasons to Become a Scientist, *American Scientist* **76** (September/October 1988) 450.

5. Quotation taken from *American Scientist* **76** (September/October 1988) 454, Reason 23. (See Note 4.)

6. For further development of this theme, see my essay A Heretical Experiment in Teaching Physics, in *Waves and Grains: Reflections on Light and Learning* (Princeton University Press, Princeton, NJ, 1998).

7. N. Campbell, *What Is Science?*, Dover, New York, 1921, p. 27.

8. Quotations are from R. Westfall, *Never At Rest: A Biography of Isaac Newton* (Cambridge University Press, London, 1980), Chapter 10, pp. 402, 406.

9. R. P. Feynman, *The Character of Physical Law*, MIT Press, Cambridge, MA, 1965, p. 13.

10. S. Fraiberg, *The Magic Years*, Charles Scribner's Sons, New York, 1959, p. ix.

11. Cited in W. I. B. Beverage, *The Art of Scientific Investigation*, Vintage, New York, 1950, p. 192.

12. A. Einstein, *The World As I See It*, The Wisdom Library, New York, 1949, p. 5.

Selected Papers by the Author

Chapter 1

1. The Vortex Tube: A Violation of the Second Law?, *European Journal of Physics* **3** (1982) 88.

The paper describes the construction of the vortex tube and gives a simple thermodynamic analysis showing that temperature separations much greater than those actually realized can be ideally generated without violating the laws of thermodynamics.

2a. Voice of the Dragon: The Rotating Corrugated Resonator, *European Journal of Physics* **10** (1989) 298.

2b. Voice of the Dragon, in *Waves and Grains: Reflections on Light and Learning*, Princeton University Press, Princeton, NJ, 1998, Chapter 14, pp. 339–354.

The paper (2a) recounts the theoretical and experimental study of another simple, yet remarkable, device activated by a rotational airflow producing intriguing musical tones in a corrugated tube spun about one end. The book chapter (2b) tells more about the impact of the paper and my resolution (through the study of air friction) of a curious discrepancy with a finding in a paper by F. S. Crawford [Singing Corrugated Pipes, *American Journal of Physics* **42** (1974) 278]. The device did not originate in Japan where I first encountered it, but, unknown to me at the time of the study, had been marketed in the United States as a "hummer."

Chapter 2

1. Musical Mastery of a Coke™ Bottle: Physical Modeling by Analogy, *The Physics Teacher* **36** (1998) 70.

A Coke bottle, although it may superficially resemble a deformed single open-ended organ pipe, can be modeled more successfully as an acoustical resonant LC circuit. The paper presents a theory for the fundamental tone of the bottle and experimental results confirming the analysis.

2. Flying High, Thinking Low? What Every Aeronaut Needs to Know, *The Physics Teacher* **36** (1998) 288.

Starting with the precarious situation of the "lawnchair pilot," the paper develops a theory for the variation of air density with altitude in three different models of the atmosphere in order to arrive at expressions for the altitude of a lighter-than-air balloon of specified dimensions.

3. Cool in the Kitchen: Radiation, Conduction, and the Newton "Hot Block" Experiment, *The Physics Teacher* **38** (2000) 82.

Beginning with an examination of the historical paper in which Isaac Newton mentions for the first (and only) time the law by which hot objects cool, the paper presents a detailed experimental study of the cooling of familiar household objects that do not obey Newton's law very well, even under circumstances where the law was expected to hold.

4. "String Theory": Equilibrium Configurations of a Helicoseir, *European Journal of Physics* **19** (1998) 379.

An additional fascinating example, not discussed in Chapter 2, of the uncommon physics of a familiar object is that of the shapes assumed by a long, thin cord or chain (like a lamp chain) held at one end and spun. This paper presents a comprehensive theoretical treatment of the dynamics of a rotating cord (in Greek: "helico" + "seir") subject to uniform gravity and centripetal force and the experimental recordings (by means of a digital camera and harmonic motion analyzer) of the equilbrium modes at various rotational frequencies.

Chapter 3

1a. Distinctive Quantum Features of Electron Intensity Correlation Interferometry. *Il Nuovo Cimento* **B97** (1987) 200.
1b. Applications of Photon Correlation Techniques to Fermions, *Photon Correlation Techniques and Applications*, edited by J. B. Abiss and A. E. Smart, Optical Society of America Proceedings Vol. 1, Optical Society of America, Washington, DC, 1988, p. 26.

The two papers discuss quantum mechanical aspects of Hanbury Brown–Twiss-type experiments uniquely characteristic of charged massive particles. Correlations attributable to the spin-statistics connection and the influence of external potentials upon these correlations show up in the coincidence count rate at two detectors, the variance in count rate at one detector, and the conditional probability of particle detection as a function of delay time.

2a. On the Feasibility of Observing Electron Antibunching in a Field-Emission Beam, *Physics Letters* A**120** (1987) 442.

2b. On the Feasibility of a Neutron Hanbury Brown–Twiss Experiment, *Physics Letters* A**132** (1988) 154.

In contrast to the belief that a short coherence time and low beam degeneracy necessarily prevent the direct observation of fermion anti-correlation in a free-particle beam, the first paper shows that electron antibunching is potentially observable with current technology. This is not the case with neutrons, however, as discussed in the second paper.

3. Gravitationally-Induced Quantum Interference Effects on Fermion Antibunching, *Physics Letters* A**122** (1987) 226.

The effect of gravity on the quantum mechanical phase of a multi-particle system influences the way in which the particles cluster in time and space. Here is an example of the influence of a gravitational potential in the absence of a gravitational force.

4. Fermion Ensembles That Manifest Statistical Bunching, *Physics Letters* A**124** (1987) 27.

Although it has long been thought that electrons manifest only anti-correlations as a result of Fermi–Dirac statistics, there are, in principle, special types of electron states associated with the two input beams of an interferometer that give rise to variety of particle correlations, including effects similar to photon bunching.

5. Second-Order Temporal and Spatial Coherence of Thermal Electrons, *Il Nuovo Cimento* B**99** (1987) 227.

Thermal or blackbody radiation has played a seminal role in the development of quantum mechanics and is one of the most thoroughly studied systems in physics. Although often considered to be the epitome of incoherent light, blackbody radiation does exhibit interference effects, as demonstrated, for example, by the Hanbury Brown–Twiss experiments with starlight. Examination of the coherence properties of a system of thermal electrons, a fermionic analog of blackbody radiation, shows the profound distinctions arising from quantum statistics, the spinorial character of the basic fields, and conservation of particle number.

6. An Aharonov–Bohm Experiment with Two Solenoids and Correlated Electrons, *Physics Letters* **A148** (1990) 154.

The "entanglement" of quantum states was regarded by Schrödinger as one of the strangest features of quantum mechanics. Here, two "back-to-back" AB experiments with momentum-correlated electrons from a single source manifest strange long-range correlations characteristic of the Einstein–Podolsky–Rosen paradox.

7. Aharonov–Bohm Effects of the Photon, *Physics Letters* **A156** (1991) 131.

Even though it has no electric charge, a photon can theoretically interact with the vector potential field outside a region of magnetic flux as a result of quantum electrodynamic processes involving the virtual production of correlated pairs of electrons and positrons.

8. Optical Manifestations of the Aharonov–Bohm effect by Ion Interferometry, *Physics Letters* **A182** (1993) 323.

All experimental tests of the AB effect to date have employed the electron, a structureless point particle. The use of charged particles with a composite internal structure, however, permits novel tests of the state dependence of the AB effect and the consistency of different quantum equations of motion. A distinctive feature of the proposed experiment is to detect the AB effect by means of quantum interference manifested in the spontaneous emission from the ion. Apart from the focus on the AB effect, this paper was one of the first to propose experiments involving ion interferometry.

9a. Electron Source Brightness and Degeneracy from Fresnel Fringes in Field Emission Point Projection Microscopy, *Journal of Vacuum Science and Technology* **A12** (1994) 542.
9b. Brighter Than a Million Suns: Electron Interferometry with Atom-Sized Sources, in *Foundations of Modern Physics: 70 Years of Matter Waves*, Editions Frontieres, Gif-sur-Yvette 1994, p. 273.
9c. The Brightest Beam in Science: New Directions in Electron Microscopy and Interferometry, *American Journal of Physics* **63** (1995) 800.

The development of ultrasharp field-emission tips which emit electrons from a region of one or a few atoms has led to startling new advances in electron microscopy and interferometry. The tips produce a strongly focused, highly coherent electron beam which can be used in a low-voltage point-projection microscope for low-energy, lensless holographic imaging. The first paper describes such a microscope for which the electron beam is sufficiently coherent that it gives rise to Fresnel fringes. From analysis of the fringe spacings and intensity are obtained the image magnification, source size, transverse coherent width, instrumental resolution, and source brightness. The second paper discusses the importance of brightness and its relation to beam intensity, coherence, and degeneracy. The third paper discusses the unique features of the point-projection microscope as a new tool for exploring quantum phenomena.

Chapter 4

1. Radiofrequency Spectroscopy of Hydrogen Fine Structure in $n = 3$, 4, 5, *Physical Review Letters* **26** (1971) 347.

The paper describes the experimental investigation of a broad range of fine-structure states of hydrogen atoms generated by electron-capture collisions of accelerated protons with gas targets. Various radiofrequency and microwave fields are employed together to suppress overlapping transitions and allow measurement of energy intervals between selected states. This was one of the first applications of the merger of atomic beam technology and radiofrequency spectroscopy.

2. Interaction of a Decaying Atom with a Linearly Polarized Oscillating Field, *Journal of Physics B: Atomic and Molecular Physics* **5** (1972) 1844.

A two-level quantum system interacting with an oscillating electromagnetic field is a fundamental quantum mechanical problem of importance to spectroscopy, but for which no exact analytical solution to the Schrödinger equation is known. The equation is commonly solved in the so-called "rotating-wave" approximation which discards antiresonant terms in the Hamiltonian—a procedure not always adequate for treatment of unstable states, particularly in the radio-frequency domain. This paper provides a more general analytical solution to the oscillating field problem that is almost indistinguishable from exact numerically integrated solutions for all cases of practical interest.

3. Observation of Fine Structure Quantum Beats Following Stepwise Excitation in Sodium D States, *Physical Review Letters* **33** (1974) 1063.

The use of two synchronously pulsed dye lasers was used to prepare sodium atoms in a linear superposition of fine-structure states. A useful feature of the experiment was to demonstrate the marked effects of light polarization on the phase of the quantum beats; this effect was employed to enhance beat contrast.

4. On the Anomalous Fine Structure in Sodium Rydberg States, *American Journal of Physics* **48** (1980) 244.

A simple quantum mechanical model, based on the virtual excitation of a core electron occurring together with the actual excitation of a valence electron, was developed to account in part for the anomalous reversal of the sodium $D_{3/2}$ and $D_{5/2}$ fine-structure levels.

5. General Theory of Laser-Induced Quantum Beats, Parts I and II, *Physical Review* **A18** (1978) 1507, 1517.

The two papers present a comprehensive theory of quantum beats generated by pulsed laser excitation. The first article is concerned with the excitation of an atom by a single laser in the absence of external fields; the second treats the sequential excitation of an atom by two lasers and the influence of an external static magnetic field on the beats. Of particular interest are the nonlinear effects arising from multiple interactions between the atom and the light during passage of an intense laser pulse, and the marked difference in quantum beat patterns between the cases of weak and strong external magnetic field.

6. The Curious Problem of Spinor Rotation, *European Journal of Physics* **1** (1980) 116.

Experimental tests are discussed that address the question of whether or not the 360° rotation of a spinor wave function can be observed.

7. The Distinguishability of 0 and 2π Rotations by Means of Quantum Interference in Atomic Fluorescence, *Journal of Physics B: Atomic and Molecular Physics* **13** (1980) 2367.

This paper describes the details of how, by observation of quantum beats from atoms coherently excited by a pulsed laser and subsequently irradiated by a radiofrequency field, one can distinguish a cyclic transition between two atomic states from no transition at all.

8a. Quantum Interference Test of Orbital Angular Momentum Eigenvalues Predicted for a Spinless Charged Particle in the Presence of Long-Range Magnetic Flux, *Physical Review Letters* **51** (1983) 1927.

8b. Angular Momentum and Rotational Properties of a Charged Particle Orbiting a Magnetic Flux Tube, *Fundamental Questions in Quantum Mechanics*, edited by L. M. Roth and A. Inomata, Gordon and Breach, New York, 1986, p. 177.

These papers discuss the curious properties of a charged particle rotating about an inaccessible magnetic field, such as that within a very long solenoid. An experimental test employing a split beam of charged particles was proposed to determine which set of angular momentum eigenvalues is relevant to describing the effects of rotation. The experiment can distinguish between particle paths that wind a different number of time around the solenoid.

9. On Measurable Distinctions Between Quantum Ensembles, in *Annals of the New York Academy of Sciences: New Techniques and Ideas in Quantum Measurement Theory*, edited by D. M. Greenberger, vol. 480, New York Academy of Science, New York, 1986, p. 292.

This paper is concerned with observable differences between quantum systems in definite, although statistically distributed, eigenstates and systems in a linear superposition of these same eigenstates.

10. Quantum Interference in the Fluorescence from Entangled Atomic States, *Physics Letters* **A149** (1990) 413.

The long-distance quantum beat effect is treated in detail and shown to be insensitive to atomic motion and to allow macroscopic atomic separation.

Chapter 5

1. Relativistic Time Dilatation of Bound Muons and the Lorentz Invariance of Charge, *American Journal of Physics* **50** (1982) 251.

This paper discusses the issues of bound-particle motion, the determination of the bound-muon lifetime, and the argument for charge invariance as a consequence of atomic neutrality.

2. Zeeman Effect in Heavy Muonic Atoms, *American Journal of Physics* **51** (1982) 605.

A muon bound to a nucleus of sufficiently large atomic number can have a classical orbit located within the nuclear interior where the electrostatic potential experienced by the muon resembles that of a harmonic oscillator. This paper discusses the energy level structure of such a muonic atom subject to electrostatic, spin–orbit, and magnetic interactions.

3. The Lifetime of the Dimuon Atom, *Il Nuovo Cimento* **D2** (1983) 848.

The theoretical existence of exotic atoms with *two* (or more) electrons in the same orbit replaced by unstable elementary particles raises interesting questions concerning the effects of quantum statistics on particle lifetime. Would the constraints of the Pauli exclusion principle speed up, slow down, or not affect at all the decay of two ground-state muons? The outcome is in some ways surprising.

Chapter 6

1. Interference Colors with Hidden Polarizers, *American Journal of Physics* **49** (1981) 881.

The paper explains the interference colors produced by birefringent cellophane with Rayleigh scattering and Brewster angle reflection serving as the "hidden" polarizers.

2. Investigation of Light Amplification by Enhanced Internal Reflection, Parts I and II, *Journal of the Optical Society of America* **75** (1983) 1732, 1739.

The theory developed in Part I and experimental tests described in Part II of light amplification by total reflection from a medium with a population inversion are shown to be in good agreement, thereby confirming the once controversial phenomenon of enhanced reflection. This phenomenon now provides a basis for all-optical telecommunication systems employing optical fibers, instead of metal wires, and optical amplification instead of amplification of electronic signals.

3a. Specular Light Scattering from a Chiral Medium, *Lettere al Nuovo Cimento* **43** (1985) 378.

3b. Reflection and Refraction at the Surface of a Chiral Medium: Comparison of Gyrotropic Constitutive Relations Invariant or Noninvariant Under a Duality Transformation, *Journal of the Optical Society of America* **A3** (1986) 830.

In the first paper, the Fresnel amplitudes for light reflected from a transparent isotropic optically active medium are derived from the two supposedly equivalent, but fundamentally different, sets of chiral material relations and shown to lead to physically distinguishable effects. The second, more comprehensive paper extends the theory to absorbing chiral media and traces the origin of the inequivalence to the imposition of electromagnetic boundary conditions. These papers provide a basis for the correct treatment of the electrodynamics and optics of chiral media, a subject that has since become of widespread interest for reasons as diverse as national defense and the search for extraterrestrial intelligent life.

4. Effects of Circular Birefringence on Light Propagation and Reflection, *American Journal of Physics* **54** (1986) 69.

This paper demonstrates, among other things, the perhaps surprising result that for reflection within an anisotropic chiral medium the angle of incidence need not equal the angle of reflection.

5. Light Reflection from a Naturally Optically Active Birefringent Medium, *Journal of the Optical Society of America* **A7** (1990) 1163.

Ordinarily, except for propagation along special directions (the optic axes) of an anisotropic optically active medium, the effects of optical activity are overwhelmed by the much stronger linear birefringence of the medium. This paper shows that the differential reflection of circularly polarized light can still be sensitive to the weak chiral interactions of the material.

6a. Experimental Method to Detect Chiral Asymmetry in Specular Light Scattering from a Naturally Optically Active Medium, *Physics Letters* **A126** (1987) 171.

6b. Experimental Configurations Employing Optical Phase Modulation to Measure Chiral Asymmetries, *Journal of the Optical Society of America* **A5** (1988) 1852.

These papers describe experimental configurations employing the photoelastic modulator to measure the difference in reflection of left and right circularly polarized light, as well as other optical manifestations of left–right asymmetry in chiral media.

7. Large Enhancement of Chiral Asymmetry in Light Reflection near Critical Angle, *Optics Communications* **74** (1989) 129.

Under conditions of total reflection the unequal scattering of left and right circularly polarized light by an optically active medium can be orders of magnitude greater than for ordinary reflection.

8. Differential Amplification of Circularly Polarised Light by Enhanced Internal Reflection from an Active Chiral Medium, *Optics Communications* **74** (1989) 134.

The paper demonstrates that right and left circularly polarized light can be selectively amplified by reflection from an optically active medium with a population inversion. Such selective amplification could make possible novel techniques in quantum information processing and cryptography.

9. Wave Propagation Through a Medium with Static and Dynamic Birefringence: Theory of the Photoelastic Modulator, *Journal of the Optical Society of America* **A7** (1990) 672.

The unusual behavior of the photoelastic modulator first exhibited in experiments designed to test the theory of light reflection from optically active materials is fully accounted for by an analysis of light propagation through a medium with nonparallel axes of static and dynamic birefringence. The paper discusses means of circumventing the effects of static birefringence and expanding the use of photoelastic modulation to novel experimental configurations.

10. Multiple Reflection from Isotropic Chiral Media and the Enhancement of Chiral Asymmetry, *Journal of Electromagnetic Waves and Applications* **6** (1992) 587.

This paper discusses the conditions under which multiple reflection between two parallel optically active surfaces can enhance the difference in reflectance of left and right circularly polarized light.

11. Chiral Reflection from a Naturally Optically Active Medium, *Optics Letters* **17** (1992) 886.

The difference in reflection of left and right circularly polarized light from a sample of naturally optically active molecules was enhanced and quantitatively observed for the first time.

Chapter 7

1. Satellite Test of Intermediate-Range Deviation from Newton's Law of Gravity, *General Relativity and Gravitation* **19** (1987) 511.

This paper analyzes the problem of a test mass subjected to the combined influence of gravity and the "fifth force" within a closed spherical shell in orbit.

2. Rotational Degeneracy Breaking of Atomic Substates: A Composite Quantum System in a Noninertial Reference Frame, *General Relativity and Gravitation* **21** (1989) 517.

It is shown in this article that the equations of motion for the center of mass and internal coordinates of an atom undergoing rotation can be separated (as in an inertial frame) and that the coupling to the rotating frame splits otherwise degenerate magnetic substates. A quantum beat experiment with atomic Rydberg states is proposed to demonstrate this rotational-level splitting.

3a. Rotationally Induced Optical Activity in Atoms, *Europhysics Letters* **9** (1989) 95.
3b. Effect of the Earth's Rotation on the Optical Properties of Atoms, *Physics Letters* **A146** (1990) 175.

A quantum mechanical derivation is given in the first paper of rotational circular birefringence in atoms; the treatment is generalized in the second paper to include other optical effects as well expected in the case of atoms on the spinning Earth.

4. Circular Birefringence of an Atom in Uniform Rotation: The Classical Perspective, *American Journal of Physics* **58** (1990) 310.

The atom in a rotating reference frame is treated by classical mechanics, and a classical interpretation of rotational optical activity is given in terms of the Coriolis force. The paper also brings out explicitly the connection between the behavior of systems in a field-free rotating reference frame and that in an inertial reference frame with a static magnetic field.

5a. Measurement of Hydrogen Hyperfine Splittings as a Test of Quantum Mechanics in a Noninertial Reference Frame, *Physics Letters* **A152** (1991) 133.

5b. Optical Activity Induced by Rotation of Atomic Spin, *Il Nuovo Cimento* **D14** (1992) 857.

It is shown that the level structure (paper 5a) and optical properties (paper 5b) of atoms with zero internal orbital angular momentum are nevertheless affected by rotation as a result of the coupling of particle spin to the angular velocity of rotation. The hydrogen hyperfine structure, which is one of the most precisely measured of all physical quantities, provides a good system for testing these predictions.

Chapter 8

1. On the Run: Unexpected Outcomes of Random Events, *The Physics Teacher* **37** (1989) 218.

This paper describes the theory of runs in elementary terms and applies it to various experiments (computer calculation of random numbers, coin selection, nuclear decay) to determine whether or not the outcomes represent a random process.

2a. Tests of Alpha-, Beta-, and Electron Capture Decays for Randomness, *Physics Letters* **A262** (1999) 265.

2b. Tests for Randomness of Spontaneous Quantum Decay, *Physical Review* **A61** (2000) 042106-1.

In these papers, different kinds of nuclear decay processes were examined experimentally and the resulting data tested for randomness by means of runs with respect to target values, parity, and sequential differences.

3. Experimental Tests for Randomness of Quantum Decay Examined as a Markov Process, *Physics Letters* **A272** (2000) 1.

In this paper, four distinct nuclear disintegration processes were recorded over a long succession of counting intervals and examined as a Markov process to ascertain whether the decay of nuclei is influenced by their past history. No such dependence was found.

Chapter 9

1. Symmetry Breaking in a Five-Dimensional Universe with Implications for the Nature of Dark Matter, *Gravity Research Foundation 2000 Essay* (GRF, Wellesley Hills, MA, 2000).

This paper makes a first attempt at constructing a gauge-field theory of gravity based on spontaneously broken spacetime symmetry of a neutral 5-vector field. (A 5-vector, in analogy to a 4-vector, has five components, two of which in this case are temporal.) In an effort to account for Newton's gravitational constant as the expectation value of a scalar field, the theory led to the prediction of low-mass vector bosons, which would have formed a degenerate Bose–Einstein gas throughout virtually the entire evolution of the universe and thereby provide a source of dark matter.

2. Cosmic Degenerate Matter: A Possible Solution to the Problem of Missing Mass, *Classical & Quantum Gravity* **18** (2001) L37.

This paper provides a theory of dark matter (in ordinary four-dimensional space–time) as a Bose–Einstein condensate of very low-mass neutral bosons arising from the spontaneous breaking of reflection symmetry of a scalar (not vector) field. An estimate of the boson mass is made by assuming a condensate density equal to the critical background density for a universe with density parameter $\Omega = 1$.

3. Coherent Degenerate Dark Matter: A Galactic Superfluid?, *Classical & Quantum Gravity* **18** (2001) L103.

The theory in the preceding article is developed further in this paper, which derives expressions for the transition temperature and condensate coherence length and estimates the mass of the scalar boson based on the density of matter in the galactic halo (rather than on the critical background density). In addition, the possibility that dark matter in the halos of rotating galaxies may exhibit superfluid properties is first proposed, and an estimate is made of the number of quantized vortices in the Andromeda Galaxy (M31), a Galaxy similar to that of the Milky Way.

4. Dark Matter as a Cosmic Bose-Einstein Condensate and Possible Superfluid, *General Relativity & Gravitation* **34** (2002) 633.

In this paper, the process of symmetry breaking and the evolution and distribution of dark matter as a degenerate Bose–Einstein gas are studied in more detail. The nonlinear Schrödinger equation for a Bose–Einstein condensate is solved, leading to the distribution of dark matter and the rotational velocity of luminous matter in a galactic halo of spherical symmetry. Comparison with the rotation curves of the

Andromeda (M31) and Triangulum (M33) Galaxies leads to an estimate of the boson mass consistent with the value obtained in the previous article (deduced from the coherence length).

Chapter 10

1. Science as a Human Endeavor, *American Journal of Physics* **53** (1985) 715.

To help students develop a more accurate and sympathetic perspective of the nature of science, the paper describes a selection of readings and topics of discussion emphasizing science as a cultural activity rather than as a methodological abstraction.

2a. Two Sides of Wonder: Philosophical Keys to the Motivation of Science Learning, *Synthèse* **80** (1989) 43.
2b. Raising Questions: Philosophical Significance of Controversy in Science, *Science and Education* **1** (1992) 163.

Much of the present chapter was adapted from these essays (particularly the first) which served as the basis of invited talks presented at the First and Second International Conferences on the History and Philosophy of Science Teaching respectively held at Florida State University, Tallahassee, Florida in 1989 and at Queen's University, Kingston, Ontario in 1992. The second paper examines several historical and contemporary scientific controversies that illustrate how different the actual practice of science is from that based on widely prevailing idealizations taught in schools.

About the Author

Mark P. Silverman received his Ph.D. in physics from Harvard University. He has been Chief Researcher at the Hitachi Advanced Research Laboratory (Tokyo), Joliot Professor of Physics at the Ecole Supérieure de Physique et Chimie (Paris), and Erskine Fellow at the University of Canterbury (Christchurch). Dr. Silverman teaches at Trinity College (Hartford).

Other books by Dr. Silverman are

And Yet It Moves: Strange Systems and Subtle Questions in Physics (Cambridge, 1993).

More Than One Mystery: Explorations in Quantum Interference (Springer, 1995).

Waves and Grains: Reflections on Light and Learning (Princeton, 1998).

Probing The Atom: Interactions of Coupled States, Fast Beams, and Loose Electrons (Princeton, 2000).

Index